Human Impacts on the Environment: Past, Present and Future

Human Impacts on the Environment: Past, Present and Future

Edited by Aaron Scott

Syrawood
PUBLISHING HOUSE

New York

Published by Syrawood Publishing House,
750 Third Avenue, 9th Floor,
New York, NY 10017, USA
www.syrawoodpublishinghouse.com

Human Impacts on the Environment: Past, Present and Future
Edited by Aaron Scott

International Standard Book Number: 978-1-68286-735-8 (Hardback)

Cataloging-in-Publication Data

Human impacts on the environment : past, present and future / edited by Aaron Scott.
    p. cm.
Includes bibliographical references and index.
ISBN 978-1-68286-735-8
1. Human beings--Effect of environment on. 2. Human ecology.
3. Nature--Effect of human beings on. I. Scott, Aaron.
GF75 .H86 2019
304.2--dc23

TABLE OF CONTENTS

PREFACE

The human impact on the environment is understood from the scientific study of anthropocene. It is a proposed epoch in earth's history characterized by significant human influence on earth's geology, ecosystems and climate. The loss of biodiversity is due to an accelerated rate of species extinction. This is witnessed through the massive decline of marine phytoplankton in the world's oceans and decrease in algal biomass. Other aspects of human impacts include the large-scale migration of species, rise in global and ocean temperatures, etc. This is caused by human overpopulation, overconsumption, intensive fishing and farming, etc. This book is a compilation of chapters that discuss the most vital concepts to understand the effects of human activities on the environment. It contains some path-breaking studies in this field. It will serve as a valuable source of reference for all experts and students.

The researches compiled throughout the book are authentic and of high quality, combining several disciplines and from very diverse regions from around the world. Drawing on the contributions of many researchers from diverse countries, the book's objective is to provide the readers with the latest achievements in the area of research. This book will surely be a source of knowledge to all interested and researching the field.

In the end, I would like to express my deep sense of gratitude to all the authors for meeting the set deadlines in completing and submitting their research chapters. I would also like to thank the publisher for the support offered to us throughout the course of the book. Finally, I extend my sincere thanks to my family for being a constant source of inspiration and encouragement.

Editor

Air pollution impacts due to petroleum extraction in the Norwegian Sea during the ACCESS aircraft campaign

P. Tuccella*,‖,¶, J. L. Thomas*, K. S. Law*, J.-C. Raut*, L. Marelle*,**, A. Roiger†, B. Weinzierl‡, H. A. C. Denier van der Gon§, H. Schlager† and T. Onishi*

Emissions from oil/gas extraction activities in the Arctic are already important in certain regions and may increase as global warming opens up new opportunities for industrial development. Emissions from oil/gas extraction are sources of air pollutants, but large uncertainties exist with regard to their amounts and composition. In this study, we focus on detailed investigation of emissions from oil/gas extraction in the Norwegian Sea combining measurements from the EU ACCESS aircraft campaign in July 2012 and regional chemical transport modeling. The goal is to (1) evaluate emissions from petroleum extraction activities and (2) investigate their impact on atmospheric composition over the Norwegian Sea. Numerical simulations include emissions for permanently operating offshore facilities from two datasets: the TNO-MACC inventory and emissions reported by Norwegian Environment Agency (NEA). It was necessary to additionally estimate primary aerosol emissions using reported emission factors since these emissions are not included in the inventories for our sites. Model runs with the TNO-MACC emissions are unable to reproduce observations close to the facilities. Runs using the NEA emissions more closely reproduce the observations although emissions from mobile facilities are missing from this inventory. Measured plumes suggest they are a significant source of pollutants, in particular NO_x and aerosols. Sensitivities to NO_x and NMVOC emissions show that, close to the platforms, O_3 is sensitive to NO_x emissions and is much less sensitive to NMVOC emissions. O_3 destruction, via reaction with NO, dominates very close to the platforms. Far from the platforms, oil/gas facility emissions result in an average daytime O_3 enhancement of +2% at the surface. Larger enhancements are predicted at noon ranging from +7% at the surface to +15% at 600 m. Black carbon is the aerosol species most strongly influenced by petroleum extraction emissions. The results highlight significant uncertainties in emissions related to petroleum extraction emissions in the Arctic.

Keywords: oil/gas emissions; oil/gas pollution; Norwegian Sea oil/gas extraction; Arctic pollution; air quality modeling

1. Introduction

Modeling and observational studies have shown that sources contributing to Arctic air pollution, namely aerosols and trace gases, are primarily located in the northern mid-latitudes (e.g. Wespes et al. 2012; Shindell et al., 2008; Jacob et al., 2010; Law et al., 2014). Previous work has highlighted the sensitivity of the Arctic climate to short-lived pollutants, such as black carbon and tropospheric ozone (O_3), from mid-latitudes (e.g., Quinn

et al., 2008; Serreze and Barry, 2011; Yang et al. 2014) and the impact of soot deposition on snow and ice on surface albedo (e.g., Hansen and Nazerenco, 2004; Flanner et al., 2007, 2009; Jiao et al., 2014). However, the dominance of remote pollution sources may be in the process of changing due to increased access to the Arctic region, associated with climate change leading, for example to reduced sea ice, and economic drivers already resulting in additional shipping and oil-gas exploration extraction

* LATMOS/IPSL, UPMC Univ. Paris 06 Sorbonne Université, UVSQ, CNRS, Paris, FR

† Deutsches Zentrum für Luft- und Raumfahrt (DLR), Institut für Physik der Atmosphäre, 10 Oberpfaffenhofen, DE

‡ University of Vienna, Faculty of Physics, Aerosol Physics and Environmental Physics, Boltzmanngasse 5, A-1090 Vienna, AT

§ TNO Climate, Air and Sustainability, Princetonlaan 6, 3584 CB Utrecht, NL

‖ NUMTECH, 6 allée Alan Turing, CS 60242, 63178 Aubiere, FR

¶ Laboratoire de Météorologie Dynamique, Ecole Polytechnique, 91128 Palaiseau, FR

** Center for International Climate and Environmental Research, Oslo, NO

Corresponding author: P. Tuccella (paolo.tuccella@lmd.polytechnique.fr)

activities in the Arctic (Stephenson et al., 2011; AMAP, 2010). It has been known for some time that certain local sources such as metal smelting are important sources of aerosols, especially sulfate (SO_4^{2-}), in the eastern Arctic (Prank et al., 2010) and more recently, other near Arctic sources have been shown to have an important impact on Arctic air pollution. In particular, Stohl et al. (2013) suggested that Russian flaring emissions associated with oil and gas extraction and seasonally varying domestic combustion emissions (Klimont et al., 2017) may contribute as much as 42% to surface Arctic annual mean black carbon (BC). This is because Arctic surface sites, such as Zeppelin, Alert and Barrow, are sensitive to surface emissions in or near to the Arctic (Hirdman et al., 2010). Inclusion of these emissions in model simulations led to improved model simulations of BC concentrations and its seasonal cycle in the Arctic (Eckhardt et al., 2015). Local emissions from shipping, related to transport of goods, tourism or fishing, have also been shown to be important sources of aerosols and O_3 (Peters et al., 2011; Eckhardt et al., 2013, Marelle et al., 2016). Ødemark et al. (2012) estimated that current petroleum activity may increase summer mean O_3 in Alaska and western Russia by 5%, and shipping may contribute to a seasonal surface O_3 change of 12% along the Norwegian coast and west coast of Greenland. Ødemark et al. (2012) estimated that the annual average radiative impact from air pollutants, normalized to mean column burden enhancements from current Arctic oil/gas and shipping activities, are similar to the global average. BC aerosols and tropospheric O_3 can lead to increased warming locally in the Arctic. Increases in petroleum extraction may have significant impacts on Arctic climate in the future given that estimates suggest that 30% of global undiscovered gas and 13% of oil is north of the Arctic Circle and mostly located offshore (Gautier et al., 2009).

From 1990 to 2004, Arctic oil production took place in western Russia (79%), Alaska (18%), and Norway (3%), whereas gas production was dominated by western Russia (96%) and Alaska (3%) (Peters et al., 2011). However, current air pollutant emissions from activities related to gas and oil extraction are very uncertain in the Arctic where very little independent data is available to validate reported estimates. For example, there is one order of magnitude difference in oil/gas emissions of NMVOCs in the Norwegian Sea in global inventories derived by Peters et al. (2011) and ECLIPSE (Evaluating the Climate and Air Quality ImPacts of Short-livEd Pollutants) (Klimont et al., 2017). Studies in other regions, such as the midwest in the United States of America (USA), also suggest that inventories underestimate emissions of CH_4 and NMVOCs from oil/gas extraction (e.g., Xiao et al., 2008; Pétron et al., 2012). Activities related to oil/gas extraction produce emissions of carbon dioxide (CO_2), methane (CH_4), nitrogen oxides ($NO + NO_2 = NO_x$), non-methane volatile organic compounds (NMVOCs), carbon monoxide (CO), sulfur dioxide (SO_2) as well as direct emissions of aerosol particulate matter (PM) containing BC and organic aerosols (OA). These emissions can lead to the production of secondary pollutants namely O_3, SO_4^{2-} and secondary organic aerosols (SOA).

Uncertainties in these emissions are due, in part, to the fact that they result from a variety of activities. For example, combustion of fossil fuel for energy production on fixed offshore production installations for heat and light generation, processing and export of hydrocarbons and the treatment and reinjection of water used in the extraction process produces emissions of CO_2, NO_x, SO_2 and CO (UK Oil and Gas, 2015, report available at http://oilandgasuk.co.uk/environment-report.cfm). Natural gas or diesel fuel is used for this purpose although the use of low NO_x turbines can reduce emissions from this source. Flaring, related to maintenance, well testing and safety procedures, and venting, either intentional or as a result of crude oil transfer procedures (e.g. to/from tankers), are also a source of air pollutants. Flaring is highly variable resulting in emissions of aerosols such as BC and trace gases such as CO_2 and NO_x whilst venting of gases leads to emissions of CH_4 and NMVOCs. In a region like the Norwegian continental shelf, combustion from turbines is an important source of CO_2 and NO_x emissions (http://www.npd.no/en/). As well as emissions from fixed facilities, air pollutants are also emitted from mobile installations, such as storage tankers and drilling rigs, which also produce emissions from fuel combustion in their turbines.

In this study, we investigate the impact of pollutant emissions from oil/gas extraction using a combination of regional modeling and analysis of data collected during flights as part of the European Union (EU) Arctic Climate Change and Society (ACCESS) aircraft campaign in July 2012. During this campaign, measurements were taken in the vicinity of extraction installations in the Norwegian Sea (Roiger et al., 2015). We note that these flights represent, to our knowledge, some of the first reported data on pollutant emissions from oil and gas extraction under summertime Arctic conditions. In this study, we use the data to evaluate available emission datasets and estimate the local and regional scale impacts of these emissions on Arctic atmospheric chemical composition and potential implications for regional air quality. We focus primarily on evaluating the impacts of petroleum extraction emissions on O_3 since only rather limited data is available with which to evaluate the model in terms of aerosols. We use a chemical transport model (Weather Research and Forecast coupled with chemistry: WRF-Chem) run at high resolution (2 km) with available point source emissions for production facilities in the Norwegian Sea region. We describe the ACCESS airborne campaign and the WRF-Chem model setup in Section 2 and the emission datasets in Section 3. In Section 4, the ACCESS measurements are used to evaluate the performance of the model simulations and to provide insights into the validity of the emission inventories that are used. The sensitivity of modeled O_3 to NO_x and NMVOC emissions is discussed in Section 5. The impact of oil/gas emissions on regional levels of O_3 and particulate matter (PM) over the Norwegian Sea is discussed in Section 6. The conclusions are given in section 7.

2. Campaign description and model setup
2.1 ACCESS campaign
An overview of the ACCESS aircraft campaign is given in Roiger et al. (2015). For completeness, we briefly describe the details of the ACCESS campaign, conducted in July 2012 using the DLR Falcon-20 aircraft based in northern Norway. The main aim was to collect data on emerging local pollution sources in the Arctic in order to evaluate their impacts on atmospheric composition and on regional air quality and climate. During the campaign, emissions from shipping and oil/gas extraction, as well as pollution plumes transported from metal smelting activities in north-west Russia and from Siberian fires were studied (see Roiger et al., 2015 for further discussion). An analysis of ACCESS flights focusing on ship emissions and their impacts can be found in Marelle et al. (2016). The Falcon-20 was equipped with trace gas (NO_x, SO_2, O_3, and CO) and aerosol instrumentation including total and non-volatile aerosol particle number concentration and accumulation mode refractory black carbon (rBC) mass mixing ratios.

During the ACCESS campaign 2 flights were dedicated to sampling emissions downwind of offshore oil/gas extraction facilities in the Norwegian Sea. The geographic range and tracks of both flights are shown in **Figure 1a**.

While the installations that were sampled are south of the Arctic Circle, they are considered to be within the geographic Arctic as defined by the Arctic Council AMAP (Arctic Monitoring and Assessment Programme) which created an area under its Arctic Environmental Protection Strategy, including the Norwegian Sea. The flights took place on 19 and 20 July 2012, in good meteorological conditions with high visibility, and were performed in close collaboration with the Norwegian oil company Statoil (www.statoil.com). The synoptic situation during the flights was dominated by a low-pressure system located to the north of Scandinavia leading to northerly winds close to the Norwegian coast on 19 July 2012. On 20 July 2012, the low-pressure system moved to the northwest, resulting in north-westerly, rather than northerly winds over the Norwegian Sea. These conditions meant that the plume samplings were not influenced by anthropogenic emissions along the Norwegian coast. Further details about the meteorological conditions are given in Roiger et al. (2015).

As shown in **Figure 1b**, the flight on 19 July 2012 was conducted as a survey to sample emissions from a large number of oil and gas production platforms (Kristin, Åsgard A and B, Heidrun and Norne), a storage condensate tanker (Åsgard C), storage tankers (Randgrid and

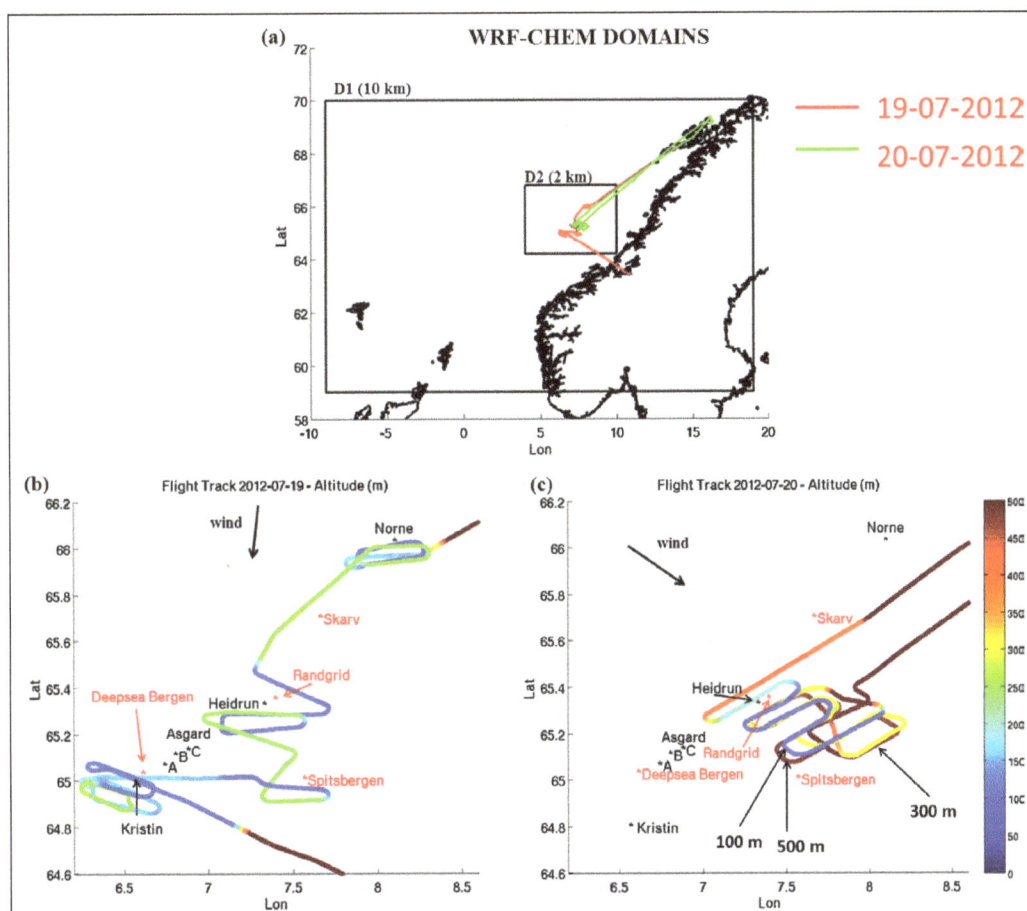

Figure 1: WRF-Chem model domains. Map of WRF-Chem model domains (D1 and D2) with DLR-Falcon flight tracks for the 19 and 20 July 2012 flights **(a)**, and maps of the aircraft altitude during the flights of 19 **(b)** and 20 July **(c)** 2012. See text for details. DOI: https://doi.org/10.1525/elementa.124.f1

Skarv), and drilling rigs (Deepsea Bergen and Transocean Spitsbergen). The flight on 20 July 2012 focused on detailed probing of emissions from the Heidrun production facility using an S-shaped track to sample the plume(s) at various intervals and altitudes (100 m, 300 m and 500 m) downwind as shown in **Figure 1c**. We note that the Norne facility was continuously flaring, on 19 and 20 July 2012, Heidrun was intermittently flaring on both days, whereas Åsgard A was not flaring on 19 July (personal communication, Statoil).

An example of data collected around the platforms is illustrated in **Figure 2**. It shows time series for NO$_x$, SO$_2$, nucleation mode aerosols and total non-volatile particle number concentrations measured on the 19 July in plumes downwind of the Åsgard A and B oil production facilities and the condensate storage tanker, Åsgard C. Moderately enhanced levels of nucleation mode aerosols and low levels of SO$_2$ were observed downwind of Åsgard A and B most likely due to release of VOCs and production of secondary organic aerosols (Roiger et al., 2015). In contrast, particulate matter emissions from the storage

tanker Åsgard C were much higher and comprised of non-volatile aerosols. Enhanced SO$_2$ in plumes from this source points to production of sulfate aerosols and enhancements in BC particles were also noted. NO$_x$ was also higher downwind of this source and other mobile sources (drilling rigs, storage tankers) compared to the fixed production facilities most likely due to combustion emissions. Emissions from drilling rigs (not shown) also contained moderate levels of SO$_2$, particles, and enhanced BC. These emissions from mobile sources are more characteristic of shipping emissions. Lower NO$_x$ and SO$_2$ downwind of the production facilities may be due to the use of natural gas for energy production. The ACCESS observations are discussed in more detail in Section 4 when they are used to evaluate the model simulations.

2.2 WRF-Chem model
The impact of emissions from offshore oil/gas installations on air quality in Norwegian Sea was simulated using the WRF-Chem model (version 3.4.1). WRF-Chem is an online model where the physical and chemical processes are fully

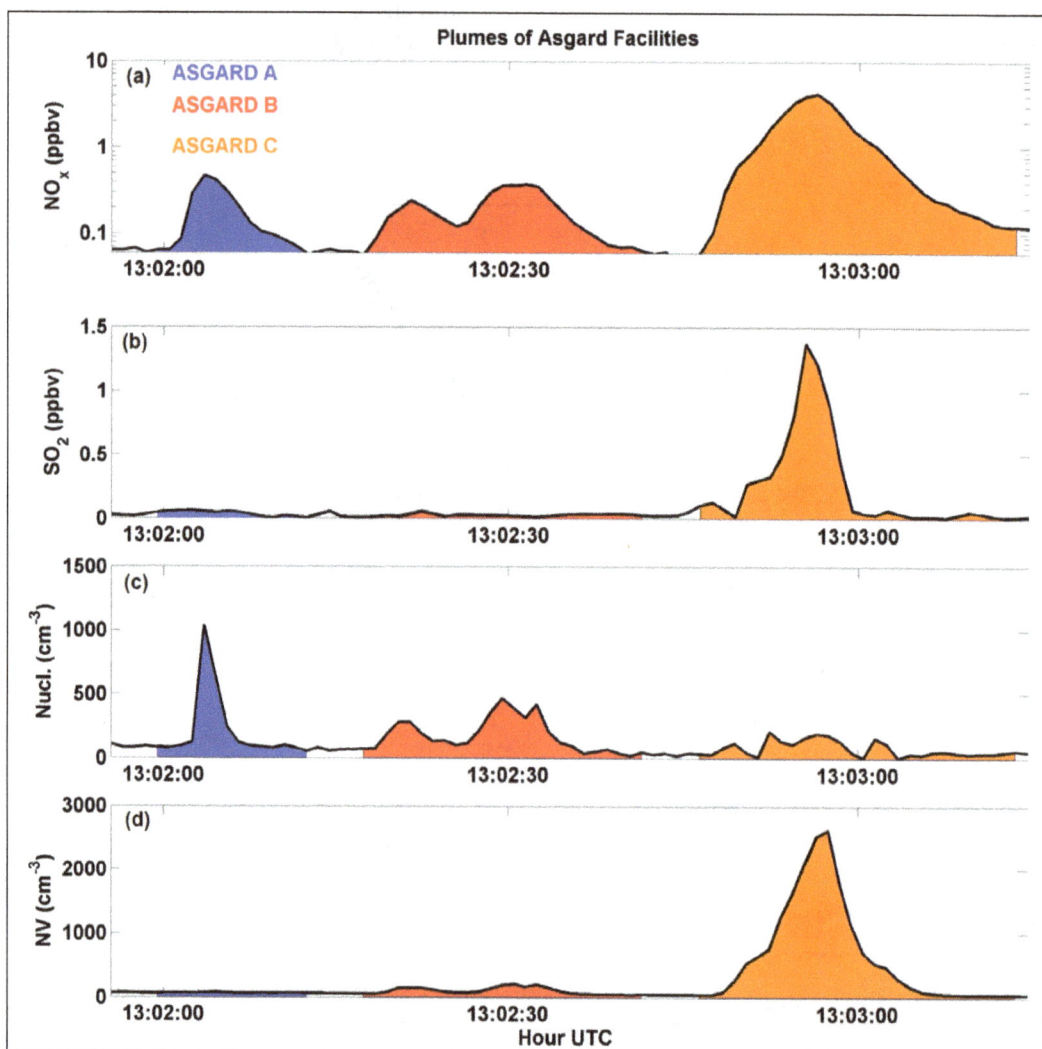

Figure 2: Observed chemical compound composition in the Asgard plume. Observed time series of NO$_x$ (ppbv) **(a)**, SO$_2$ (ppbv) **(b)**, concentration of nucleation mode particles (cm^{-3}) **(c)**, total non-volatile (NV) particle concentration (cm^{-3}) measured in the plumes of Asgard facilities during the flight of 19 July 2012. DOI: https://doi.org/10.1525/elementa.124.f2

consistent (Grell at al., 2005). The model was run with two 1-way nested domains centered on the Norwegian Sea (see **Figure 1a**) denoted as Domain 1 (D1) and Domain 2 (D2), respectively. The larger outer Domain 1 has 130 × 130 cells and a horizontal resolution of 10 km while the inner nested Domain 2 has 151 × 151 cells with resolution of 2 km. The vertical grid has 33 eta levels up to 50 hPa. The physical and chemical parameterizations used in this work are listed in **Table 1**. Nudging, using meteorological analyses from the European Centre for Medium-range Weather Forecasts (ECMWF) of temperature, water vapor mixing ratio and wind speed, was applied every 6 hours above the planetary boundary layer (PBL) in the outer Domain 1. The choice of this model setup gives the best agreement between meteorological variables and observations collected during ACCESS campaign. The comparison of observed and predicted meteorology is shown in **Figure 3** and will be discussed in the Section 4.

The model was run with the RACM-ESRL gas phase chemical mechanism (Kim et al., 2009), an updated version of the Regional Atmospheric Chemistry Mechanism (RACM) (Stockwell et al., 1997). RACM is a chemical mechanism designed for air pollution studies and includes a full range of photolysis and gas-phase reactions. Aerosol particle dynamics is simulated using the Modal Aerosol and Dynamics for Europe (MADE) scheme (Ackermann et al., 1998). MADE uses three overlapping log-normally distributed modes: Aitken, accumulation and coarse. The

Table 1: WRF-Chem model physical and chemical parameterizations. DOI: https://doi.org/10.1525/elementa.124.t1

Physical process	WRF-Chem Options/Parameterizations
Cloud Microphysics	Morrison (Morrison et al., 2009)
Cumulus	New Grell (G3) (update version of Grell and Devenyi, 2002)
Shortwave radiation	RRTM (Iacono et al., 2008)
Longwave radiation	RRTM (Iacono et al., 2008)
PBL	MYNN (Nakanishi and Niino, 2006)
Surface Layer	Monin-Obukhov
Surface	NOAH LSM (Chen and Dudhia, 2001)
Gas-Phase chemistry	RACM-ESRL (Stockwell et al., 1997; Ahmadov et al., 2012)
Aerosol chemistry	MADE/SOA-VBS (Ackermann et al. 1998; Ahmadov et al., 2012)

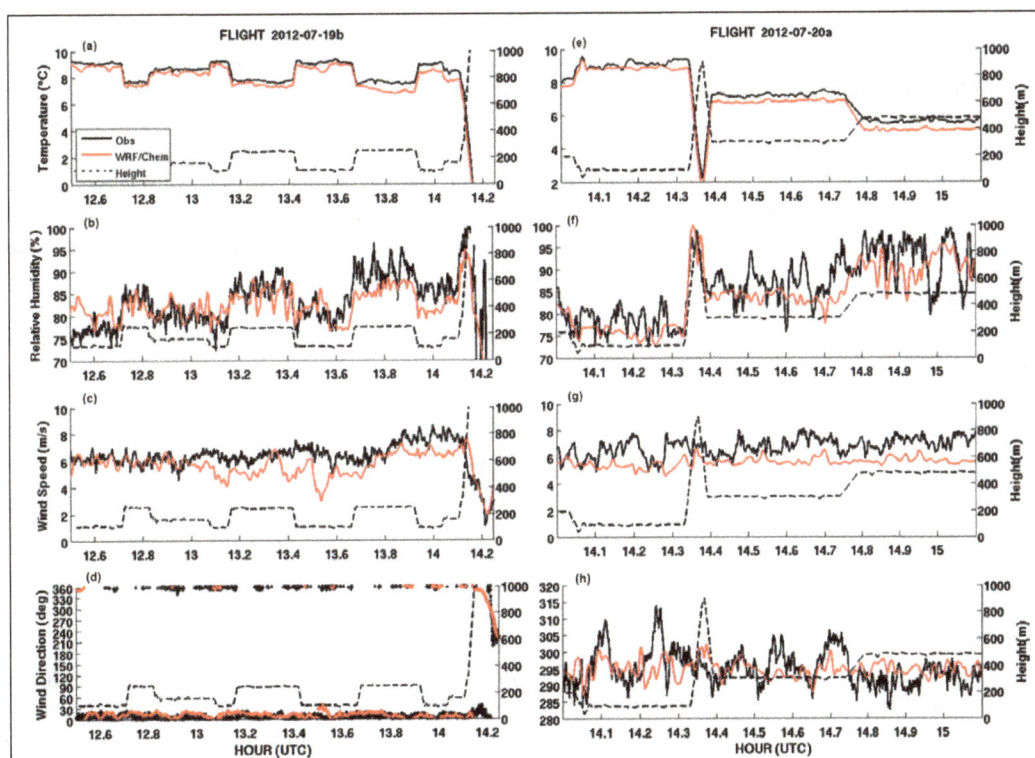

Figure 3: Comparison between observed and modeled meteorology. Time series of observed (black) and simulated (CTRL run, red) temperature **(a, e)**, relative humidity **(b, f)**, wind speed **(c, g)** and wind direction **(d, h)** on 19 (left) and 20 (right) July 2012. DOI: https://doi.org/10.1525/elementa.124.f3

species treated in the Aitken and accumulation modes include inorganic ions, primary particulate matter with aerodynamic diameters less than 2.5 μm ($PM_{2.5}$) that also includes the fine fraction of dust and sea salt, BC, particulate organic matter (POM), SOA, and aerosol water. Unspeciated PM_{10} with aerodynamic diameters less than 10 μm, dust, and sea salt are treated in the coarse mode. SOA production is based on the volatility basis set approach (Ahmadov et al., 2012). Photolysis rates are simulated using the Fast-J scheme (Wild et al., 2000). Dry deposition velocities for trace gases are calculated according to Erisman et al. (1994), while for aerosols the parameterization of Wesely and Hicks (2000) is used. Dry deposition velocities for condensable organic vapors are assumed to be 25% of calculated deposition velocity for HNO_3. Cloud chemistry is treated using the scheme of Walcek and Taylor (1986). Wet deposition from convective updrafts and large scale precipitation is included in the model runs. Aerosol feedbacks on radiation and clouds are not included in these simulations. The tendency terms in the continuity equation are diagnosed following Wong et al. (2009).

WRF-Chem was run over both domains from 14 to 20 July 2012, with 14 to 16 July considered as spin-up for the chemistry. A series of 30 h simulations were performed on each day starting at 00 UTC. The run in the outer Domain 1 was initialized with initial and boundary meteorological conditions provided by ECMWF 6-hourly analyses, at a resolution of $0.125° × 0.125°$. Chemical boundary conditions were provided using the output of the global Model for Ozone and Related Chemical Tracers (MOZART) (Emmons et al., 2010) every 6 hours. In order to reproduce background O_3 concentrations outside of plumes, MOZART predicted O_3 values that are used for initial and boundary conditions were multiplied by a factor of 1.1. For the inner Domain 2, initial meteorological conditions as well as meteorological and chemical boundary conditions were taken from the Domain 1 simulations. Due to resolution, the cumulus parameterization was only used in the outer Domain 1. The emissions and simulations performed as part of this study are described in the next section.

3. Emissions and model simulations

In this study, we ran WRF-Chem using available emissions from activities related to oil and gas extraction in the Norwegian Sea. The model was run with anthropogenic emissions taken from TNO-MACC for 2009 (Kuenen et al., 2014) for the larger coarse resolution (Domain 1) simulations, also used as boundary conditions for the high-resolution (2 km) inner Domain 2 simulations. In Domain 2, we used publically reported point source emissions from the Norwegian Environment Agency (NEA) (http://www.norskeutslipp.no/en/Offshore-industry/?SectorID=700) for 2012. We used these emission estimates because the data is provided at high resolution for specific facilities, in contrast to global inventories like Peters et al. (2011) or ECLIPSE (Klimont et al., 2016) that are only available at resolutions larger than 50 km. Nevertheless, it is important to note that emissions for certain storage tankers (Randgrid, Skarv) and drilling rigs

(Deepsea Bergen, Transocean Spitzbergen), associated with the offshore industry, are not taken in account in the NEA or TNO-MACC inventories because they are mobile emission sources and/or their emissions are not available. Therefore, in this paper we focus on permanently installed facilities (Heidrun, Norne, Åsgard (A + B + C), Kristin) that were sampled by the ACCESS flights. Note that we used annual average TNO-MACC shipping emissions in all runs (Domains 1 and 2).

The TNO-MACC inventory is a gridded high-resolution (7 × 7 km) regional inventory for Europe providing total annual emissions of NO_x, SO_2, total NMVOC, CH_4, NH_3, CO, and primary aerosols for different Selected Nomenclature for Air Pollutants (SNAP) source sectors (Vestreng, 2003). Emissions are split into area and point sources. Examples of area sources are residential combustion, agriculture and transport sectors, whereas the point sources include power plants, refineries, and major industries such as iron and steel plants (Denier van der Gon et al., 2010). The 2009 emissions for oil and gas extraction facilities in the Norwegian Sea are reported as point sources (SNAP sector 5) and are given in **Table 2**. The TNO-MACC inventory only takes into account emissions of NO_x and NMVOCs from the Åsgard complex, Heidrun, and Norne. It does not report emissions of SO_2, BC, primary PM, or OC from these facilities.

The NEA emissions are provided as point sources for the Åsgard complex (total A + B + C), Heidrun, Norne and Kristin (see **Table 2**). Again, drilling rigs and storage tankers are not included. In the case of the Åsgard complex, made up of 2 production facilities (A, B) and a condensate storage tanker (C), separate emissions for Åsgard A, B, and C were estimated based on ACCESS measurements collected close to and downwind of these facilities on the 19 July 2012 (**Figure 2**). Emissions from each individual installation are given by the product of total emission multiplied by the ratio of the area under each single peak to the total area. The distance between these platforms is sufficient such that mixing of plumes originating from different platforms is unlikely. We make the assumption that plumes from different facilities are independent, and do not consider mixing between plumes. Therefore, the areas below the observed plume peaks shown for NO_x, SO_2 and particles in **Figure 2** are proportional to the emission for each given compound. In the case of NO_x, emissions were split into 6%, 12%, and 82% for Åsgard A, B, and C, respectively. SO_2 was assigned only to Åsgard C because observed SO_2 in Åsgard A and B plumes was under the detection limit of 25 pptv (Roiger et al., 2015). As noted earlier, Åsgard A and B are production facilities emitting negligible SO_2 and non-volatile particles, but a large fraction of nucleation mode particles which could be SOA formed from the nucleation of NMVOCs emitted as a result of venting. In contrast, Åsgard C is a tanker emitting a large amount of SO_2 and non-volatile particles, and a smaller fraction of nucleation mode particles. Therefore, NMVOC emissions were divided by using the observed peak area of nucleation particles as a proxy. Based on these observations, 30%, 47%, and 23% of total NMVOC were attributed to Åsgard A, B, and C, respectively.

Table 2: Total annual emissions from facilities in the Norwegian Sea from the TNO-MACC inventory and reported by the Norwegian Environment Agency (NEA). Aerosol emissions are not included in the inventories and were additionally estimated for the NEA inventory as described in Section 3. DOI: https://doi.org/10.1525/elementa.124.t2

	NO_x (t/y)		NMVOC (t/y)		SO_2 (t/y)		PM (t/y)		EC (t/y)		OC (t/y)	
	TNO	NEA	TNO	NEA	TNO	NEA	TNO	NEA[a]	TNO	NEA[a]	TNO	NEA[a]
Kristin	–	180	57	23	–	4	–	4	–	1	–	1
Åsgard (total)	63	2284	8780	5739	–	239	–	115	–	14	–	15
Åsgard A	–	145	–	1747	–	–	–	3	–	1	–	1
Åsgard B	–	270	–	2675	–	–	–	6	–	2	–	2
Åsgard C	–	1869	–	1317	–	239	–	106	–	11	–	12
Heidrun	61	1775	550	255	–	8	–	38	–	10	–	12
Norne	33	658	560	325	–	5	–	14	–	4	–	4

a) Not taken from NEA but additionally estimated.

The NEA emissions do not include aerosols. Therefore emissions of PM, BC and POC were estimated using emission factors for Norwegian oil/gas operations for 2004 reported by Peters et al. (2011). Emitted aerosol mass was calculated by assuming that it is proportional to NO_x emissions using the ratio between the emission factor of a given aerosol compound and that of NO_x. The values obtained under this assumption are reported in **Table 2**.

In this study, we have used the most relevant emissions estimates for the facilities that were the focus of the measurement campaign. These emissions are available from NEA for the year 2012, which corresponds to the year of the measurement campaign. For regional background emissions, we use the TNO-MACC inventory, which is most recently available for 2009. We also compare model runs using TNO-MACC emissions with the more up to date 2012 NEA emissions for comparison. Several notable differences are apparent between the two inventories used in this study as shown in **Table 2**. TNO-MACC NO_x emissions for the facilities are about 30 times smaller than those from NEA while TNO-MACC NMVOC emissions are about a factor 1.5–2.0 higher than NEA. NEA reporting, which is available yearly, reveals that there were some changes due to inter-annual variability in emissions between 2009 and 2012, especially for NO_x. For example, Heidrun and Norne NO_x emissions were lower by 9% and 37%, and NMVOC emissions by 22% and 2%, respectively in 2012 compared to 2009. In contrast, 2012 emissions from Asgard complex were higher than 2009 by 16% and 2% for NO_x and NMVOC, respectively. These differences in annual amounts, which reflect inter-annual variability in the emissions, are not large enough to account for the differences between TNO-MACC and NEA reported in **Table 2**.

The horizontal and vertical distribution of the emissions, their time variability, NMVOC speciation and aggregation in WRF-Chem model species follows Tuccella et al. (2012). Following this method emissions are distributed over WRF-Chem vertical levels depending on the SNAP

sector (Vestreng, 2003). In particular for oil/gas sector (SNAP 5), 90% of emissions are distributed within the first 90 m and 10% between 90 to 170 m. Temporal variability is calculated using generic monthly, weekly, daily and hourly factors derived from Schaap et al. (2005). Following Tuccella et al. (2012), NMVOC emissions are speciated using UK data (Passant, 2002) and aggregated into WRF-Chem species following the reactivity weighting factor principle (Middleton et al., 1990). Following Passant (2002), NMVOC emissions are split into 88% alkanes, 5% alkenes, 0.6% aromatics and 7% are unassigned, i.e. the fraction of compounds in the speciation database whose assignment to a lumped species is difficult or arbitrary. Primary aerosol mass emissions are distributed 10% into Aitken mode and 90% into accumulation mode.

In the reference control run (CTRL), the model was run for the inner Domain 2 using reported NEA emissions and additionally estimated aerosol emissions for the oil/gas production facilities. All other emissions, such as shipping, were from TNO-MACC emissions which were also used in the outer Domain 1. This included shipping emissions and, in the case of Domain 1, all sectors for anthropogenic emissions along the Norwegian coast. For comparison purposes, we also performed a run over Domain 2 using the TNO-MACC emissions, referred to as TNO. In this case, we only used available emissions of NO_x and NMVOCs, i.e. without any estimated aerosol or SO_2 emissions. Based on the differences noted earlier between TNO-MACC and NEA NO_x and NMVOC emission inventories, we examine the sensitivity of model results to levels these emissions in Section 5. The details of all model runs, including the sensitivity tests, are given in **Table 3**.

4. Model evaluation and interpretation of the ACCESS aircraft campaign measurements

In this section, model simulations are evaluated against the ACCESS aircraft campaign data. Results are also used to interpret the measurements collected downwind from the platforms. We focus on the CTRL run simulations in

Table 3: Description of model runs used to study the impact of oil/gas emissions in the high resolution domain (D2). DOI: https://doi.org/10.1525/elementa.124.t3

Run Name	Description
TNO	Run with TNO emissions (see Table 2)
CTRL	Run with NEA emissions (see Table 2) and estimated aerosol emissions using the emission factors in Peters et al. (2011)
T1	No emissions
T2	CTRL with NO_x facilities emissions doubled
T3	CTRL with NO_x facilities emissions reduced by a factor 2
T4	CTRL with NMVOC emissions increased 5 times

Domain 2 performed at high resolution with NEA and estimated aerosol emissions in order to assess whether the model is able to reproduce pollution plumes from oil/gas production facilities. The results discussed in this section are representative of model behavior in the region close to the facilities and up to 10–30 km downwind.

Correct prediction of meteorology in WRF-Chem is the first step in studying platform emissions where simulation of wind speed and direction is of crucial importance for the correct prediction of plume position and extent. **Figure 3** compares observed and simulated time series of temperature, relative humidity, wind speed and direction along the flight tracks. WRF-Chem underestimates the observed temperature by about 0.5°C and observed relative humidity by 2% on average. Wind speed is underpredicted with a mean bias of ~1 m/s, while the error in the wind direction is on average a few degrees. The errors in modeled temperature and relative humidity may affect the prediction of chemical rates and aerosol formation, whereas discrepancies in simulating wind fields may lead to discrepancies in plume transport, including plume location, pollutant concentrations, and plume dilution, as discussed in the following sections.

Figure 4 shows maps of predicted NO_x and O_3 mixing ratios as well as particle number concentration from the CTRL run at 100 and 300 m on 20 July downwind of Heidrun compared with the aircraft measurements along the flight tracks. We show results for 20 July 2012 because the comparison using a map view on 19 July is more complex due to the sampling of several facilities at several altitudes during this flight (see **Figures 1b** and **1c**). In general, the model captures the overall distribution of the NO_x plumes at all altitudes, but has a tendency to disperse and dilute the peaks over several grid cells, leading to an underestimation of concentrations in some plumes. The absence of emissions in the NEA inventory for the Randgrid storage tanker may also explain part of this underestimation. The simulated width of the plumes is larger than observed which may also be attributed to model resolution. These differences are produced by small differences in simulated wind speeds and wind directions. For example, an error of 1 m/s in predicted wind speed leads to a transport discrepancy of about 8 km over 2 hours, which is roughly the distance found between modeled and measured plumes further downwind from the platforms. Background O_3 is overestimated by +1.5 ppbv (about +6%, the same bias is

also found on 19 July). WRF-Chem captures lower O_3 in the plumes but, in general, overestimates the observations. One explanation is that O_3 has not undergone sufficient titration due to the missing NO_x sources from the Randgrid ship. Another reason is the dilution of NO_x in the model, due to model resolution, leads to less efficient titration of O_3. The shift, spread and dilution of the plumes is also found in simulated aerosol particle number. In general, the model predicts higher aerosol numbers than observed as shown in **Figures 4e** and **4f**.

The model was also run using NO_x and NMVOC point source emissions from the TNO-MACC inventory for fixed oil/gas extraction facilities in Domain 2 (run TNO). Results are shown in the Supplementary Material (Figure S1). As noted in Section 3, TNO-MACC NO_x emissions are much lower than the NEA estimates resulting modeled NO_x that is 10–100 times lower than the observations. This, together with the fact that NMVOC emissions are higher in TNO-MACC, leads to an overestimation of O_3 in plumes by about 5 ppbv. These results indicate that model simulations using TNO-MACC emissions, albeit at high resolution (7 km), are unable to represent the composition of plumes resulting from permanent oil/gas extraction installations in the Norwegian Sea.

Figure 5 shows the observed and predicted NO_x time series extracted from the CTRL simulations along the flights on 19 and 20 July. The different facilities were identified using FLEXPART-WRF runs (not shown) (Brioude et al. 2013) and indicate a clear separation between the different plumes. WRF-Chem reproduces the observed peaks of NO_x on 19 July 2012 from Åsgard C, Heidrun and Norne although NO_x mixing ratios downwind of Åsgard C are overestimated by the model. This bias arises from the distribution of the NO_x emissions between Åsgard installations, which was based on the ACCESS observations as described in Section 3. Since, during the measurement period, Åsgard A was on a low production cycle with only one reinjection compressor running (Statoil, pers. communication), the attribution of emissions to Åsgard C could be overestimated. In contrast, the model does not show a systematic bias in reproducing the Norne peaks, which was the only facility that was constantly flaring on 19 July due to a plant trip earlier that day. This may also partly explain the overestimation of NO_x peaks downwind of the Heidrun platform on 19 July. Indeed, the NEA emissions are based on total annual amounts, and

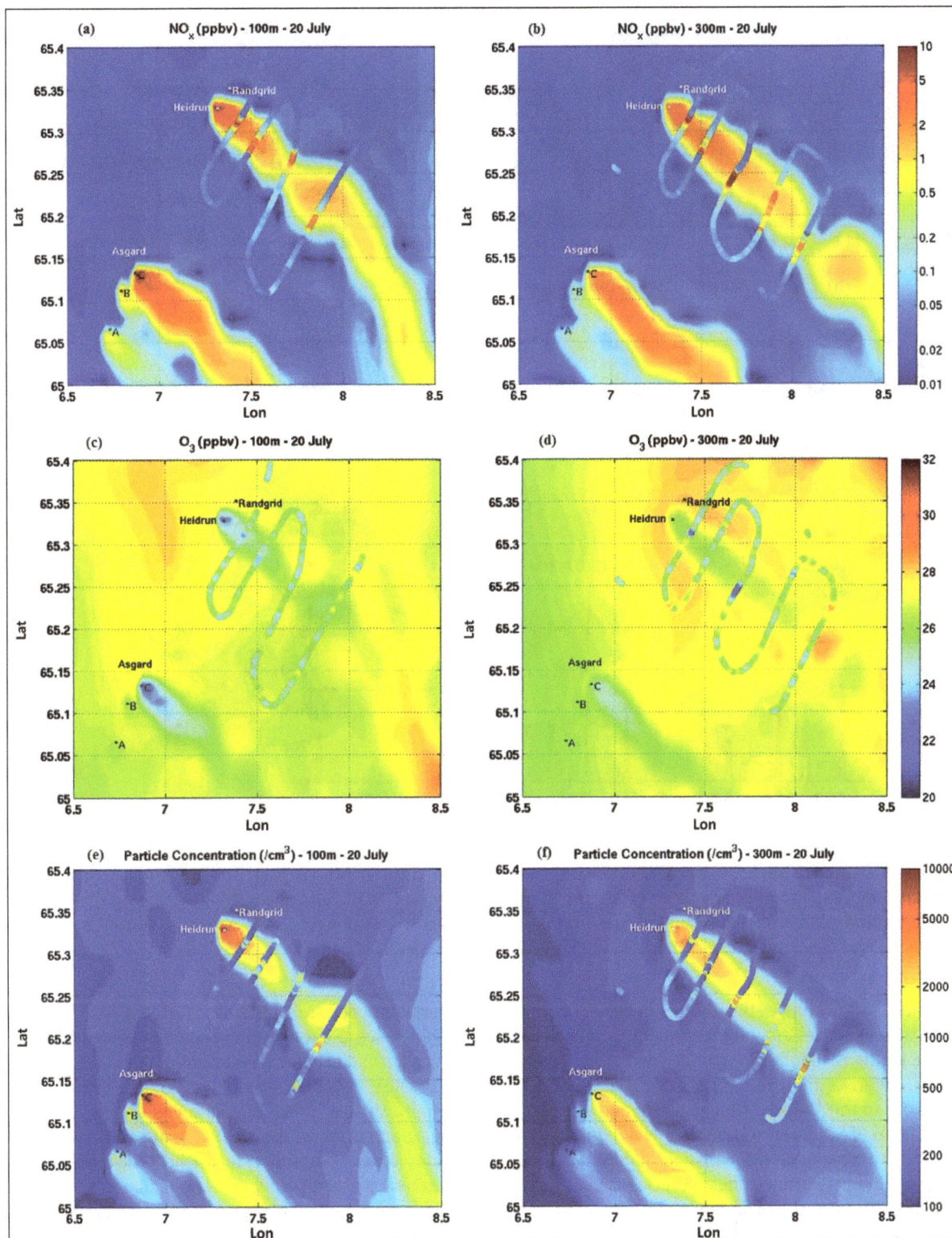

Figure 4: Comparison between observed and modeled NO$_x$ and O$_3$ on 20 July 2012. Observed (track) and simulated (CTRL run, color shading) NO$_x$ **(a, b)** and O$_3$ **(c, d)** mixing ratios (ppbv), and particle number concentrations (particle/cm^3) **(e, f)** at 100 (left) and 300 (right) m at 14 UTC on 20 July 2012. DOI: https://doi.org/10.1525/ elementa.124.f4

therefore, the emissions implemented in the model are representative of average combustion, flaring and venting activity. Other factors, such as the modeled winds, plume dispersion and lack of emissions from certain drilling ships and storage tankers also impact the model results, as discussed earlier.

Figure 6 shows the comparison between observed and modeled time series of aerosol number concentration.

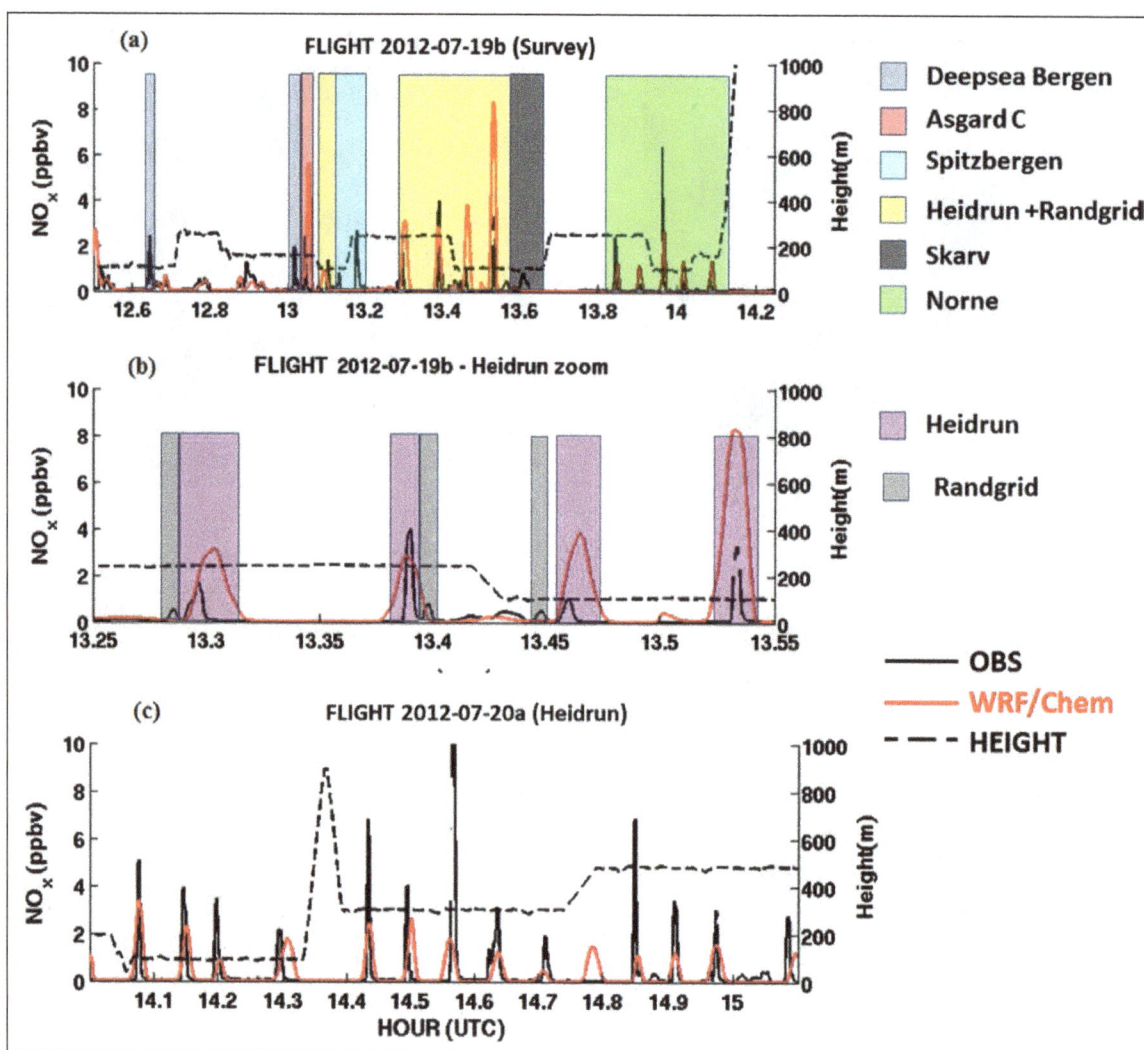

Figure 5: Comparison between observed and modeled NO$_x$ and O$_3$ on 19 July 2012. Observed (black) and simulated (CTRL run red) time series of NO$_x$ on 19 **(a)** and 20 **(c)** July 2012. The panel **(b)** is a zoom on the Heidrun and Randrid plume of the 19 July. Model simulations included emissions from the Asgard, Heidrun and Norne facilities (see text for details). DOI: https://doi.org/10.1525/elementa.124.f5

The model overestimates the number of aerosol particles in Åsgard C and Heidrun plumes on 19 July, whereas the Norne plume is simulated reasonably well. The Åsgard C bias likely arises from the attribution of all Åsgard SO$_2$ emissions to the Åsgard C storage tanker and subsequent production of SO$_4^{2-}$ aerosols. Aerosol emissions for other facilities were based on NO$_x$ emissions following Peters et al. (2011) but may include direct particle emissions of, for example, BC, as a result of flaring or combustion as well as production of SOA from NMVOC oxidation as a result of venting. Thus, Heidrun particle numbers may be overestimated because this facility was only intermittently flaring, therefore releasing fewer aerosols into the atmosphere.

The model results for aerosol particle number may also be affected by assumptions made about the emissions and representation of aerosol dynamics in the model. For example, aerosol mass emission in the model is assumed to be distributed 10% in Aitken mode and 90% in the accumulation mode. Also, the standard deviation of the

lognormal modes are constant in the aerosol model and may not be representative of the real particle size distribution. Unfortunately, these assumptions could not be verified because measurements of aerosol size distributions were not available. A quantitative analysis of the impact of these assumptions on the results discussed above would require a series of sensitivity tests where the modeled aerosol emission size distribution and standard deviation of the lognormal modes are varied, but this is not the aim of our work. A proper emission size distribution deduced on the basis of the observations could improve the simulation of aerosol particle number concentration but would not eliminate the bias because the correct emission size distribution is a small factor in the prediction of the size distribution, as shown previously by Elleman and Covert (2010). Using a constant standard deviation of lognormal modes could have an impact on the aerosol size distribution especially when the nucleation and growth of aerosol particles is strong (Makkonen et al., 2009) and can

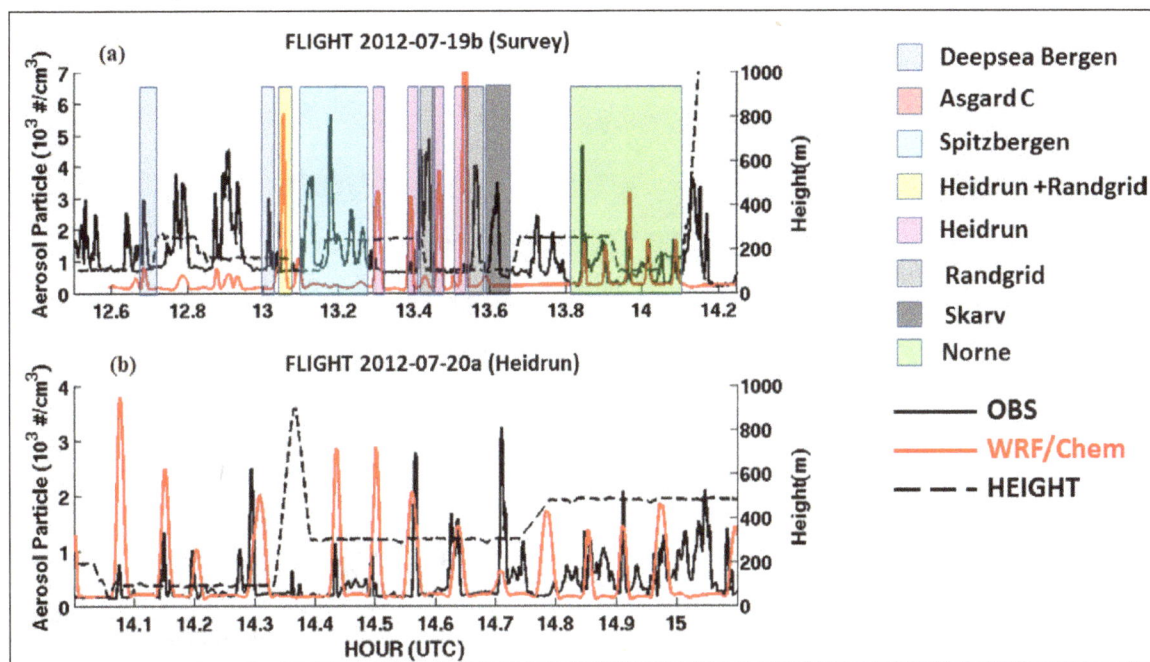

Figure 6: Comparison between observed and modeled particle concentration on 19 and 20 July. Observed (black) and simulated (CTRL run, red) time series of particle concentration on 19 (upper panel) and 20 (bottom) July 2012. Model simulations included emissions from the Asgard, Heidrun and Norne facilities (see text for details). DOI: https://doi.org/10.1525/elementa.124.f6

also affect aerosol loads. For example, Brock et al. (2016) showed that models with constant the standard deviations have larger errors in predicted aerosol optical depths.

Whilst the model results for NO_x and O_3 agree qualitatively with the observations along the flight tracks, such a quantitative evaluation of model performance using point by point comparisons between the model and observations is affected by dilution of concentrations within the model grid and differences in the location of modeled and observed peaks. To overcome this, we also compared pollutant concentrations averaged over modeled and observed plumes. More precisely, the observations were averaged over the time interval where mixing ratios were larger than background concentrations. The model results were averaged over the predicted plumes close to the observed plumes, where modeled values were larger than background mixing ratios. We only used data from 19 July because it is not possible to separate the Heidrun and Randgrid plumes during the flight on 20 July 2012. **Figure 7** shows scatter plots of NO_x and O_3 concentrations averaged over observed and predicted plumes. NO_x is simulated with a correlation of 0.70, a linear regression slope of 1.21, and a bias within a factor of 2. These results show more clearly that WRF-Chem tends to overestimate the values of NO_x close to the facilities, and underestimate NO_x farther away, especially for Heidrun.

O_3 is reproduced less well in the plumes with a correlation of 0.45 although modeled O_3 shows negligible bias for Åsgard C and Heidrun plumes (+0.3%) with respect the observations. The negative O_3 bias in the Heidrun plume is due to biases at low altitudes (less than 100 m)

at distances of less than 7 km from the platform where biases reach −10%. Moving away from the installation the model bias is less than 2%. In contrast, simulated O_3 in the Norne plumes is high biased by 1.5 ppbv (+5%). This bias increases further away from the installation ranging from +4% close to up to +8% far from platform. The differences found in the O_3 biases in the predicted plumes are related to different relative amounts of NO_x and NMVOC emitted from different processes. In addition, as discussed previously, many of these processes are intermittent whereas the emission profiles used in the simulations are representative of an average activity. Moreover, significant uncertainties arise from deriving temporal variability in the emissions from an inventory reporting annual totals which are broken down into hourly or daily emissions using generic time profiles. The sensitivity of O_3 to NO_x and NMVOC emissions is investigated in Sections 5 and 6. Model results are also affected by errors in the simulation of wind speed and direction and uncertainties in the emissions although results obtained for simulated NO_x and aerosol particle concentrations suggest that the variations due to errors in simulated wind field are much less than those arising from using the NEA based or TNO-MACC inventories.

5. Sensitivity of modeled O₃ to oil/gas platform emissions

As discussed in Section 4, simulation of modeled O_3 perturbations due to petroleum extraction emissions is dependent on accurate knowledge about these emissions which are highly variable and linked to the various activi-

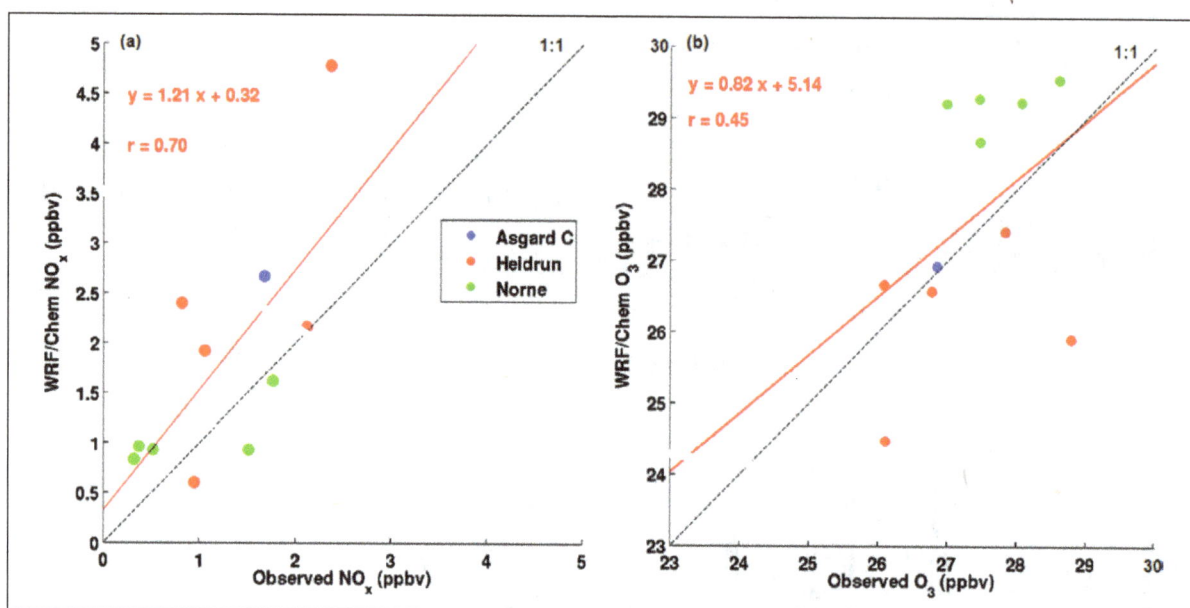

Figure 7: Scatter plot of observed and modelled NO$_x$ and O$_3$. Scatter plots of observed and modeled (CTRL run) NO$_x$ **(a)** and O$_3$ **(b)** mixing ratio in the plumes on 19 July 2012. The red lines represent the best least-square linear fit, regression line, 1:1 line, and Pearson's correlation coefficient are also shown. See text for details. DOI: https://doi.org/10.1525/elementa.124.f7

ties on the fixed installations. To investigate the impact of these uncertainties on O$_3$ close to the facilities, we examine model sensitivity to NO$_x$ and NMVOC emissions by performing a series of runs listed in **Table 3**. Examination of model sensitivity is also driven by the evaluation of model results against the ACCESS observations. In the first sensitivity test (T1) emissions of NO$_x$ and NMVOC were switched off in the CTRL run over Domain 2. For the second (T2) and third (T3) tests we increased (decreased) NO$_x$ emissions by a factor of 2. We also explored the sensitivity of model results to NMVOC emissions since there is a difference of a factor of 1.5 to 2 between NEA and TNO-MACC. In this last test (T4), NMVOC emissions were increased by a factor of 5.

The results are reported in **Table 4** which provides O$_3$ concentrations averaged over the modeled plumes as a function of distance from the facilities and altitude. Switching off the platform emissions (run T1) almost always leads to higher O$_3$ concentrations in the plumes compared to the CTRL run, especially close to Heidrun where the enhancement is 3.3 ppbv (+14%) and is the result of less titration by NO$_x$. Further downwind at 36 km from Heidrun lower emissions results in slightly less O$_3$ as O$_3$ production switches from a NO$_x$-saturated to a NO$_x$-limited regime. These results suggest that these oil/gas emissions are having a significant impact locally on Arctic boundary layer O$_3$ in the southern Norwegian Sea. This point is investigated further in the next section. Increasing NO$_x$ emissions by a factor of 2 (run T2) results in lower O$_3$ in all plumes especially few kilometers downwind the facilities due to increased O$_3$ titration. O$_3$ is lower by –6.3% (–1.7 ppbv) for the Åsgard C plume, and reaches 12% (–2.9 ppbv) and –3.5% (–1 ppbv) in the Heidrun and Norne plumes, respectively. Model results are less sensitive to decreasing NO$_x$ emissions by a factor of 2 (run T3) with,

in this case, O$_3$ enhancements that are roughly half those obtained in run T2. For example, O$_3$ increases by 6.5% (1.7 pbbv) in the plume close to the Heidrun platform. The run T4 with higher NMVOC emissions produces only limited changes in O$_3$, the largest enhancement 2.6% (0.7 ppbv) being in the Åsgard C plume. These results illustrate the non-linear behavior of O$_3$ chemistry in remote areas and suggest that modeled O$_3$ is in a NO$_x$-satured regime in the plumes up to about 20 km downwind of the platforms. Our results suggest that O$_3$ production from oil/gas extraction activities in Norwegian Sea is more sensitive to NO$_x$ emissions than NMVOC emissions especially close to the platforms. This sensitivity is generally lower moving away from the platforms (see Section 6). While it is difficult to verify whether the NEA or TNO-MACC NMVOC emissions are more likely to be correct, runs using the TNO-MACC emissions are unable to represent the pollution from oil/gas platforms in the Norwegian Sea, as noted previously.

6. Regional impact of oil/gas emissions on the Norwegian Sea

The discussion so far has been limited to evaluation of model simulations, including the sensitivity to emissions up to 20–30 km downwind of the facilities. Here, we assess local and regional impacts due to oil/gas extraction emissions on O$_3$ and aerosol distributions in southern Norwegian Sea during July 2012 over scales several 100 km. We examine model results averaged over areas of about 100 km^2 which are comparable to the size of grid cells in global models. Differences between the CTRL run and the run where all O$_3$ and aerosol precursor emissions were switched off (T1) are used to investigate the contribution of oil/gas extraction to regional pollution. Sensitivities to NO$_x$ and NMVOC emissions are also examined. Model

Table 4: Simulated average O_3 in CTRL run and sensitivity runs (see Table 3). Modeled ozone is averaged over observed plumes along the flight track of 19 July. The units are in ppbv, and the percentages indicate the variation with respect to the CTRL run. The distance from the platforms and altitude where WRF-Chem O_3 is averaged are also reported. DOI: https://doi.org/10.1525/elementa.124.t4

	Distance (km)	Altitude (m)	CTRL (ppbv)	T1 (ppbv)	T2 (ppbv)	T3 (ppbv)	T4 (ppbv)
ÅSGARD C	13	150	26.9	28.6 (+6.3%)	25.2 (−6.3%)	27.9 (+3.7%)	27.6 (+2.6%)
	7	95	24.5	27.8 (+14%)	21.6 (−12%)	26.1 (+6.5%)	24.5 (−)
	7	240	26.6	28.1 (+5.6%)	25.3 (−4.9%)	27.4 (+3.0%)	26.7 (+0.4%)
HEIDRUN	13	95	25.9	27.5 (+6.2%)	24.2 (−6.6%)	26.8 (+3.5%)	26.0 (+0.4%)
	17	240	26.9	27.6 (+3.8%)	25.3 (−4.9%)	27.2 (+2.3%)	26.6 (−)
	36	95	27.4	27.2 (−0.7%)	26.9 (−1.8%)	27.6 (+0.7%)	27.5 (+0.4%)
	1.7	95	28.7	29.6 (+3.1%)	27.7 (−3.5%)	29.1 (+1.4%)	28.7 (−)
	3	240	29.2	29.7 (+1.7%)	28.7 (−1.7%)	29.5 (+1.0%)	29.2 (−)
NORNE	7.3	150	29.2	29.7 (+1.7%)	28.5 (−2.4%)	29.5 (+1.0%)	29.3 (+0.3%)
	10	95	29.3	29.7 (+1.4%)	28.6 (−2.4%)	29.6 (+1.0%)	29.4 (+0.3%)
	10	240	29.5	29.9 (+1.4%)	29.0 (−1.7%)	29.8 (+1.0%)	29.6 (+0.3%)

1. Run with NEA emissions (see Table 2).
2. No emissions.
3. CTRL with NO_x facilities emissions doubled.
4. CTRL with NO_x facilities emissions reduced by a factor 2.
5. CTRL with NMVOC emissions increased 5 times.

diagnosed net photochemical O_3 production rates together with vertical mixing rates are also used to examine processes influencing O_3 distributions.

6.1 Ozone

Figure 8 shows the contribution of oil/gas extraction emissions to surface daytime average O_3, noontime O_3 at 600 m, and daytime average O_3 burdens in the PBL expressed as partial 1000 m columns on the 19 and 20 July. The spatially averaged enhancements downwind of the facilities in the areas A_1 and A_2 shown on **Figure 8** are given in **Tables 5** and **6**. The average impact of oil/gas emissions on O_3 is negative up to 40–50 km downwind of the platforms due to the release of large amounts of NO_x emissions from the facilities. It then becomes positive further downwind from the platforms where lower NO_x concentrations favor photochemical O_3 production. Average daytime O_3 enhancements in areas A_1 and A_2 at the surface (lowest model layer) with respect to background concentrations of 25–30 ppbv are about +2% (+0.6 ppbv) but locally reach +5–7% (+2 ppbv). On both days O_3 enhancements at noon are up to +4 pbbv (+15%) at 600 m and +2 pbbv (+8%) at surface (not shown). The change in the partial column is about +2% (about 2×10^{19} particles/m^2) or 0.07 DU. This result is comparable to the Arctic mean annual change of 0.05 DU due to petroleum extraction activities reported by

Odemark et al. (2012) associated with an average radiative forcing of +1.3 mW/m^2.

In order to better understand which chemical and dynamical mechanisms are controlling the local O_3 budget during the summertime Arctic boundary layer the terms of the O_3 continuity equation were analyzed. These include horizontal and vertical advection, net photochemical production rates, and vertical turbulent mixing. In WRF-Chem, dry deposition is included in the vertical mixing term since dry deposition velocity is the boundary condition at surface for the flux of a species. In both the CTRL and T1 runs, background O_3 is dominated by the advection terms with net photochemical production and vertical mixing rates that are 10 times smaller than the advection term. However, O_3 responses to oil/gas emissions are driven by changes in net photochemical production and turbulent mixing plus deposition and differences in the advection terms are not important and are not correlated with emission perturbations. **Figure 9** shows cross sections of the differences (CTRL-T1) in the average daytime net O_3 photochemical production and net vertical mixing terms (both in ppbv/h) extracted along lines C_1 and C_2 shown in **Figure 8**. C_1 is representative of the processes close to the platforms (few kilometers), whereas C_2 represents the behavior far, about 180 km, from the facilities where O_3 column changes are a maximum on 20 July.

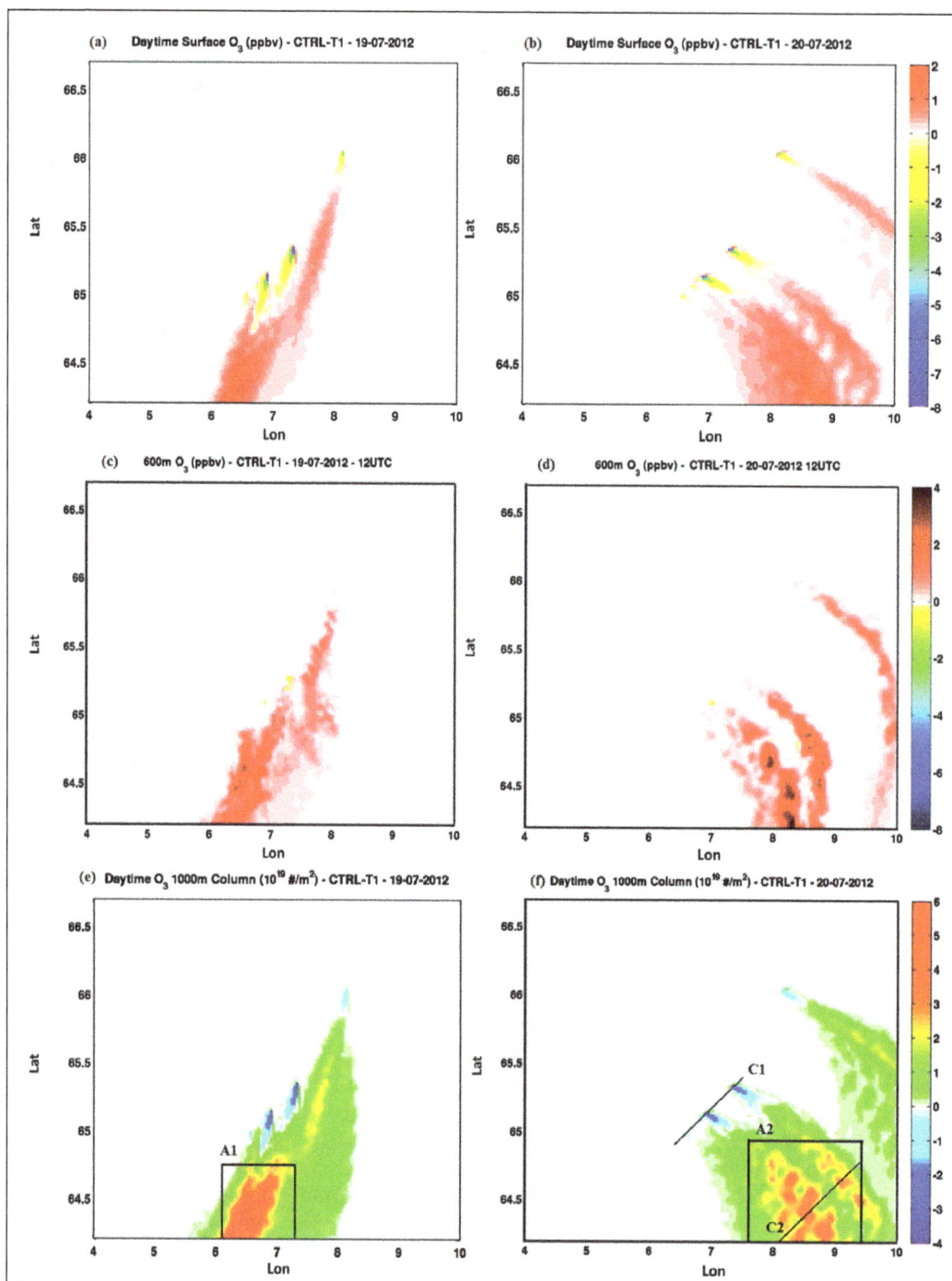

Figure 8: Contribution of oil/gas emissions to O₃ budget. Contribution of oil/gas emissions (calculated as the difference between control run (CTRL) and (T1, no emissions) to surface average day time O₃ **(a, b)**, 600 m O₃ at 12 UTC **(c, d)**, and partial 1000 m column **(e, f)** of average day time ozone on 19 (left) and 20 July 2012. A1 and A2 are the areas used to calculate the average enhancements reported in Tables 5 and 6. C1 and C2 represent the lines used to plot vertical cross sections in Figure 9. DOI: https://doi.org/10.1525/elementa.124.f8

Near the facilities, O₃ is destroyed by photochemistry from the surface up to about 300–400 m. In contrast, O₃ is produced at higher altitudes (300–700 m) but this production is small (0.1 ppbv/h) compared to the destruction rate at lower altitudes of more than 70 ppbv/h locally. In the model, O₃ destroyed near the surface is rapidly replaced by O₃ from higher altitudes by vertical mixing. In **Figure 9**, we note that the difference CTRL-T1 in the vertical

mixing rate is negative (downwards) between 200 and 800 m and positive from surface up to 150 m. Net photochemical destruction at the surface is larger than vertical mixing rate, therefore, the net impact of oil/gas emissions on O₃ is negative very close to the platforms. Downwind of the facilities a different behavior is observed. According to the model, O₃ is photochemically produced in the boundary layer reaching maximum average values of up to

Table 5: Contribution from oil/gas extraction emissions to surface O_3 and aerosols. The values are calculated as day-time average for ozone in the area A_1 and A_2 (see Figure 8), and diurnal average for aerosols in the areas A_3 and A_4 (see Figure 10). DOI: https://doi.org/10.1525/elementa.124.t5

Component	19-07-2012	20-07-2012
O_3 (ppbv)	0.6 (2.1%)	0.6 (2.4%)
$PM_{2.5}$ (ng/m³)	20 (3.7%)	10 (1.5%)
POM (ng/m³)	3.9 (8.1%)	2.0 (4.6%)
BC (ng/m³)	2.2 (48%)	1.2 (16%)
Primary $PM_{2.5}$ (ng/m³)	7.8 (11.1%)	4.2 (4.5%)
SO_4^{2-} (ng/m³)	3.9 (2%)	1.8 (0.70%)
SOA (ng/m³)	1.2 (12%)	0.56 (8.8%)

Table 6: As Table 5, but for partial 0–1000 m columns. DOI: https://doi.org/10.1525/elementa.124.t6

Component	19-07-2012	20-07-2012
O_3 (10^{19} #/m²)	2.5 (2.3%)	2.0 (1.9%)
$PM_{2.5}$ (µg/m²)	17 (3.0%)	9.4 (1.2%)
POM (µg/m²)	2.9 (4.7%)	1.6 (2.5%)
BC (µg/m²)	1.8 (35%)	0.94 (15%)
Primary $PM_{2.5}$ (µg/m²)	5.9 (6.7%)	3.3 (2.5%)
SO_4^{2-} (µg/m²)	4.6 (2.1%)	2.6 (0.75%)
SOA (µg/m²)	1.3 (6.9%)	0.67 (4.4%)

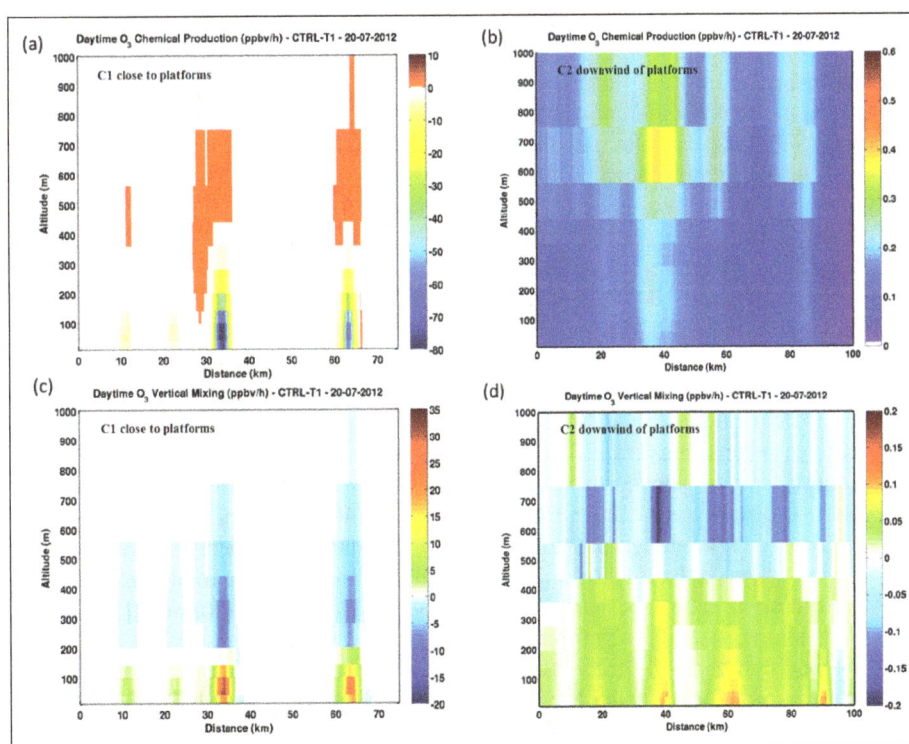

Figure 9: O_3 photochemical production and vertical mixing terms. Cross sections daytime average of the absolute differences (CTRL-T1) in net photochemical O_3 production **(a, c)** and net vertical mixing **(b, d)**. The cross sections are extracted along the lines C1 (left panels) and C2 (right panels) displayed in Figure 8. C1 is representative of the processes close to the facilities, whereas C2 represent the processes far from platforms. DOI: https://doi.org/10.1525/elementa.124.f9

+0.6 ppbv/h at 550–750 m. The difference CTRL-T1 in the vertical mixing tendencies is negative at these altitudes, whereas it is positive close to the surface suggesting that surface O_3 far from platforms is influenced by O_3 transported downwards to the surface by turbulent mixing.

Given that platform emissions are highly uncertain, the sensitivity tests discussed in the previous section can also be used to shed light on the sensitivity of modeled O_3 to NO_x and NMVOC emissions downwind from the facilities. Sensitivity tests show that regional budget of O_3 calculated in areas A_1 and A_2 is sensitive to both NO_x and NMVOC emissions. When NO_x emissions are doubled (T2), surface O_3 within A_1 (19 July) is lower by up to 2 ppbv in plumes downwind of Åsgard and Heidrun, and a weak enhancement of few tenths of ppbv is simulated in the Norne plume. O_3 decreases by up to 1 ppbv since in the run T3 there is not enough NO_x to form O_3. In the sensitivity test T4 where NMVOC emissions were increased by a factor of 5, O_3 is enhanced up to 2 ppbv but only downwind of the Åsgard complex, which are the facilities with the largest release of NMVOCs (see **Table 2**). Changes in the partial columns show a clearer pattern than results at the surface. Increasing or decreasing NO_x (runs T2 and T3) produce similar increases/decreases in O_3 burdens in A1 and A2 of 1–2% while increasing NMVOC emissions by a factor of 5 produces an increase of 2–3% in Asgard C plume. This illustrates that O_3 is less sensitive to NO_x emissions downwind of the platforms and in the case of NMVOC emissions it depends there being sufficient emissions for a particular source.

It is interesting to note differences in terms of amplitude of the response of O_3 to emission variations between our work and other studies examining the impacts of petroleum extraction emissions in other regions of the world. Studies conducted under very different conditions over land in winter in the Uintah Basin (Utah) and Upper Green River Basin (UGRB) (Wyoming) show a wide range of sensitivities in O_3 to emissions. Edwards et al. (2013) and Ahmadov et al. (2015) using box and 3-D models, respectively, showed that O_3 production in Uintah Basin region is sensitive to both NO_x and NMVOC emissions due to the fact that ozone production is at the crossover point between NO_x sensitive and VOC sensitive regimes. Differences between O_3 production in the Uintah Basin and the Norwegian Sea could be due many factors. For example, total NMVOC emissions released in Utah are 30 times larger than those over the Norwegian Sea (for a comparison see **Table 2** of Ahmadov et al. (2015)). Carter and Seinfeld (2012) examined O_3 formation in the UGRB oil/gas production region at two different sites in 2008 and 2011. They showed that, in one case, O_3 production was sensitive to NO_x emissions, whereas in another case it was sensitive to NMVOC emissions. In these studies, net O_3 production was larger than reported here over the Norwegian Sea. However, it should be noted that the studies in Uintah and UGRB basins were conducted under very different atmospheric conditions compared to our study. Significant O_3 production during wintertime episodes was characterized by cold, stagnant conditions, with very shallow boundary layers and snow cover on the ground, while our study was conducted in the near Arctic summer in the

Norwegian marine boundary layer. We may expect that emission sensitivities would be different under conditions such as the Arctic winter or spring when temperatures are much colder and boundary layers are shallower. The presence of sea-ice may also play a role since it can affect pollutant concentrations by, for example, modifying of surface albedo, photolysis rates and deposition velocities.

6.2 Particulate matter

Figure 10 displays the contribution of oil/gas emissions to the daytime average $PM_{2.5}$ mass at the surface and to burdens expressed as partial 1 km columns, calculated as the difference between the CTRL run and the run with emissions switched off (T1). The average contributions calculated far from the facilities in the areas A_3 and A_4 (see **Figure 10**) are provided in **Tables 5** and **6**. The largest amount of $PM_{2.5}$, up to 100 ng/m³, predicted by the model is located close to the platforms.

On 19 July 2012, $PM_{2.5}$ shows relatively small average enhancements (around +3%) in A_3. The average contribution to POM is about +8% at surface and about 5% in the partial 1000 m column. Black carbon is the aerosol species that changes most when oil/gas emissions are included in the simulations with changes at the surface estimated to be +2.2 ng/m³ (+48%) with increases in the column burden of +1.8 μg/m² (+35%). It should be noted that the largest contribution to BC mass is from Åsgard C and Heidrun, the facilities with higher (estimated) BC emissions (see **Table 2**). The average contribution to sulfate is +3.9 ng/m³ (+2%) at surface and +2.1 μg/m² (+2%) in the column, respectively. Predicted sulfate is dominated by Åsgard C emissions which is a condensate storage tanker releasing large amounts of SO_2. Surface and column changes of SOA are estimated to be +12% and 6.9% respectively but SOA mass is very sensitive to NMVOC emissions. Results from the sensitivity test T4 with increased NMVOC emissions produce surface SOA increases of up to 7 ng/m³ close to the facilities with respect to the CTRL run, with an average enhancement in A_3 of 1.0 ng/m³ (+7%). Nitrate and ammonium present negligible enhancements. Unspeciated $PM_{2.5}$ increases of 7.8 ng/m³ (11%) at surface and 5.9 μg/m² (7%) in the column. With regard to the vertical distribution of the aerosol (not shown), $PM_{2.5}$ mass change (CTRL-T1) is largest from surface up to about 800 m. Finally, it is interesting to assess the relative contribution of each aerosol species to total mass. The largest contribution to $PM_{2.5}$ is given by SO_4^{2-} (40%). The contribution of POC and unspeciated $PM_{2.5}$ are 10% and 5% respectively, whereas black carbon and SOA are 2% of the total mass.

Aerosol mass enhancements on 20 July, during the flight downwind of Heidrun, are 2–3 times less than those calculated on 19 July (see **Tables 5** and **6**). One factor that may explain these differences is that during 20 July the wind speed in A4 is about 1 m/s larger than in A3 with respect previous day. Moreover the wind direction switched to northwesterly in early morning, whereas it was northerly all day on 19 July. This illustrates the sensitivity of the model results to different meteorological conditions. Specifically, a change in wind speed and direction leads to aerosol mass loads that differ at least by a factor 2.

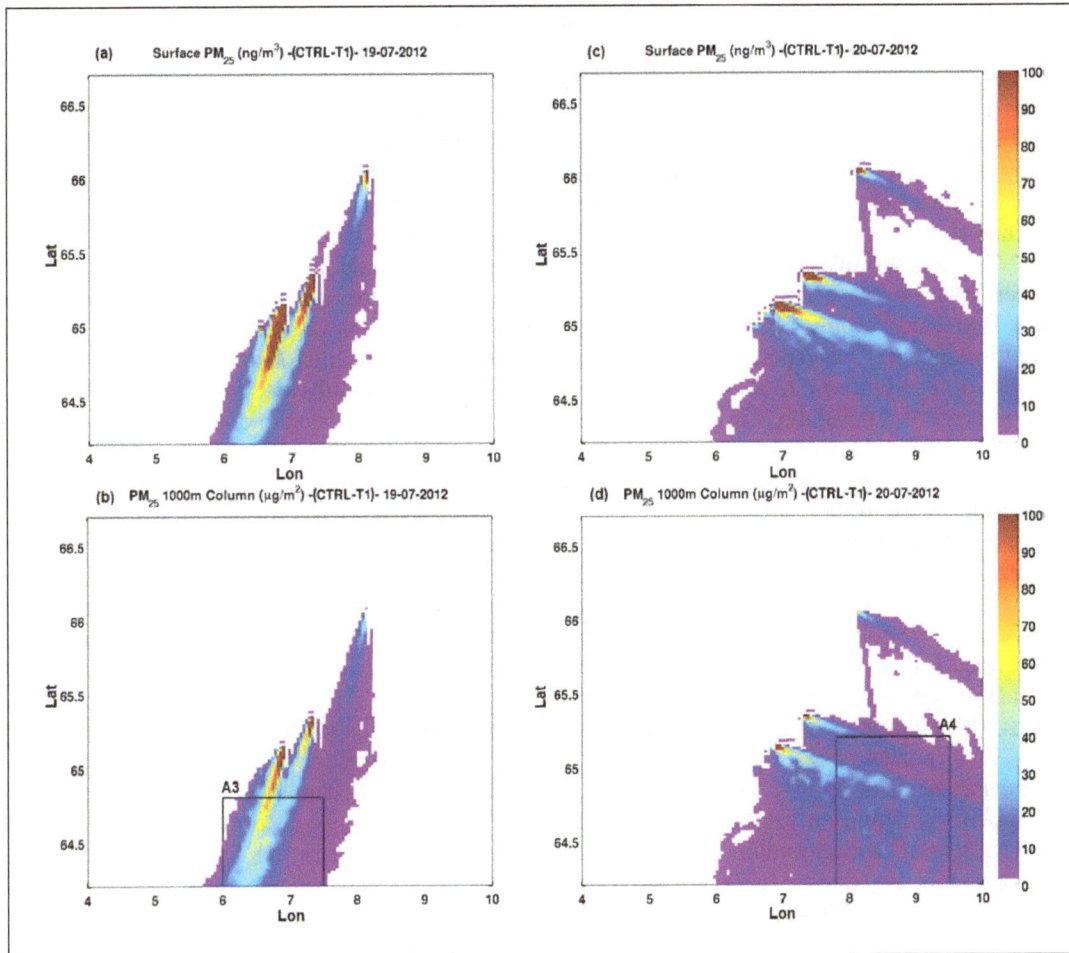

Figure 10: Contribution of oil/gas emissions to PM$_{2.5}$ budget. Contribution of oil/gas emissions (calculated as the difference between control run (CTRL) and (T1, no emissions) to surface PM$_{2.5}$ **(a, c)** and partial 1000 m column **(b, d)** of average diurnal PM$_{2.5}$ on 19 (left) and 20 July 2012. A3 and A4 are the areas used to calculate the average enhancements reported in Tables 5 and 6. DOI: https://doi.org/10.1525/elementa.124.f10

The differences in the columnar aerosol burden between the two days may be explained analyzing the day-to-day variability of wind speed and direction persistence in the two areas A3 and A4. The variability in the wind speed and direction results in more dispersed pollution plumes on the 20 July leading to reduced pollutant concentrations compared to 19 July.

7. Summary and Conclusions

In this work we have presented a study to assess the impact of oil/gas emissions on air pollution during the ACCESS aircraft campaign over the Norwegian Sea in July 2012. The campaign included sampling of pollution plumes from a range of facilities including production platforms, storage tankers, and drilling rigs. Numerical simulations were conducted with the WRF-Chem model using available point source emissions, and model results were evaluated against data collected during the campaign. The sensitivity of O$_3$ and aerosols to oil/gas emissions was investigated.

Emissions from activities related to oil and gas extraction are highly uncertain and represent a significant source of uncertainty in the assessment of local pollution impacts in the present-day and future Arctic or sub-Arctic regions. In this study, we used two emission estimates (NEA and

TNO-MACC) to examine local and regional impacts on atmospheric composition over the Norwegian Sea in July 2012. Large differences were found between the two inventories. NO$_x$ emissions included in TNO-MACC are much smaller compared to Norwegian emissions, by a factor 20–30, whereas the TNO-MACC NMVOC emissions are 1.5–2.0 times larger than the Norwegian NEA data. It is important to note that both inventories do not include aerosol emissions. The emissions of PM, BC and primary OC for NEA were estimated using emission factors for Norwegian Arctic reported by Peters et al. (2011).

Results from a control simulation run with NEA emissions were evaluated using the aircraft measurements. The model was run over a high-resolution domain covering the region of the ACCESS flights allowing assessment of model behavior up to 30 km downwind the facilities. WRF-Chem is able to capture the overall distribution of the plumes of NO$_x$, O$_3$, and aerosol particle number concentrations at several altitudes. Nevertheless, predicted plumes are diluted over several grid points, and modeled peaks are not precisely at the position observed. These differences are attributable to the model grid resolution (7 km) and to uncertainties in simulated winds. Simulations with TNO emissions do not reproduce observed NO$_x$ and O$_3$ with

NO_x being 10–100 times lower than the observations, and O_3 within plumes overestimated by 5 ppbv.

Modeled average NO_x and O_3 concentrations in plumes are reproduced with a correlation coefficient of 0.70 and 0.45, respectively. The bias in predicted O_3 is sensitive to differences in the relative amounts of NO_x and NMVOC emissions, and depends on operational procedures, such as flaring, at different facilities which can be very intermittent. Predicted NO_x and aerosol particles are overestimated in the plumes downwind of facilities that were not flaring, whereas WRF-Chem captures better enhancements downwind of installations that reported normal activity. The emissions are representative of average activity and do not take in account the variations in daily activity of the facilities. The impact of uncertainties in emissions on O_3 levels in the plumes was investigated using sensitivity tests, showing that close to the platforms O_3 is sensitive to NO_x rather than NMVOC emissions. Further downwind, results are sensitive to both NO_x and NMVOC emissions.

In order to assess the wider impact of oil/gas emissions on O_3 and aerosols we compared NEA simulation runs to runs performed without emissions. We find that both aerosols and O_3 are enhanced up to 50 km downwind of the platforms with average daytime enhancements in O_3 of up to +5–7% (+2 ppbv) locally at the surface and larger noontime enhancements of up to 4 ppbv at around 600 m. This represents a significant regional enhancement in O_3 above background concentrations of 25–30 ppbv. These enhancements are due to a switch from net photochemical destruction close to the platforms to net photochemical production accompanied by vertical mixing of O_3 downwind. In terms of aerosols, the largest changes are found in black carbon which increases 2.2 ng/m^3 (+48%) at the surface and +1.8 µg/m^2 (+35%) in the column due to oil/gas extraction emissions. Primary $PM_{2.5}$, to which $SO4^{2-}$ makes the largest contribution, is enhanced by 7.8 ng/m^3 (11%) at the surface and 5.9 µg/m^2 (7%) in the column.

Our results indicate the necessity to improve estimates of oil/gas extractions emissions, including their temporal variability and spatial location. In addition, the impact of oil/gas emissions found in this study is likely to be underestimated because emissions from mobile platforms such as storage tankers and drilling rigs are not taken in account in either the NEA or TNO-MACC inventories. At the same time more detailed shipping emissions need to be included. Another significant source of uncertainty is associated with the fact that oil/gas extraction aerosol emissions are not included in current inventories. It is also interesting to note that the Norwegian emissions from oil and gas exploration (SNAP 5) in the TNO-MACC inventory are based on the official Norwegian emissions reporting to EMEP (available at http://www.ceip.at) as reported in 2011. The NEA emissions are apparently not included in the Norwegian reporting to EMEP, at least not at the time of construction of the TNO-MACC inventory by Kuenen et al. (2014). Our analysis suggests that the TNO-MACC inventory does not fully represent the Norwegian offshore Oil and Gas air pollutant emissions. We suggest that a correction in the TNO-MACC data could be made for the

emissions from oil and gas exploration in the Norwegian Sea based on the NEA data and information but an additional effort will be needed for the PM emissions which were also lacking in NEA.

We note also that, in this study, the impact of petroleum extraction emissions on pollutant concentrations has been calculated based on a rather short time period covering the ACCESS campaign. More accurate regional assessment would require simulations on time scales of several months, which is not the focus of the present study. Impacts may also be larger at other times of year, in particular, in winter or spring when it is expected that boundary layer depths are shallower and different chemical processes may operate due less sunlight, lower temperatures, etc.. Future exploration in the Arctic may also take place in regions with sea-ice or snow which could also have an impact on the chemical and dynamical processing of these emissions.

The results obtained in this study point to the need for further work to assess emerging air quality and climate impacts due to oil/gas emissions in the Arctic and demonstrate that improvements in current emission inventories and knowledge of processes are required in order to take in account all the various types of emissions associated with oil/gas extraction and production. For this purpose further intensive campaigns including measurements of NMVOCs, aerosol size distributions and composition, are desirable in the Arctic oil/gas extraction regions, such as northern Russia, where substantial oil/gas extraction is already occurring. In this study we have focused on the local scale impacts of these emissions. In the future, more accurate and exhaustive evaluation of oil/gas emissions impacts and their radiative forcing are desirable on regional scales.

Acknowledgements

The authors are grateful to the DLR flight department for their support during campaign, and to the Statoil company for providing the information about the operational activity of the facilities.

Funding information

The work presented here was funded from the European Union under Grant Agreement n° 265863-ACCESS (http://www.access-eu.org) within the Ocean of Tomorrow call of the European Commission Seventh Framework Programme. This work benefited from access to IDRIS HPC resources (GENCI: Grand Équipement National de Calcul Intensif, allocations 2014-017141, 2015-017141 and 2016-017141) and the IPSL mesoscale computing center (CICLAD: Calcul Intensif pour le Climat, l'Atmosphère et la Dynamique). French authors also acknowledge funding from the CNRS Chantier Arctique PARCS (Pollution in the Arctic System) project.

Competing interests

The authors have no competing interests to declare.

Author contributions
- Performed model simulations: PT
- Contributed to conception and design: PT, JLT, KSL
- Contributed to acquisition of data: JLT, KSL, JCR, LM, AR, BW, HS, TO
- Contributed to analysis and interpretation of data: PT, JLT, KSL, JCR, HACDG
- Drafted and/or revised the article: PT, JLT, KSL, JCR, AR, HACDG
- Approved the submitted version for publication: PT, JLT, KSL, JCR, LM, AR, BW, HACDG, HS, TO

References

Ackermann, IJ, Hass, H, Memmesheimer, Ebel, A, Binkowski, FS and Shankar, U 1998 Modal aerosol dynamics model for Europe: Development and first allpications, *Atmos. Environ.*, **32**(17): 2981–2999. DOI: https://doi.org/10.1016/S1352-2310(98)00006-5

Ahmadov, R, McKeen, SA, Robinson, A, Bahreini, R, Middlebrook, A, de Gouw, J, Meagher, J, Hsie, E, Edgerton, E, Shaw, S and Trainer, M 2012 A volatility basis set model for summertime secondary organic aerosols over the eastern United States in 2006, *J. Geophys. Res.*, **117**(D06): 301. DOI: https://doi.org/10.1029/2011JD016831

Ahmadov, R, McKeen, S, Trainer, M, Banta, R, Brewer, A, Brown, S, Edwards, PM, de Gouw, JA, Frost, GJ, Gilman, J, Helmig, D, Johnson, B, Karion, A, Koss, A, Langford, A, Lerner, B, Olson, J, Oltmans, S, Peischl, J, Pétron, G, Pichugina, Y, Roberts, JM, Ryerson, T, Schnell, R, Senff, C, Sweeney, C, Thompson, C, Veres, PR, Warneke, C, Wild, R, Williams, EJ, Yuan, B and Zamora, R 2015 Understanding high wintertime ozone pollution events in an oil- and natural gas-producing region of the western US, *Atmos. Chem. Phys*, **15**: 411–429. DOI: https://doi.org/10.5194/acp-15-411-2015

AMAP: Assessment 2007 Oil and Gas in the Arctic: Effects and Potential effects, Arctic Monitoring and Assessment Program (AMAP), Oslo, Norway, 2010.

Brock, CA, Wagner, NL, Anderson, BE, Beyersdorf, A, Campuzano-Jost, P, Day, DA, Diskin, GS, Gordon, TD, Jimenez, JL, Lack, DA, Liao, J, Markovic, MZ, Middlebrook, AM, Perring, AE, Richardson, MS, Schwarz, JP, Welti, A, Ziemba, LD and Murphy, DM 2016 Aerosol optical properties in the southeastern United States in summer – Part 2: Sensitivity of aerosol optical depth to relative humidity and aerosol parameters, *Atmos. Chem. Phys*, **16**: 5009–5019. DOI: https://doi.org/10.5194/acp-16-5009-2016

Carter, WPL and Seinfeld, JH 2012 Winter ozone formation and VOC incremental reactivities in the Upper Green River Basin of Wyoming, *Atmos. Environ*, **50**: 255–266. DOI: https://doi.org/10.1016/j.atmosenv.2011.12.025

Chen, F and Dudhia, J 2001 Coupling an advanced land-surface/hydrology model with the Penn State/NCAR MM5 modeling system, Part I: Model description

and implementation, *Mon. Weather Rev,* **129**: 569–585. DOI: https://doi.org/10.1175/1520-0493

Denier van der Gon, HAC, Visschedijk, A, Van der Brugh, H and Dröge, R 2010 A high resolution European emission database for the year 2005, a contribution to the UBA-project PAREST: Particle Reduction Strategies, TNO report TNO-034-UT-2010-01895_RPT-ML, Utrecht.

Eckhardt, S, Hermansen, O, Grythe, H, Fiebig, M, Stebel, K, Cassiani, M, Baecklund, A and Stohl, A 2013 The influence of cruise ship emissions on air pollution in Svalbard – a harbinger of a more polluted Arctic? *Atmos. Chem. Phys*, **13**: 8401–8409. DOI: https://doi.org/10.5194/acp-13-8401-2013

Edwards, PM, Young, CJ, Aikin, K, deGouw, J, Dubé, WP, Geiger, F, Gilman, J, Helmig, D, Holloway, JS, Kercher, J, Lerner, B, Martin, R, McLaren, R, Parrish, DD, Peischl, J, Roberts, JM, Ryerson, TB, Thornton, J, Warneke, C, Williams, EJ and Brown, SS 2013 Ozone photochemistry in an oil and natural gas extraction region during winter: simulations of a snow-free season in the Uintah Basin, Utah, *Atmos. Chem. Phys*, **13**: 8955–8971. DOI: https://doi.org/10.5194/acp-13-8955-2013

Elleman, RA and Covert, DS 2010 Aerosol size distribution modeling with the Community Multiscale Air Quality modeling system in the Pacific Northwest: 3. Size distribution of particles emitted into a mesoscale model, *J. Geophys. Res*, **115**: D03204. DOI: https://doi.org/10.1029/2009JD012401

Emmons, LK, Walters, S, Hess, PG, Lamarque, J-F, Pfister, GG, Fillmore, D, Granier, C, Guenther, A, Kinnison, D, Laepple, T, Orlando, J, Tie, X, Tyndall, G, Wiedinmyer, C, Baughcum, SL and Kloster, S 2010 Description and evaluation of the Model for Ozone and Related chemical Tracers, version 4 (MOZART-4), *Geosci. Model Dev*, **3**: 43–67. DOI: https://doi.org/10.5194/gmd-3-43-2010

Erisman, JW, Vanpul, A and Wyers, P 1994 Parametrization of surfaceresistance for the quantification of atmospheric deposition of acidifying pollutants and ozone, *Atmos. Environ*, **28**(16): 2595–2607. DOI: https://doi.org/10.1016/1352-2310(94)90433-2

Flanner, MG, Zender, CS, Hess, PG, Mahowald, NM, Painter, TH, Ramanathan, V and Rasch, PJ 2009 Springtime warming and reduced snow cover from carbonaceous particles, *Atmos. Chem. Phys*, **9**: 2481–2497. DOI: https://doi.org/10.5194/acp-9-2481-2009

Flanner, MG, Zender CS, Randerson, JT and Rasch, PJ 2007 Present-day climate forcing and response from black carbon in snow, *J. Geophys. Res*, **112**(D11): 202. DOI: https://doi.org/10.1029/2006JD008003

Gautier, DL, Bird, KJ, Charpentier, RR, Grantz, A, Houseknecht, DW, Klett, TR, Moore, TE, Pitman, JK, Schenk, CJ, Schuenemeyer, JH, Sorensen, K, Tennyson, ME, Valin, ZC and Wandrey, CJ 2009 Assessment of undiscovered oil and gas in the Arctic, *Science*, **324**: 1175. DOI: https://doi.org/10.1126/science.1169467

Grell, GA and Devenyi, D 2002 A generalized approach to

parameterizing convection combining ensemble and data assimilation techniques, *Geophys. Res. Lett*, **29**(14): p. 38. DOI: https://doi.org/10.1029/2002GL015311

Grell, GA, Peckham, SE, Schmitz, R, McKeen, SA, Frost, G, Skamarock, WC and **Eder, B** 2005 Fully coupled 'online' chemistry in the WRF model. *Atmos. Environ*, **39**: 6957–6976. DOI: https://doi.org/10.1016/j.atmosenv.2005.04.027

Hansen, J and **Nazarenko, L** 2004 Soot climate forcing via snow and ice albedos, *P. Natl. Acad. Sci. USA*, **101**: 423–428. DOI: https://doi.org/10.1073/pnas.2237157100

Hirdman, D, Sodemann, H, Eckhardt, S, Burkhart, JF, Jefferson, A, Mefford, T, Quinn, PK, Sharma, S, Ström, J and **Stohl, A** 2010 Source identification of short-lived air pollutants in the Arctic using statistical analysis of measurement data and particle dispersion model output, *Atmos. Chem. Phys*, **10**: 669–693. DOI: https://doi.org/10.5194/acp-10-669-2010

Iacono, MJ, Delamere, JS, Mlawer, EJ, Shephard, MW, Clough, SA and **Collins, WD** 2008 Radiative forcing by long-lived greenhouse gases: Calculations with the AER radiative transfer models, *J. Geophys. Res*, **113**(D13): 103. DOI: https://doi.org/10.1029/2008JD009944

Jacob, DJ, Crawford, JH, Maring, H, Clarke, AD, Dibb, JE, Emmons, LK, Ferrare, RA, Hostetler, CA, Russell, PB, Singh, HB, Thompson, AM, Shaw, GE, McCauley, E, Pederson, JR and **Fisher, JA** 2010 The Arctic Research of the Composition of the Troposphere from Aircraft and Satellites (ARCTAS) mission: design, execution, and first results, *Atmos. Chem. Phys*, **10**: 5191–5212. DOI: https://doi.org/10.5194/acp-10-5191-2010

Jiao, C, Flanner, MG, Balkanski, Y, Bauer, SE, Bellouin, N, Berntsen, TK, Bian, H, Carslaw, KS, Chin, M, De Luca, N, Diehl, T, Ghan, SJ, Iversen, T, Kirkevåg, A, Koch, D, Liu, X, Mann, GW, Penner, JE, Pitari, G, Schulz, M, Seland, Ø, Skeie, RB, Steenrod, SD, Stier, P, Takemura, T, Tsigaridis, K, van Noije, T, Yun, Y and **Zhang, K** 2014 An AeroCom assessment of black carbon in Arctic snow and sea ice, *Atmos. Chem. Phys*, **14**: 2399–2417. DOI: https://doi.org/10.5194/acp-14-2399-2014

Kim, SW, Heckel, A, Frost, GJ, Richter, A, Gleason, J, Burrows, JP, McKeen, S, Hsie, EY, Granier, C and **Trainer, M** 2009 NO_2 columns in the western United States observed from space and simulated by a regional chemistry model and their implications for NO_x emissions, *J. Geophys. Res. Atmos*, **114**: D11301. DOI: https://doi.org/10.1029/2008JD011343

Klimont, Z, Hoglund, L, Heyes, Ch, Rafaj, P, Schoepp, W, Cofala, J, Borken-Kleefeld, J, Purohit, P, Kupianen, K, Winiwarter, W, Amann, M, Zhao, B, Wand, SX, Bertok, I and **Sander, R** 2016 Global scenarios of air pollution and methane: 1990–2050, in preparation.

Kuenen, JJP, Visschedijk, AJH, Jozwicka, M and

Denier van der Gon, HAC 2014 TNO-MACC_II emission inventory: a multi-year (2003–2009) consistent high-resolution European emission inventory for air quality modelling, *Atmos. Chem. Phys. Discuss*, **14**: 5837–5869. DOI: https://doi.org/10.5194/acpd-14-5837-2014

Law, KS, Stohl, A, Quinn, PK, Brock, C, Burkhart, J, Paris, J-D, Ancellet, G, Singh, HB, Roiger, A, Schlager, H, Dibb, J, Jacob, DJ, Arnold, SR, Pelon, J and **Thomas, JL** 2014 Arctic Air Pollution: New Insights From POLARCAT-I PY, *Bull. Amer. Meteor. Soc.* DOI: https://doi.org/10.1175/BAMS-D-13-00017.1

Makkonen, R, Asmi, A, Korhonen, H, Kokkola, H, Järvenoja, S, Räisänen, P, Lehtinen, KEJ, Laaksonen, A, Kerminen, V-M, Järvinen, H, Lohmann, U, Bennartz, R, Feichter, J and **Kulmala, M** 2009 Sensitivity of aerosol concentrations and cloud properties to nucleation and secondary organic distribution in ECHAM5-HAM global circulation model, *Atmos. Chem. Phys*, **9**: 1747–1766. DOI: https://doi.org/10.5194/acp-9-1747-2009

Marelle, L, Thomas, JL, Raut, J-C, Law, KS, Jalkanen, J-P, Johansson, L, Roiger, A, Schlager, H, Kim, J, Reiter, A and **Weinzierl, B** 2016 Air quality and radiative impacts of Arctic shipping emissions in the summertime in northern Norway: from the local to the regional scale, *Atmos. Chem. Phys*, **16**: 2359–2379. DOI: https://doi.org/10.5194/acp-16-2359-2016

Middleton, P, Stockwell, WR and **Carter, WP** 1990 Aggregation and analysis of volatile organic compound emissions for regional modelling, *Atmos. Environ*, **24**: 1107–1133. DOI: https://doi.org/10.1016/0960-1686(90)90077-Z

Morrison, H, Thompson, G and **Tatarskii, V** 2009 Impact of cloud microphysics on the development of trailing stratiform precipitation in a simulated squall line: comparison of one- and two-moment scheme, *Mon. Weather Rev*, **137**: 991–1007. DOI: https://doi.org/10.1175/2008MWR2556.1

Nakanishi, M and **Niino, H** 2006 An improved Mellor-Yamada Level-3 Model: its numerical stability and application to a regional prediction of advection fog, *Bound.-Lay. Meteorol*, **119**: 397–407. DOI: https://doi.org/10.1007/s10546-005-9030-8

Ødemark, K, Dalsøren, SB, Samset, BH, Berntsen, TK, Fuglestvedt, JS and **Myhre, G** 2012 Short-lived climate forcers from current shipping and petroleum activities in the Arctic, *Atmos. Chem. Phys*, **12**: 1979–1993. DOI: https://doi.org/10.5194/acp-12-1979-2012

Passant, N 2002 Speciation of UK emissions of NMVOC, AEAT/ENV/0545, AEA Technology, London.

Peters, GP, Nilssen, TB, Lindholt, L, Eide, MS, Glomsrød, S, Eide, LI and **Fuglestvedt, JS** 2011 Future emissions from shipping and petroleum activities in the Arctic, *Atmos. Chem. Phys*, **11**: 5305–5320. DOI: https://doi.org/10.5194/acp-11-5305-2011

Pétron, G, Frost, G, Miller, BR, Hirsch, AI, Montzka, AE, Karion, A, Trainer, M, Sweeney, C, Andrews, AE, Miller, L, Kofler, J, Bar-Ilan, A, Dlugokencky, EJ, Patrik, L, Moore Jr, CT, Ryerson, TB, Siso, C, Kolodzey, W, Lang, PM, Conway, T, Novelli, P, Masarie, K, Hall, B, Guenther, D, Kitzis, DF, Miller, J, Welsh, D, Wolfe, D, Neff, W and Tans, P 2012 Hydrocarbon emissions characterization in the Colorado Front Range: A pilot study, *J. Geophys. Res*, **117**. DOI: https://doi.org/10.1029/2011JD016360

Prank, M, Sofiev, M, Denier van der Gon, HAC, Kaasik, M, Ruuskanen, TM and Kukkonen, J 2010 A refinement of the emission data for Kola Peninsula based on inverse dispersion modelling. *Atmos. Chem. Phys*, **10**: 10849–10865. DOI: https://doi.org/10.5194/acp-10-10849-2010

Quinn, PK, Bates, TS, Baum, E, Doubleday, N, Fiore, AM, Flanner, M, Fridlind, A, Garrett, TJ, Koch, D, Menon, S, Shindell, D, Stohl, A and Warren, SG 2008 Short-lived pollutants in the Arctic: their climate impact and possible mitigation strategies, *Atmos. Chem. Phys*, **8**: 1723–1735. DOI: https://doi.org/10.5194/acp-8-1723-2008

Roiger, A, Thomas, J-L, Schlager, H, Law, K, Kim, J, Schafler, A, Marelle, L, Raut, J-C, Weinzierl, B, Reiter, A, Rose, M, Dahlkotter, F, Minikin, A, Krisch, I, Scheibe, M, Stock, P, Baumann, R, Bouarar, I, Clerbeaux, C, George, M, Onishi, T and Flemming, J 2015 Quantifyng emerging local anthropogenic emissions in the Arctic region: the ACCESS aircraft campaign experiment, *Bull. Amer. Meteor. Soc*. DOI: https://doi.org/10.1175/BAMS-D-13-00169.1

Schaap, M, Roemer, M, Sauter, F, Boersen, G, Timmermans, R and Builtjes, PJH 2005 LOTOS-EUROS: Documentation, TNO report B&O-A, 2005–297, Apeldoorn.

Serreze, MC and Barry, RC 2011 Processes and impacts of Arctic amplification: a research synthesis, *Global Planet. Change*, **77**: 85–96. DOI: https://doi.org/10.1016/j.gloplacha.2011.03.004

Shindell, DT, Chin, M, Dentener, F, Doherty, RM, Faluvegi, G, Fiore, AM, Hess, P, Koch, DM, MacKenzie, IA, Sanderson, MG, Schultz, MG, Schulz, M, Stevenson, DS, Teich, H, Textor, C, Wild, O, Bergmann, DJ, Bey, I, Bian, H, Cuvelier, C, Duncan, BN, Folberth, G, Horowitz, LW, Jonson, J, Kaminski, JW, Marmer, E, Park, R, Pringle, KJ, Schroeder, S, Szopa, S, Takemura, T, Zeng, G, Keating, TJ and Zuber, A 2008 A multi-model assessment of pollution transport to the Arctic, *Atmos. Chem. Phys*, **8**: 5353–5372. DOI: https://doi.org/10.5194/acp-8-5353-2008

Stephenson, S, Smith, L and Agnew, J 2011 Divergent long-term trajectories of human access to the Arctic, *Nature Climate Change*, **1**: 156–160. DOI: https://doi.org/10.1038/nclimate1120.

Stockwell, WR, Kirchner, F, Kuhn, M and Seefeld, S 1997 A new mechanism for regional atmospheric chemistry modeling, *J. Geophys. Res*, **102**(D22): 25,847–25,879. DOI: https://doi.org/10.1029/97JD00849.

Stohl, A, Klimont, Z, Eckhardt, S, Kupiainen, K, Shevchenko, VP, Kopeikin, VM and Novigatsky, AN 2013 Black carbon in the Arctic: the underestimated role of gas flaring and residential combustion emissions, *Atmos. Chem. Phys*, **13**: 8833–8855. DOI: https://doi.org/10.5194/acp-13-8833-2013.

Tuccella, P, Curci, G, Visconti, G, Bessagnet, B, Menut, L and Park, RJ 2012 Modeling of gas and aerosol with WRF-Chem over Europe: Evaluation and sensitivity study, *J. Geophys. Res*, **117**(D03): 303. DOI: https://doi.org/10.1029/2011JD016302.

Vestreng, V 2003 Review and revision: Emission data reported to CLRTAP, EMEP/MSC-W Note 1/2003, 134 pp, *Norw. Meteorol. Inst, Oslo*. [Available at http://emep.int/publ/reports/2003/mscw_note_1_2003.pdf.]

Walcek, CJ and Taylor, GR 1986 A theoretical method for computing vertical distribution of acidity and sulfate production within clouds, *J. Atmos. Sci*, **43**: 339–355. DOI: https://doi.org/10.1175/1520-0469

Wesely, ML and Hicks, BB 2000 A review of the current status of knowledge on dry deposition, *Atmos. Environ*, **34**(12–14): 2261–2281. DOI: https://doi.org/10.1016/S1352-2310(99)00467-7

Wespes, C, Emmons, L, Edwards, DP, Hannigan, J, Hurtmans, D, Saunois, M, Coheur, P-F, Clerbaux, C, Coffey, MT, Batchelor, RL, Lindenmaier, R, Strong, K, Weinheimer, AJ, Nowak, JB, Ryerson, TB, Crounse, JD and Wennberg, PO 2012 Analysis of ozone and nitric acid in spring and summer Arctic pollution using aircraft, ground-based, satellite observations and MOZART-4 model: source attribution and partitioning, *Atmos. Chem. Phys*, **12**: 237–259. DOI: https://doi.org/10.5194/acp-12-237-2012

Wild, O, Zhu, X and Prather, MJ 2000 Fast-J: Accurate simulation of in- and below cloud photolysis in tropospheric chemical models, *J. Atmos. Chem*, **37**: 245–282. DOI: https://doi.org/10.1023/A:1006415919030

Wong, DN, Barth, M, Skamarock, W, Grell, G and Worden, J A Budget of the Summertime Ozone Anomaly Above Southern United States using WRF-Chem, AGU Fall Meeting, San Francisco, CA, USA, 14–18 December 2009.

Xiao, Y, Logan, JA, Jacob, DJ, Hudman, RC, Yantosca, R and Blake, DR 2008 Global budget of ethane and regional constraints on US sources, *J. Geophys. Res*, **113**(D21): 306. DOI: https://doi.org/10.1029/2007JD009415

Yang, Q, Bitz, CM and Doherty, SJ 2014 Offsetting effects of aerosols on Arctic and global climate in the late 20th century, *Atmos. Chem. Phys*, **14**: 3969–3975. DOI: https://doi.org/10.5194/acp-14-3969-2014

2

Quantification of urban atmospheric boundary layer greenhouse gas dry mole fraction enhancements in the dormant season: Results from the Indianapolis Flux Experiment (INFLUX)

Natasha L. Miles[*], Scott J. Richardson[*], Thomas Lauvaux[*], Kenneth J. Davis[*], Nikolay V. Balashov[*], Aijun Deng[*], Jocelyn C. Turnbull[†,‡], Colm Sweeney[‡], Kevin R. Gurney[§], Risa Patarasuk[§], Igor Razlivanov[§], Maria Obiminda L. Cambaliza[‖,¶] and Paul B. Shepson[‖]

We assess the detectability of city emissions via a tower-based greenhouse gas (GHG) network, as part of the Indianapolis Flux (INFLUX) experiment. By examining afternoon-averaged results from a network of carbon dioxide (CO_2), methane (CH_4), and carbon monoxide (CO) mole fraction measurements in Indianapolis, Indiana for 2011–2013, we quantify spatial and temporal patterns in urban atmospheric GHG dry mole fractions. The platform for these measurements is twelve communications towers spread across the metropolitan region, ranging in height from 39 to 136 m above ground level, and instrumented with cavity ring-down spectrometers. Nine of the sites were deployed as of January 2013 and data from these sites are the focus of this paper. A background site, chosen such that it is on the predominantly upwind side of the city, is utilized to quantify enhancements caused by urban emissions. Afternoon averaged mole fractions are studied because this is the time of day during which the height of the boundary layer is most steady in time and the area that influences the tower measurements is likely to be largest. Additionally, atmospheric transport models have better performance in simulating the daytime convective boundary layer compared to the nighttime boundary layer. Averaged from January through April of 2013, the mean urban dormant-season enhancements range from 0.3 ppm CO_2 at the site 24 km typically downwind of the edge of the city (Site 09) to 1.4 ppm at the site at the downwind edge of the city (Site 02) to 2.9 ppm at the downtown site (Site 03). When the wind is aligned such that the sites are downwind of the urban area, the enhancements are increased, to 1.6 ppm at Site 09, and 3.3 ppm at Site 02. Differences in sampling height affect the reported urban enhancement by up to 50%, but the overall spatial pattern remains similar. The time interval over which the afternoon data are averaged alters the calculated urban enhancement by an average of 0.4 ppm. The CO_2 observations are compared to CO_2 mole fractions simulated using a mesoscale atmospheric model and an emissions inventory for Indianapolis. The observed and modeled CO_2 enhancements are highly correlated ($r^2 = 0.94$), but the modeled enhancements prior to inversion average 53% of those measured at the towers. Following the inversion, the enhancements follow the observations closely, as expected. The CH_4 urban enhancement ranges from 5 ppb at the site 10 km predominantly downwind of the city (Site 13) to 21 ppb at the site near the landfill (Site 10), and for CO ranges from 6 ppb at the site 24 km downwind of the edge of the city (Site 09) to 29 ppb at the downtown site (Site 03). Overall, these observations show that a dense network of urban GHG measurements yield a detectable urban signal, well-suited as input to an urban inversion system given appropriate attention to sampling time, sampling altitude and quantification of background conditions.

Keywords: urban; greenhouse gas; carbon dioxide; methane; tower; in-situ

[*] Department of Meteorology, The Pennsylvania State University, University Park, Pennsylvania, US

[†] National Isotope Centre, GNS Science, Lower Hutt, NZ

[‡] National Oceanic and Atmospheric Administration/University of Colorado, Boulder, Colorado, US

[§] Arizona State University, Tempe, Arizona, US

[‖] Purdue University, West Lafayette, Indiana, US

[¶] Ateneo de Manila University, Katipunan Ave, Quezon City, Metro Manila, Philippines 1108, PH

Corresponding author: Natasha L. Miles (nmiles@psu.edu)

1 Introduction

Atmospheric greenhouse gas (GHG) mole fractions continue to rise rapidly (currently at about 2.5 ppm/year), primarily in response to anthropogenic emissions from fossil fuel consumption (IPCC 2014). Of these anthropogenic emissions, about 70% originate from urban areas (IEA 2008). Climate change mitigation will require reductions of GHG emissions, thus the ability to quantify urban GHG emissions is essential for assessing the effectiveness of mitigation efforts.

Quantification of anthropogenic GHG emissions is traditionally accomplished via "bottom-up" accounting or inventory methods (e.g. Marland et al., 1985; Andres et al., 1999). Interest in evaluating emissions at regional scales has motivated the development of spatially-distributed (CDIAC, Andres et al., 1999) and more temporally-resolved CO_2 emissions products (Vulcan, Gurney et al., 2009; Hestia, Gurney et al., 2012). A broadly utilized air quality emissions product, the Emission Database for Global Atmospheric Research, (EDGAR, European Commission JRC/PBL, 2013) provides a global assessment of spatially-resolved CH_4 (as well as CO_2 and other GHG) emissions. Most recently, a number of global products have used night lights and other remote sensing techniques to develop spatially-distributed emissions estimates (Oda and Maksyutov 2011; Rayner et al., 2010), sometimes including an uncertainty assessment (Asefi-Najafabady et al., 2014). All of these products use the same large-scale data utilized in national inventory products, but take a variety of approaches to distribute these emissions in space and time.

Inventory approaches are rich in information about sectoral emissions and spatial distribution, but challenging to assemble and maintain over time, and vulnerable to systematic errors (Marland and Boden, 1993; Turnbull et al., 2015). For the purposes of evaluating the effectiveness of voluntary or enforced mitigation efforts to reduce GHG emissions, independent assessment of anthropogenic GHG emissions is critical (Pacala et al., 2010; Nisbet and Weiss, 2010; Ciais et al, 2010; Durant et al., 2011). It is not yet clear what degree (resolution, precision, accuracy) of independent verification will be required. Regulations, however, are likely to be applied by sector (e.g., manufacturing sources, power generation sources, mobile sources), and thus highly resolved, accurate and precise emissions estimates from urban areas would be ideal for evaluation of emissions inventories and mitigation progress.

Atmospheric methods can potentially provide an independent assessment of emissions for cities. Depending on the objectives (trend detection, interannual variability, whole-city emissions, spatially resolved fluxes), different approaches are more or less suitable. Total emissions from an urban area have been obtained via an aircraft-based mass balance approach, comparing background and downwind mole fractions (Mays et al., 2009; Cambaliza et al., 2014; Cambaliza et al., 2015). The temporal coverage is, however, limited with aircraft, and downwind measurements alone provide little information about spatial patterns of fluxes within an urban region. Sensitivity analyses showed that the aircraft-based estimates of CO_2 and CH_4 emissions are most dependent upon determination of the appropriate background mole fraction (Cambaliza et al., 2015).

A number of experiments have been initiated in an attempt to demonstrate quantification of urban GHG emissions using tower- or building-based atmospheric approaches. McKain et al. (2012) compared simulated CO_2 mole fractions to five observational sites located in and around Salt Lake City, Utah. They argued that the similarity between observed and simulated CO_2 suggested that urban inversions are possible. The temporal duration of the study was limited to four 3–5 week time periods in 2006, and the primary focus was on the amplitude of the diurnal cycle in the region. McKain et al. (2014) solved for methane (CH_4) emissions from the city of Boston using a network of five tower and building-based observations. This study optimized whole-city emissions and did not evaluate spatial structure of emissions within the city. These studies have shown promise in quantifying whole city emissions but have not yet demonstrated the ability to resolve emissions in space, and have had limited ability to explore the sensitivity of their findings to the layout of their observational networks. Extensive networks of greenhouse gas measurements have also been implemented in Paris, France (Bréon et al., 2015), and Los Angeles, California (Verhulst et al., 2017).

Lauvaux et al. (2013) implemented a simplified version of an inversion approach to determine, in real-time, changes in the CO_2 emissions for the city of Davos, Switzerland, using an atmospheric transport model and two CO_2 measurement sites. This simple approach provided information about temporal changes in GHG emissions, but did not quantify total emissions. Further, a single measurement site is sensitive to changes in the spatial distribution within an urban region that is not representative of the whole-city emissions.

Finally, spatially- and temporally-resolved GHG emissions can be quantified with frequent, spatially-distributed measurements of GHG mole fractions merged with an atmospheric transport model and a method of solving for those fluxes most consistent with the measured and modeled GHG mole fractions. This atmospheric inversion approach has been used successfully to determine spatially- and temporally-resolved emissions consistent with agricultural inventory results in the U.S. Upper Midwest (Schuh et al., 2013; Lauvaux et al., 2012), and has been applied to Indianapolis (Lauvaux et al., 2016).

The first step towards implementing the atmospheric inversion approach at high resolution is assessing the detectability of the city emission flux via the tower-based GHG network, and documenting the spatial and temporal patterns. Here we present results from a dense network of highly-calibrated GHG sensors deployed in an urban area, as part of the Indianapolis Flux (INFLUX) experiment. This study provides a description of multi-species variability of atmospheric GHGs along with the high spatial and temporal resolution we expect to be needed to fully characterize and quantify urban emissions across space, time, and economic sectors in a large metropolitan area. We assess the detectability of city emissions via the tower-based GHG network, and quantify the spatial and temporal patterns in atmospheric GHG mole fractions associated with the urban emissions. We further compare the observed CO_2 mole fraction enhancements across the city to those predicted by a numerical modeling system that includes an inventory-based emissions estimate and an atmospheric transport model. This comparison tests the degree to which the observed CO_2 enhancements are similar to those expected from prior knowledge of emissions and atmospheric transport. Finally, we examine the sensitivity of the CO_2 results to variability in sampling height and time.

2 Methods

2.1 Study site

The study site is Indianapolis, Indiana, a medium-sized city in the Midwestern U.S. The population of Marion county, encompassing the majority of the urban area, for 2013 is 928,000 (U.S. Census Bureau; http://quickfacts.census.gov). According to the Vulcan national carbon dioxide emissions inventory (Gurney et al., 2009), the fossil fuel CO_2 emissions of Marion county are 4.3 MtC for 2013 (2014 release). Indianapolis is relatively isolated from other metropolitan areas, and agriculture is the predominant land cover type surrounding the city, except to the south, which is considerably forested (**Figure 1**). The terrain is relatively flat. The Hestia bottom-up fossil fuel CO_2 high-resolution inventory product (Gurney et al., 2012) is available for Indianapolis (Marion County) and the eight surrounding counties, providing a spatially and temporally resolved prior for top-down methods in order to evaluate and improve uncertainties in both inventories and inversions, a primary goal of INFLUX.

A map of INFLUX ground-based measurement sites and the city of Indianapolis is shown in **Figure 2**. The location, deployment date and measurements of the INFLUX sites are listed in **Table 1**, as are the known nearby sources of CO_2, CH_4, and CO. The predominant wind direction during the dormant season is from the southwest, although it varies considerably (**Figure 3**). The Harding Street Power Plant, contributing 28% of the CO_2 emissions of the city in 2002 (Gurney et al., 2012), is located in the southwest sector of the city. Between 2011 – 2013, the average monthly net electricity generation of the Harding Street Power Plant is 325,100 MWH (U.S. Energy Information Administration, 2016). During the study period, its primary fuel source is coal, but as of March 2016, its conversion to a natural gas facility was complete. There are several smaller power plants in the area as well: The Noblesville Station Power Plant (45,800 MWH average monthly generation; U.S. Energy Information Administration, 2016), 6 km to the north of Site 08, operates on steam generated from the hot exhaust of three combustion turbines fueled by natural gas. The C.C. Perry Power Plant (800 MWH mean monthly generation) is 2 km to the south of Site 03, and is primarily coal-fired during the study period but switched to natural gas in May 2014. The Eagle Valley Power Plant (109,400 MWH mean monthly generation for the period January – August 2011; 33,200 MWH mean monthly generation for September 2011 – December 2013; U.S. Energy Information Administration, 2016) is located 10 km to the south of Site 01 and is coal-fired with plans to be converted to a natural gas facility. Landfills and wastewater treatment plants are also indicated on the map (**Figure 2**). The South Side Landfill is located 6 km to the west of Site 10, and contributes 37% of the CH_4 emissions of the city (Cambaliza et al., 2015). The in-situ measurement sites were chosen such that Site 01 is the background site and Site 02 on the downwind edge of the city when the wind is from the predominant southwesterly direction. Site 09 is further downwind of the urban area, but depending on the wind direction, is another potential background site.

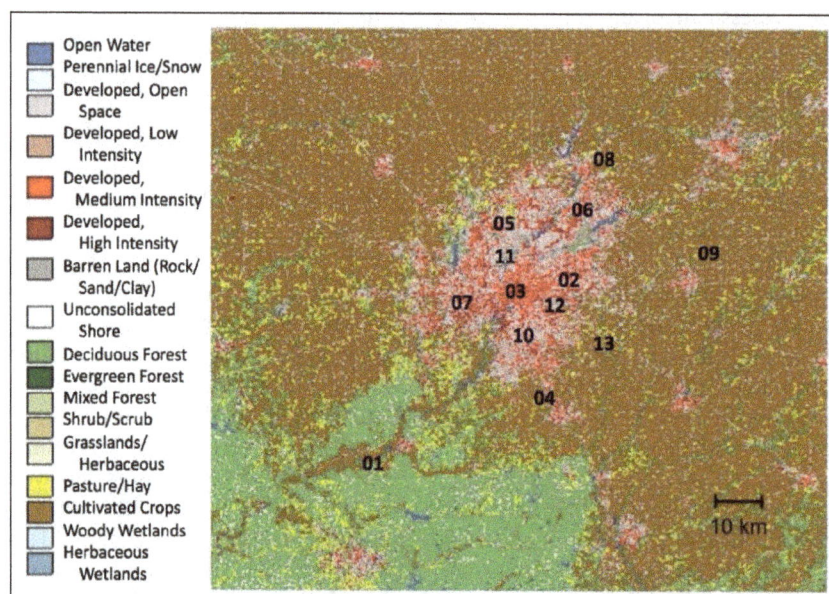

Figure 1: Land cover types for Indianapolis and the surrounding area. (National Land Cover Database 2011; Jin et al., 2013). The numbers 01–13 indicate tower site locations as listed in Table 1. DOI: https://doi.org/10.1525/elementa.127.f1

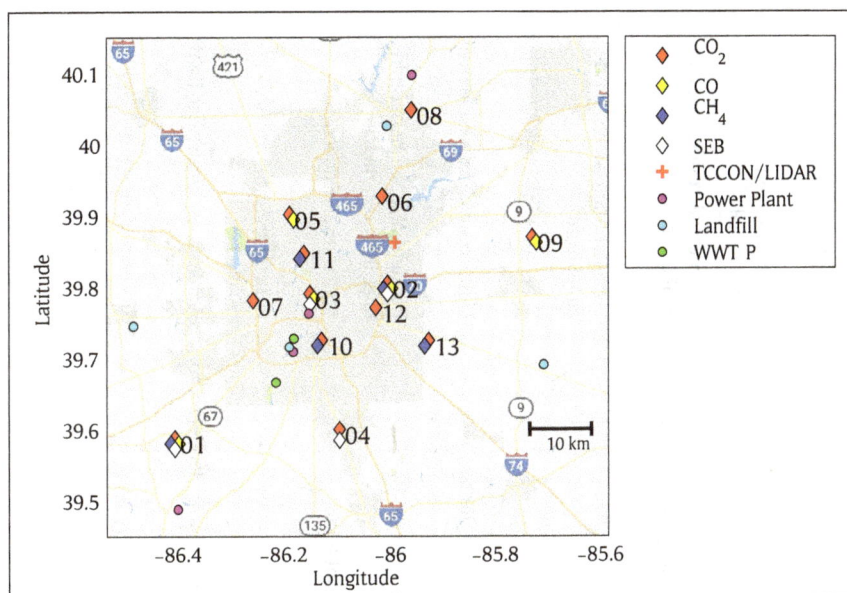

Figure 2: Map of the Indianapolis, IN, region, with INFLUX sites (as of January 2013) shown. The numbers 01–13 indicate tower site locations as listed in Table 1. The color of the marker represents the measurements at each site: red for CO_2, yellow for CO + NOAA flasks, blue for CH_4, and white for surface energy balance (SEB) fluxes. A NASA (Total Carbon Observing Network Fourier Transform Infrared) TCCON FTIR spectrometer was deployed from August – December 2012 at the site denoted by a red cross; a scanning Doppler lidar was deployed at the same site beginning in April 2013. Power plants, landfills, and wastewater treatment plants (WWTP) are indicated as well. The largest power plant is Harding Street and the largest landfill is South Side; both of these are located to the west of Site 10. Background road map: Google maps, www.google.com/maps. DOI: https://doi.org/10.1525/elementa.127.f2

Site 10 is closest to the primary power plant and landfill for the city. Site 03 is located near the downtown area, about 2 km from the center, and adjacent to a junction of two major interstate highways. Site 04 and Site 08, in particular, are 20 – 30 km from downtown, but in suburban/commercial areas of Indianapolis, and have light to medium urban development. The remaining sites are distributed around the city. Site 12 was deployed for only six months; the instrument was then relocated to a different site. The distance between each of the site-pairs (**Figure 2**) varies from 4 km (Site 02 and Site 12) to 66 km (Site 01 and Site 09).

2.2 Instrumentation

The INFLUX in-situ observation network includes twelve sites measuring CO_2 dry mole fractions. A subset of five sites additionally measure CO dry mole fraction, and a different subset of five sites additionally measure CH_4 dry mole fraction. In November 2014, four sites were upgraded from CO_2 only to CO_2 and CH_4 measurements. Measurements at two sites began in September 2010, seven sites were operational by August 2012, and nine of the sites were deployed as of January 2013 and data from these sites are the focus of this paper. The full network of twelve sites was deployed by July 2013. CO_2, CH_4, and CO dry mole fractions are measured with wavelength-scanned cavity ring down spectroscopic (CRDS) instruments (Picarro, Inc., models G2301, G2302, G2401, and G1301).

The instruments are deployed at the base of existing communications towers, with sampling tubes installed as high as possible on each tower (**Table 1**). Five of the tower measurement heights are greater than 100 m AGL, four are about 40 m AGL, and the remainder of the tower measurement heights are between 54 and 87 m AGL. Except for Site 03, the mean building height within the 1-km² area surrounding each of the towers is less than 6 m AGL and the measurements are thus expected to be above the roughness sublayer (typically 2 – 5 times the building height) most of the time. Site 03 is the closest of the INFLUX towers to the urban center, but it is about 2 km north of downtown. The tallest building in downtown Indianapolis is the Salesforce tower which is 247 m AGL and the remainder of the 20 tallest buildings are 79 – 162 m AGL (https://www.emporis.com/statistics/tallest-buildings/city/101039/indianapolis-in-usa). The buildings over 1 – 2 stories tall within a 300 m radius of the Site 03 tower are three Indiana University buildings about 70 – 150 m to the southwest which are 25 – 29 m tall and the Stutz Business Center 250 m to the southeast which is 21 m tall. The measurements may thus be within the roughness sublayer when the wind is from the southeast or southwest. The predominant landcover in the 1 km² area surrounding each tower is listed in **Table 1**. Sites 01, 05, 08, 09, and 11 have wooded landcover in the surrounding 1 km² area. Of these towers, Site 01, 05, 09, and 11 are all greater than 120 m AGL. Site 08, with about 10% wooded landcover, is 41 m AGL, and thus may at times be within the roughness sublayer. Sites 01, 02, and 03 also include measurements at 10 m AGL and one or two intermediate levels. Tubing for levels not being sampled is continuously purged in order to eliminate long residence times for the air in the tubing. The samples at all sites

Table 1: Details of INFLUX in-situ tower sites. Measurements are listed for the period of focus for this paper (2013). The CO_2 only sites were upgraded to measure both CO_2 and CH_4 in November 2014. DOI: https://doi.org/10.1525/elementa.127.t1

Site	Measurements	Installation date	Lat (deg N)	Long (deg W)	Sample height(s) (m AGL)	Known nearby sources	Predominant land-cover (in 1 km² surrounding tower)
*Site 01 – Background	CO_2/CO/CH_4/Flasks	9/2010	39.5805	86.4207	10/40/121	Power plant 10 km to the S	Wooded, sparse residential, agriculture
*Site 02 – Downwind edge	CO_2/CO/CH_4/Flasks	9/2010	39.7978	86.0183	10/40/136	I-70 200 m to the N	Residential, light commercial
*Site 03 – Downtown	CO_2/CO/Flasks	6/2012	39.7833	86.1651	10/20/40/54	I65–S 10 m to the SW; I65–N 40 m to the NE; power plant 2 km to the S	Commercial, residential, 2 km N of city center
*Site 04 – South Side	CO_2	8/2012	39.5927	86.0991	60		Light commercial, residential, agriculture
*Site 05 – NorthWest corner	CO_2/CO/Flasks	3/2012	39.8949	86.2028	125		Wooded, residential, apartment buildings
Site 06 – NorthEast corner	CO_2	7/2013	39.9201	86.0280	39	I-69 500 m to the NW; I-465 4 km to the SW	Light commercial, residential
*Site 07 – West Side	CO_2	3/2012	39.7739	86.2724	58	I465 200 m to the W	Residential, apartment buildings, light commercial
Site 08 – NorthEast	CO_2/CO/CH_4	5/2013	40.0411	85.9734	41	Power plant 6 km to the N	Agricultural, wooded
*Site 09 – Downwind/Background	CO_2/CO/Flasks	3/2012	39.8627N	85.7448W	10/40/70/130		Agricultural, golf course, wooded
*Site 10 – Downtown – South	CO_2/CH_4	3/2012	39.7181N	86.1436W	40	Landfill and power plant 6 km to the W	Warehouses, residential, light commercial
*Site 11 – Downtown – North	CO_2/CH_4	4/2013	39.8403N	86.1763W	130		Residential, university buildings, wooded, athletic fields, pond
*Site 12 – decommissioned 4/2013	CO_2	8/2012	39.7637N	86.0403W	40	4-lane roads 400m to the W and 1 km to the N; I465 800 m to the E	Residential, light commercial
*Site 13 – South East	CO_2/CH_4	4/2013	39.7173N	85.9417W	87		Agriculture, residential

*Results from these towers are presented in this manuscript.

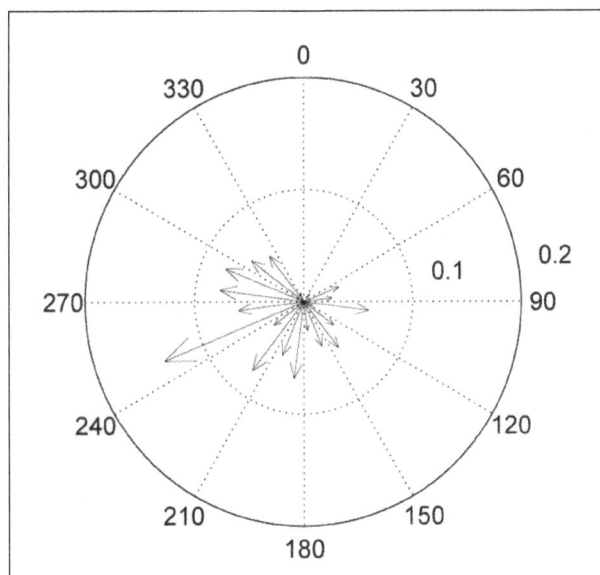

Figure 3: Probability distribution function of afternoon-averaged near-surface wind direction for 1 January − 30 April 2013. Wind direction is measured at the Indianapolis International airport (http://cdo.ncdc. noaa.gov/qclcd/QCLCD). Arrows point in the direction of origin of the afternoon-averaged mean winds. The radial length of the arrow denotes the fractional probability of wind from that direction. Wind directions are not reported for periods in which the wind speed is less than 1.6 ms⁻¹. DOI: https://doi.org/10.1525/elementa.127.f3

measuring CO have been dried since installation, to water vapor levels less than 0.6% at Site 02 and less than 0.2% at the other sites. As of late May 2013, the incoming sample air at all INFLUX sites is dried. Details of the air sampling systems at the INFLUX tower sites are described in Richardson et al. (2016).

Flow rates are approximately 240 cc min⁻¹ for the G2301, G2302, and G2401 instruments and approximately 140 cc min⁻¹ for the G1301 instruments. The measurement times are adjusted to reflect the residence time in the tubing (3–9 min for the top levels). For sites measuring at multiple heights, the 10-m and intermediate levels are each measured for 10 min of each hour, and the top level is sampled for the remainder of the hour. Four minutes of data are ignored after each transition between measurement levels and to/from field calibration gases, in order to flush the sample line.

The inter-laboratory compatibility goals set by the Global Atmosphere Watch program of the World Meteorological Organization are ±0.1 ppm CO_2 in the Northern Hemisphere and ±0.05 ppm CO_2 in the Southern Hemisphere, ±2 ppb CH_4, and ±2 ppb CO in background conditions and ±5 ppb CO in urban environments (GAW Report No. 229; 2016). Here we use the term compatibility, as advised in the GAW Report No. 229 (2016), to describe the difference between two measurements, rather than the absolute accuracy of those measurements. The specific compatibility requirements for urban environments, based on this study, are discussed in Section 4. The calibration protocol for the INFLUX sites is described in Richardson et al. (2016). Prior to deployment and following any manufacturer repairs, the instruments are calibrated for slope and offset in the laboratory using 3 to 5 NOAA-calibrated tanks, and at each site, one or two

NOAA-calibrated tanks are sampled daily for 10 min as field offset calibration points.

Six sites include co-located flask measurements (Turnbull et al., 2012) taken in the afternoon (1400–1600 LST), with comparisons yielding mean differences of 0.18 ± 0.55 ppm CO_2, 0.6 ± 5.0 ppb CH_4, and −6 ± 4 ppb CO for the period May 2011 − June 2016 (Richardson et al., 2016). Additionally, round robin tests with three NOAA-calibrated tanks were performed at all INFLUX tower sites, yielding network averaged errors of −0.09 ± 0.11 ppm CO_2, 0.2 ± 0.4 ppb CH_4 and 0 ± 2 ppb CO in the November 2013 tests (Richardson et al., 2016). Taking the magnitude of the largest of these results as the uncertainty bound, the compatibility of the values reported in this paper are 0.18 ppm CO_2, 0.6 ppb for CH_4, and 6 ppb for CO.

2.3 Numerical modeling system: Hestia and WRF-FDDA-LPDM

Hestia (Zhou and Gurney, 2010; Gurney et al., 2012), a building-level resolution inventory product for the Indianapolis area, is used as an estimate of anthropogenic CO_2 emissions. Hestia combines several datasets such as energy consumption, traffic data, industrial productivity, and electricity generation from the power plant, with models such as a building energy model. The Hestia product covers Marion county and the other eight surrounding counties and includes diurnal and seasonal variability to compute hourly emissions for any day of the year for a variety of economic sectors at the building/street scale. The CO_2 emissions are available for eight different sectors of economic activity: airport, commercial, industrial, mobility (on-road vehicles), nonroad (vehicles), residential, utility, and railroad. The 2014 release of Hestia describing emissions from 2013 is used in this paper.

The Weather Research Forecasting model (WRF version 3.5.1) modeling system uses a Four-Dimensional Data Assimilation (FDDA) technique, originally developed and tested for the Fifth-Generation Penn State/NCAR Mesoscale Model (Stauffer and Seaman 1994, Deng et al. 2004) and implemented into WRF (Deng et al. 2009) assimilating the meteorological measurements from WMO surface stations as well as vertical profiles from radiosondes. The WRF-FDDA system has been used to produce optimal dynamic analyses for air quality applications (Rogers et al. 2013), and used over the city of Davos, Switzerland, in a project to quantify urban emissions of CO_2 (Lauvaux et al., 2013). It has also been used for an aircraft-based estimate of total methane emissions from the Barnett Shale region (Karion et al. 2015). The simulation domain for the current study encompasses Indianapolis and the surrounding area in a nested mode at 9km, 3km, and 1km resolutions, with the domains covering 900 × 900 km, 297 × 297 km, and 87 × 87 km, respectively. The atmospheric boundary layer scheme used is the Mellor-Yamada-Nakanishi-Niino (MYNN) 2.5 scheme (Nakanishi and Niino, 2004) coupled to the simple urban scheme within the Noah land surface model (Chen and Dudhia, 2001). The atmospheric vertical column was described by 60 levels, with 40 levels in the lower 2 km, the first level being at about 6 m above ground. We use 3-hourly North America Regional Reanalysis (NARR) analyses at 40 × 40-km resolution for the initial conditions and lateral boundary conditions for all WRF simulation. The NARR analyses were downloaded from the Research Data Archive maintained by the Computational and Information Systems Laboratory at the National Center for Atmospheric Research. The influence functions, representing the relationship between mole fractions at the tower locations and their related flux footprints at the surface, were simulated at 1-km resolution over the inner model domain with the Lagrangian Particle Dispersion Model (LDPM) (Uliasz, 1994; Lauvaux et al., 2012). 6300 particles are released incrementally at equal intervals over one-hour periods at the inlet heights at each of the towers. Inputs to the LPDM include mean winds (u,v,w), potential temperature, and turbulent kinetic energy from WRF-FDDA-CO_2 system. Multiplying the influence functions for afternoon hours (1700 – 2100 UTC) during the period 1 January – 30 April 2013 by the total emissions from Hestia (using the afternoon average for each day), we obtain expected mean CO_2 dry mole fraction at the towers for the period.

2.4 Wind measurements

The wind data used in this study are measured at the Indianapolis International Airport (KIND), outside the southwest corner of the city. The data are part of the Integrated Surface Dataset (ISD) (https://www.ncdc.noaa.gov/isd). The weather station at the airport uses the Automated Surface Observing System (ASOS). The complete description of ASOS type stations is available at http://www.nws.noaa.gov/asos/pdfs/aum-toc.pdf. The accuracy of wind speed is ±1.0 ms^{-1} or 5% (whichever is greater) and the accuracy of wind direction is 5 degrees when wind speed is ≥ 2.6 ms^{-1}. Wind directions

are not reported for periods in which the wind speed is less than 1.6 ms^{-1}. The height of the wind instrument is about 10 m AGL. The wind data reported in ISD are the wind data at a single point in time recorded within the last 10 minutes of an hour.

2.5 Analyses

Prior to determining the enhancement in urban CO_2, CO and CH_4 dry mole fractions, we first identify well-mixed, steady-state atmospheric conditions. Well-mixed conditions are more tractable for interpretation and for comparison to mesoscale atmospheric model simulations. Furthermore, the rapid morning growth of the convective ABL causes rapid changes in mole fraction caused by entrainment, potentially masking spatial differences caused by surface fluxes. Well-mixed daytime conditions also alleviate sensitivity to nearby point sources. Here we use the term "steady state" to describe conditions under which the boundary layer depth and greenhouse gas mole fractions are not changing quickly. Composited diurnal cycles of CO_2 in July at the WLEF tower in Wisconsin indicate that the atmosphere is generally well-mixed between 1700 and 2100 UTC (1200–1600 LST) (Bakwin et al., 1998, Figure 1d). For the majority of the analyses in this paper, we consider the afternoon average to be the average over the period 17:00:00–20:59:59 UTC, which for brevity, we refer to as 1700–2100 UTC (1200–1600 LST). In Section 3.2.1, we quantify the effect of variable CO_2 mole fraction in the afternoon by considering different time periods, including a time-lagged version.

We choose one site to serve as a background, upwind boundary condition. The dry mole fractions observed at this site are subtracted from all other sites' mole fractions to isolate the enhancement in mole fraction caused by emissions within the city. The results leading to this choice are described in Section 3.1.

We use temporal averaging to quantify the mole fraction enhancements that result from urban emissions. Afternoon averages and a 15-day running average are both examined over the entire three-year record of measurement. A four-month average (January through April 2013) during the dormant season is used to quantify the long-term mole fraction enhancements. This period is chosen to take advantage of the large number of observation sites available and to avoid complications caused by biogenic fluxes that exist during summer months. In the dormant season Turnbull et al. (2015) show that the total CO_2 is an appropriate proxy for fossil-fuel CO_2, at least for Indianapolis and with a local background site. The four-month average mole fractions are also examined as a function of wind direction to quantify variability in the enhancement caused by changing winds. We also compare the four-month average observed CO_2 mole fraction enhancements at each tower site to the mole fraction enhancements simulated by the numerical modeling system.

Finally, we examine the sensitivity of our CO_2 results to variability in maximum sampling altitude and time of day used for comparisons. Long-term differences in CO_2 mole fraction as a function of height are studied at three towers where multi-level measurements were collected. We use those long-term vertical differences to estimate the mole fractions we would expect across the network if all CO_2 measurements were collected at the same altitude above ground. Similarly, a number of different definitions of well-mixed, steady-state ABL mole fractions are used to determine sensitivity of our results to that choice.

3 Results
3.1 Background sites
Next we evaluate the suitability of Site 01 and Site 09 as background sites by considering the difference between the CO_2 mole fraction measured at each site for each afternoon hour and the minimum mole fraction across the INFLUX tower network measured at the same hour for the period 1 January – 30 April 2013. In **Figure 4a**, the cumulative fraction of afternoon hours of observed CO_2 mole fraction enhancement above a given level is shown. The ideal background site would measure the lowest CO_2 mole fraction at all times (in the dormant season), within the measurement noise. Of course, the perfect background site does not exist, as this would require the wind to always originate from the predominant wind direction and that there were no local sources near the background site. For 43% of the afternoon hours Site 01 measures within 0.2 ppm of the lowest CO_2 amongst the INFLUX towers. Site 09 is less often most appropriate as a background site, but

not drastically so. For 39% of the afternoon hours, Site 09 measures within 0.2 ppm of the lowest. In comparison, the other INFLUX sites measure within 0.2 ppm of the lowest site between 0 and 19% of the afternoon hours.

When categorized into subsets during which the wind is from the southeast, southwest and northwest quadrants (**Figure 4b**), Site 09 is further from the lowest value when the wind is from the urban area (i.e., from the southwest) and Site 01 shows evidence of a source(s) to the southeast, most likely attributable to the Eagle Valley Power Plant, 10 km to the south. During this period, the wind comes from the northeast less than 8% of the time; thus the CO_2 enhancement from that direction is not considered. In general, the best background choices are Site 01 in general, Site 01 (when the wind is from the SW or NW), and Site 09 (when the wind is from the SE or NW). For each of these cases, the site in question is within 0.2 ppm of the lowest INFLUX site 42–47% of the afternoon hours. We consider Site 01 as the background site for the purpose of comparison in this paper. As will be shown in Section 3.2, the mean dormant-season afternoon difference between the CO_2 measured at Site 01 and Site 09 is small, 0.3 ppm. Thus, in terms of the time-averaged spatial results presented in this paper, choosing Site 01 as the only background site is not likely to significantly affect the results. However, if considering the temporal variability of mole fraction enhancements, the choice of background may play a more important role.

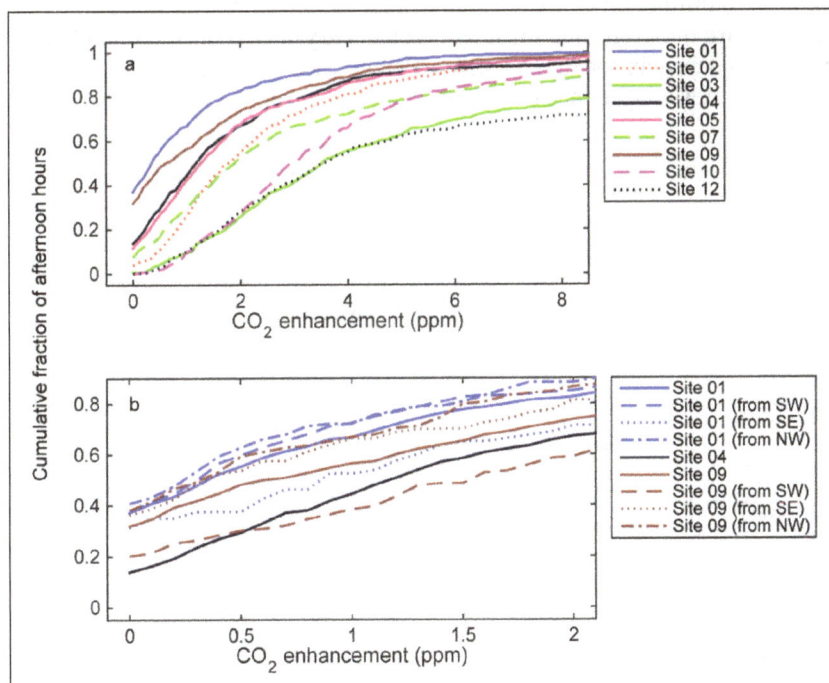

Figure 4: Cumulative fraction of afternoon hours of observed CO_2 mole fraction enhancement above a given level. a) CO_2 enhancement for all sites. Here enhancement is the difference between each site and the INFLUX network minimum for that hour. The averaging period is 1 January – 30 April 2013. Site details are listed in Table 1. Sites 01 and 09 are considered potential background sites. **b)** As in a), but for Site 01 and 09 when the wind is from the southeast (90–180°), from the southwest (180–270°), and from the northwest (270–360°). During this period, the wind comes from the northeast less than 8% of the time; thus the CO_2 enhancement from that direction is not considered. The results for Site 04 averaged over all wind directions are shown for comparison. DOI: https://doi.org/10.1525/elementa.127.f4

3.2 Urban greenhouse gas mole fractions: temporal and spatial cycles

3.2.1 Daytime dormant-season CO_2 dry mole fraction

As an example of the daily afternoon-averaged (1700–2100 UTC, 1200–1600 LST) CO_2, shown in **Figure 5a** are four weeks of data (1 – 28 January 2013). It is apparent that Site 01 generally measures the lowest during the period, except for 24 January for which Site 09 measures the lowest. The highest peak is measured at Site 03 on 8 January, but throughout the period there are days for which five different sites (Sites 02, 04, 07, 10, and 12) measure the highest CO_2. There is a period of four days (19 – 22 January) during which the CO_2 is consistently low at all of the sites. The wind speed measured at the Indianapolis International Airport (**Figure 5d**) is persistently high during this period compared to the rest of the four weeks, consistent with increased mixing of the boundary layer air, and thus lower mole fractions.

Shown in **Figure 6a** is the time series of daily afternoon-averaged CO_2 at INFLUX sites for a period of three years (1 January 2011 – 31 December 2013). Variability at various time scales is apparent. To illustrate the large variability in mole fractions, the two-sigma range (95%) of the daily CO_2 values throughout the measurement period is within 374–418 ppm (44 ppm range) for Site 02. The urban enhancement of CO_2, defined here as the difference between the afternoon-averaged CO_2 measured at a particular site and that measured at the background site (Site 01), is relatively small compared to the range of CO_2 values measured and it is difficult to distinguish between urban and background sites in **Figure 6a**. In general, the urban enhancement observed varies depending on the emissions and the weather conditions (e.g., wind speed and boundary

layer depth). 90% of the dormant season (1 January – 30 April 2013) afternoon-averaged CO_2 enhancements above the background site (Site 01) for Site 02 are between –2.41 and 7.00 ppm. For Site 03, 90% of the enhancements are between –0.34 and 9.64 ppm CO_2. In terms of detectability requirements, we instead consider the magnitude of the differences. On 90% of afternoons, the magnitude of the differences between Site 02 and Site 01 is greater than 0.47 ppm, while the compatibility of the measurement is 0.18 ppm CO_2.

In order to visualize the difference between the urban sites from the background sites as a function of time, further averaging is necessary. The daily CO_2 mole fractions, smoothed with a 15-day running mean filter, are shown in **Figure 7a**. In general, with this degree of averaging, the CO_2 shows coherent fluctuations across all of the sites, dominated presumably by variations in the hemispheric flux variations and synoptic-scale transport, rather than by the urban effects. Temporal variability is apparent at multiple scales: synoptic, seasonal, and inter-annual. On the synoptic scale of several days, weather patterns change, leading to differences in boundary layer depth, wind speed and direction and solar radiation, etc., and consequently, the CO_2 is observed to change coherently across all sites. Typical seasonal patterns of hemispheric growing-season CO_2 drawdown and dormant-season respiration are apparent as well. The seasonal minimum and maximum, determined by evaluating a 61-day running mean, occurs about August 1 and December 15, respectively. The growing-season CO_2 drawdown varies considerably amongst the observed years; the seasonal amplitude (defined as the difference between the dormant-season maximum and the previous growing season minimum) is 26/20/33 ppm CO_2

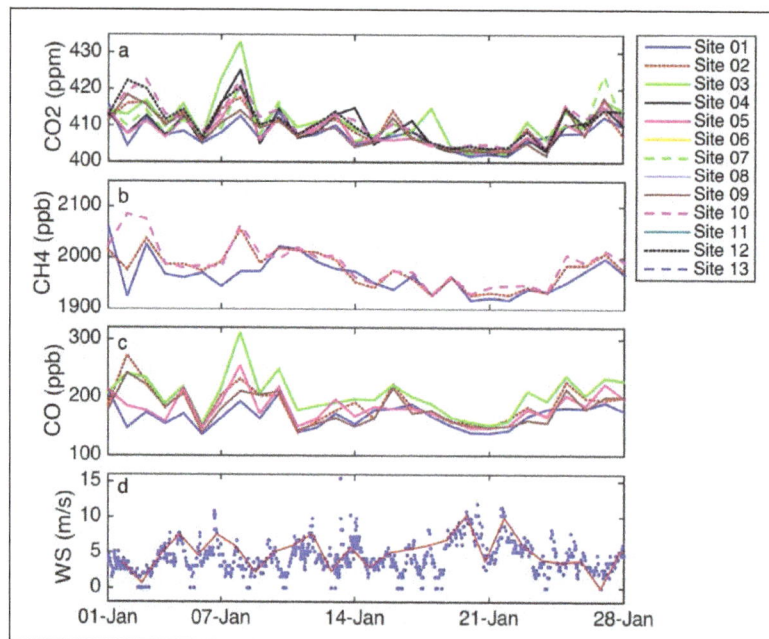

Figure 5: Afternoon-averaged daily mole fractions and wind speed for 1 – 28 January 2013. a) CO_2, **b)** CH_4 and **c)** CO. The data from the tallest measurement height at each tower is used; the measurement heights range from 39 to 136 m AGL (Table 1). Other site details are listed in Table 1 as well. **d)** Wind speeds (WS) measured at the Indianapolis airport for all hours (blue dots) and afternoon-averaged (red line). DOI: https://doi.org/10.1525/elementa.127.f5

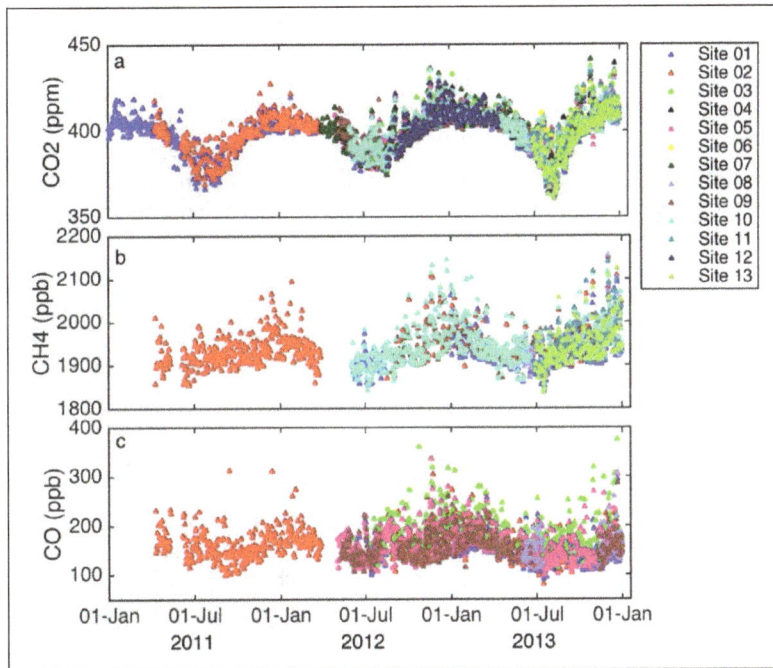

Figure 6: Afternoon-averaged daily mole fractions for the INFLUX tower sites for January 2011 – December 2013. a) CO_2, **b)** CH_4, and **c)** CO. The data from the tallest measurement height at each tower is used; the measurement heights range from 39 to 136 m AGL (Table 1). Other site details are listed in Table 1 as well. DOI: https://doi.org/10.1525/elementa.127.f6

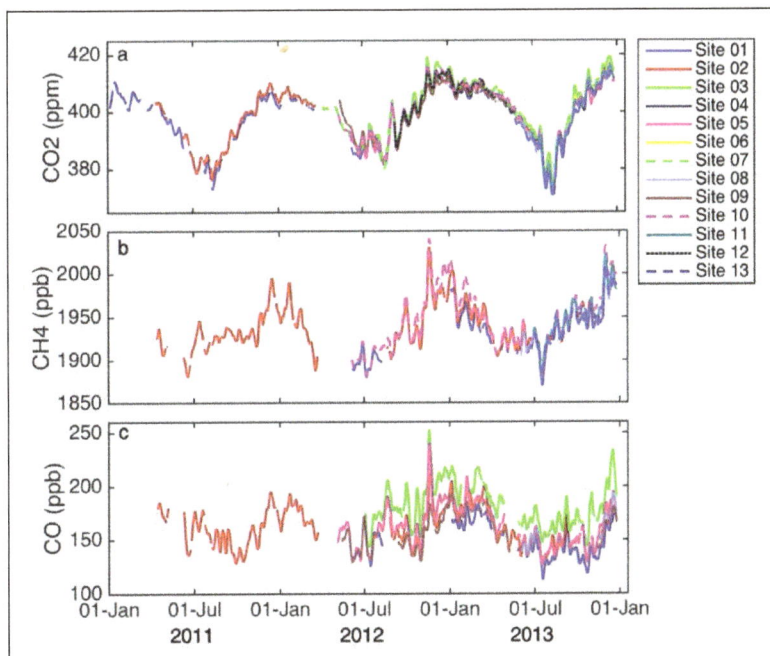

Figure 7: Afternoon-averaged daily mole fractions for January 2011 – December 2013, with 15-day smoothing applied. a) CO_2, **b)** CH_4, and **c)** CO. The data from the tallest measurement height at each tower is used; the measurement heights range from 39 to 136 m AGL (Table 1). Other site details are listed in Table 1 as well. DOI: https://doi.org/10.1525/elementa.127.f7

for the years 2011/2012/2013, respectively, at Site 01. This pattern does not vary appreciably among the sites; at Site 02, the seasonal amplitudes are 26 and 31 ppm for 2011 and 2013, respectively (2012 is not available). The decreased drawdown in 2012 is visible even in the unsmoothed afternoon-averaged CO_2 data (**Figure 6a**) and may be correlated with drought conditions observed that year; while the climatic average monthly precipitation in Indianapolis for May – July is 11.7 cm, only 3.1 cm was measured in 2012 (http://www.crh.noaa.gov).

Averaging over a period of four months (1 January – 30 April 2013) yields a clear spatial pattern induced by the city in the CO_2 signals. Dormant-season time-averaged CO_2 enhancements for each site above the background site (Site 01) are shown in **Figure 8** and listed in **Table 2**. The downtown site (Site 03) measures the largest mean CO_2, 2.9 ppm higher than the background site, whereas Site 02 measures 1.4 ppm larger than the background site. Site 09 measures only 0.3 ppm larger than the background site; this site only occasionally captures the urban plume and there is not a large constant local source of CO_2. The other sites fall between these extremes.

In the above analysis we have used the same averaging interval (1700–2100 UTC) for all the sites. In reality, the CO_2 dry mole fraction changes at the background site while the air mass advects to the downwind sites. In order to quantify this effect, we consider different definitions of well-mixed, steady-state conditions, including a time-lagged version. Shown in **Figure 9** is the observed time-averaged afternoon CO_2 dry mole fraction above background, averaged over different periods of the day (1700–2100 UTC, 2000–2300 UTC, and 2200–2300 UTC). The difference between the result using the averaging interval of 2000–2300 UTC and that using 1700–2100

UTC (the default averaging interval) is +0.2 ppm (ranging from 0.0 to 0.4 ppm) and the difference between using 2200–2300 UTC compared to using the default averaging interval is +0.4 ppm (ranging from 0.2 to 0.9 ppm), where both values are averaged across all sites. These differences in the enhancements above background are attributable to site-to-site differences in the timing of the dilution of the accumulation of emissions in the stable nocturnal boundary layer by the convective growth of the ABL. Rural sites (e.g., Site 01) exhibit a delay in the growth of the ABL relative to the urban sites.

The distance between Site 01 and the other sites is 27–66 km (**Figure 2**). Median near-surface afternoon wind speeds at the Indianapolis Airport are 5.3 ± 2.6 ms^{-1}. Thus a reasonable amount of time for air masses to traverse the distance between Site 01 and Site 02 (42 km) is 1.5–4.3 hr, for example. The actual range of transit times is much larger; in calm winds, an air mass starting at Site 01 in the beginning of the afternoon does not even reach the downwind sites during the same afternoon. But to approximate the effect on the CO_2 dry mole fraction above background, we consider a time-lagged case, subtracting the CO_2 at 1900–2000 UTC at Site 01 from the other sites' CO_2 three hours later at 2200–2300 UTC. The difference

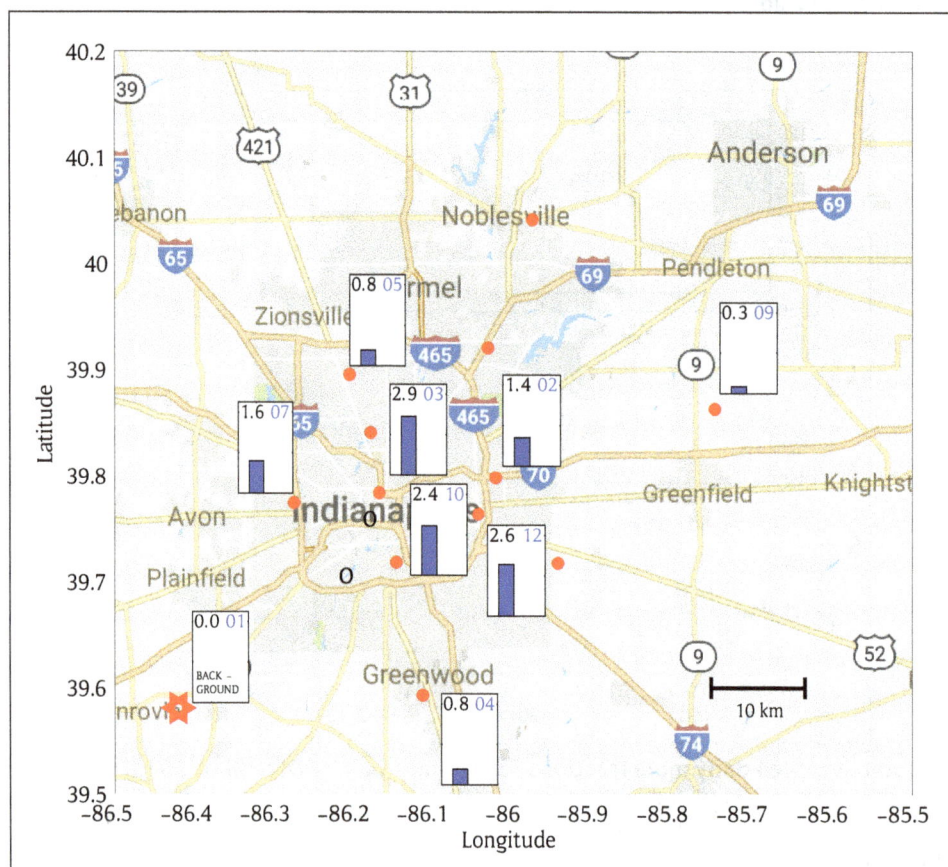

Figure 8: Observed time-averaged afternoon CO_2 dry mole fraction above background (Site 01). The averaging period is 1 January – 30 April 2013. The site number is shown in the upper right corner of each plot and the average observed CO_2 above background (in ppm) in the upper left corner. The full y-axis scale for the plots is 4.5 ppm CO_2. Red filled circles indicate the locations of the sites, with the background Site 01 indicated by a red star. Time frame is chosen to minimize biogenic signals and maximize the number of sites with available data. Sites with less than 75% data availability during the selected time period are excluded. DOI: https://doi.org/10.1525/elementa.127.f8

Table 2: Observed time-averaged CO_2 mole fraction above background (Site 01) for INFLUX tower sites, in order from least to greatest observed urban enhancement. The averaging period is 1 January – 30 April 2013. DOI: https://doi.org/10.1525/elementa.127.t2

	Site 01	Site 09	Site 04	Site 05	Site 02	Site 07	Site 12	Site 10	Site 03
Maximum sampling height (m AGL)	121	130	60	125	136	58	40	40	54
Average CO_2 above background at maximum sampling height (ppm)	–	0.3	0.8	0.8	1.4	1.6	2.5	2.5	2.9
Approximated average CO_2 above background at 40 m AGL (ppm)	–	0.3[1]	0.5 – 0.7[1,2]	0.8 – 1.5[1,2]	2.2	1.3 – 1.4[1,2]	2.1	2.1	3.5

[1]Gradient measured at Site 01 is used to approximate the gradient at this site.

[1,2]The range of values shown originates from using the gradients measured at Site 01 and Site 02 (Figure 14). Note that the background value at Site 01 is 0.4 ppm higher at 40 m AGL compared to 121 m AGL.

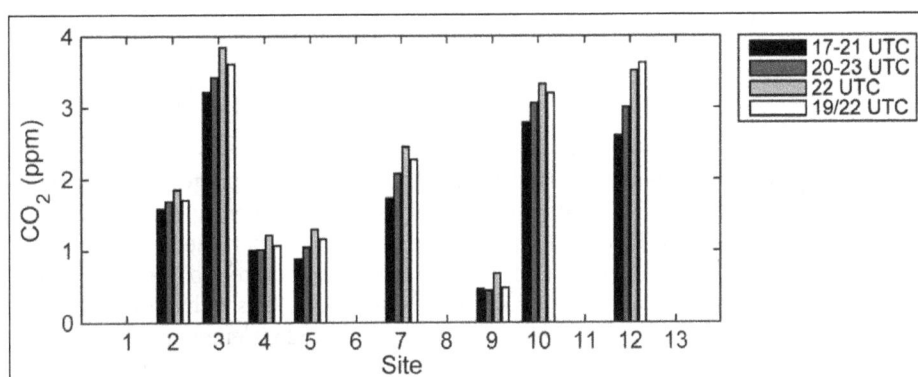

Figure 9: Observed time-averaged afternoon CO_2 dry mole fraction above background (Site 01). The averaging period is 1 January – 30 April 2013. Bars indicate different daily averaging intervals (black: 1700–2100 UTC (1200–1600 LST), dark gray: 2000–2300 UTC (1500–1800 LST), light gray: 2200–2300 UTC (1700–1800 LST)), and a time-lagged version in which the CO_2 dry mole fraction at Site 01 at 1900–2000 UTC (1400–1500 LST) is subtracted from the other sites' CO_2 dry mole fraction at 2200–2300 UTC (1700–1800 LST) (white bar). Sites 06, 08, 11, and 13 were not yet deployed during this period. DOI: https://doi.org/10.1525/elementa.127.f9

between this result and that using 1700–2100 UTC as the averaging interval, averaged across all sites, is +0.4 ppm (ranging from 0.0 to 1.0 ppm at the different sites). Overall, the time averaging choice has a significant impact (0.4 ppm is 25% of the enhancement averaged over all of the sites of 1.6 ppm), but the spatial pattern of the urban enhancement is similar in the tested cases.

3.2.2 Daytime dormant-season CH_4 dry mole fraction

The daytime afternoon-averaged CH_4 mole fraction is shown in **Figure 5b** for 1 – 28 January 2013. As for CO_2, Site 01 most often measures the lowest CH_4 (for the three sites with available data during this period). Site 10, near the large city landfill, measures the highest CH_4 and the period 19 – 22 January measures low CH_4 at all of the sites.

When we examine the entire three-year period, the variability is large and the urban effect is difficult to discern (**Figure 6b**); the two-sigma range of the CH_4 measurements at Site 02 is within 1873-2046 ppb CH_4 (range = 173 ppb). Smoothing with a 15-day filter, there

is temporal variability at various scales and the coherence between the sites is clear (**Figure 7b**), as for CO_2. Seasonal amplitudes are 80 and 86 ppb for the years 2012 and 2013, respectively, at Site 01 (2011 is not available), and 42 and 86 ppb for the years 2011 and 2013 at Site 02 (2012 is not available). The synoptic-scale amplitudes are a larger fraction of the seasonal signal, compared to CO_2. The seasonal cycle is also shifted compared to the seasonal cycle of CO_2, with the maximum occurring around November 15 and the minimum around August 15. That minimum corresponds to the time of year for maximum OH, the dominant CH_4 sink, as seen for a range of hydrocarbons in the northern hemisphere (Swanson et al., 2003).

The urban signal is detectable in the CH_4 signal, after averaging each site for the period 1 January – 30 April 2013 (**Figure 10**). The time-averaged CH_4 mole fraction above the background site varies from 5 ppb for Site 13 (southeast of the city, in an agricultural area) to 21 ppb for Site 10 (typically downwind of the South Side Landfill).

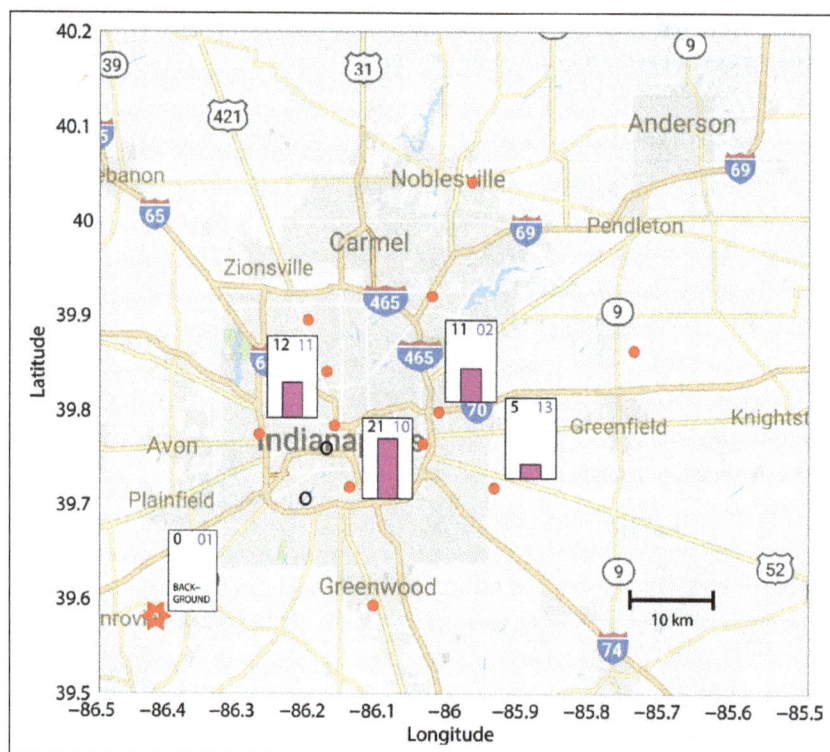

Figure 10: Observed time-averaged afternoon CH$_4$ dry mole fraction above background (Site 01). The averaging period is 1 October – 31 December 2013. The site number is shown in the upper right corner of each plot (in blue) and the average observed CH$_4$ above background (in ppb) in the upper left corner. The full y-axis scale for the plots is 28 ppb CH$_4$. Red filled circles indicate the locations of the sites, with the background Site 01 indicated by a red star. Time frame was chosen to maximize the number of sites with available data. Sites with less than 75% data availability during the selected time period were excluded. DOI: https://doi.org/10.1525/elementa.127.f10

In terms of the range of signals, 90% of the dormant season (1 January – 30 April 2013) afternoon-averaged CH$_4$ enhancements above the background site (Site 01) for Site 02 are between −13.3 and 35.5 ppb. On 90% of afternoons, the magnitude of the differences between Site 02 and Site 01 is greater than 2.2 ppb CH$_4$.

3.2.3 Daytime dormant-season CO dry mole fraction

Examining the CO mole fraction measured at the INFLUX sites for 1 – 28 January 2013 (**Figure 5c**), there are similarities with the CO$_2$ and CH$_4$ results. Site 01 again measures the lowest mole fractions. There is a large peak particularly at the downtown Site 03 on January 8 (Site 03 measures 119 ppb higher CO than Site 01 on this afternoon) and a period of overall decreased mole fractions on 19 – 22 January (the range amongst the INFLUX sites is only 15 ppb). Over the three-year period, the CO variability from day to day is large compared to the seasonal cycle (**Figure 6c**), consistent with the much shorter photochemical lifetime for CO, compared with the other two gases (Mao and Talbot, 2004). As discussed by Jobson et al. (1999) there is a well-defined (inverse) relationship between lifetime of trace gases and their atmospheric variability. The range of CO values measured at Site 02 is 114–227 ppb for 2-sigma (95%) of the values. The smoothed daily afternoon-averaged CO mole fractions are shown in **Figure 7c**. The seasonal maximum occurs around December 15, but the minimum is too variable

to determine a specific date range applicable for all three years.

The time-averaged (1 January – 30 April 2013) CO above the background site varies from 5 ppb for Site 09 to 29 ppb for the downtown Site 03 (**Figure 11**). 90% of the dormant-season afternoon-averaged CO enhancements above the background site (Site 01) for Site 02 are between −13 and 53 ppb. For Site 03, the enhancements are larger, with 90% being between 6 and 79 ppb CO. In terms of detectability requirements, 90% of afternoons exhibit the magnitudes of the differences between Site 02 and Site 01 is greater than 2 ppb. Comparatively, for Site 03 90% of the magnitudes of differences are greater than 11 ppb.

3.2.4 Urban mole fractions as a function of wind direction

Mole fraction differences across the city change, as expected, when segregating them as a function of wind direction. When only considering wind directions from the southwest, Site 02 is downwind of the city, and the average CO$_2$ mole fraction enhancement is up to 3.3 ppm (**Figure 12a**). When the wind is aligned such that Site 09 captures the urban plume (i.e., from the southwest), Site 09 measures up to 1.6 ppm CO$_2$ larger than background (**Figure 12b**). An exception is that the CO$_2$ difference between Sites 09 and 01 is almost 2 ppm when the wind is between 165° and 180° (**Figure 12b**), but with Site 01 higher; this difference is likely attributable to the (coal-fired) Eagle Valley Power Plant that is 10 km south of Site 01.

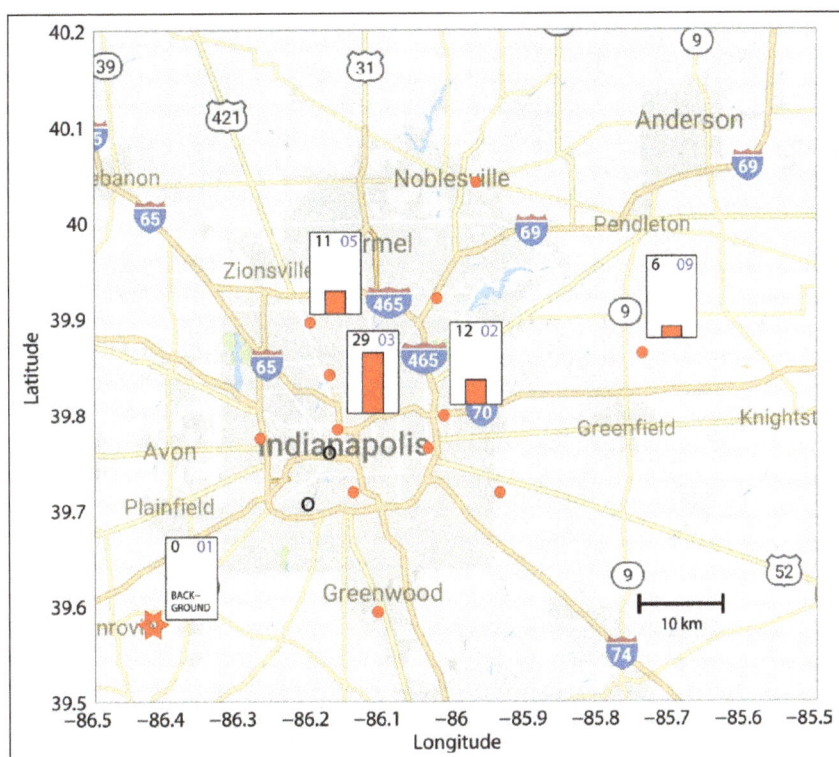

Figure 11: Observed time-averaged afternoon CO dry mole fraction above background (Site 01). The averaging period is 1 January – 30 April 2013. The site number is shown in the upper right corner of each plot (in blue) and the average observed CO above background (in ppb) in the upper left corner. The full y-axis scale for the plots is 40 ppb CO. Red filled circles indicate the locations of the sites, with the background Site 01 indicated by a red star. Time frame was chosen to maximize the number of sites with available data. Sites with less than 75% data availability during the selected time period were excluded. DOI: https://doi.org/10.1525/elementa.127.f11

The CO urban enhancement is up to 34 ppb at Site 02 and 20 ppb at Site 09 when the wind is aligned such that the sites are downwind of the city (**Figure 12c and d**). The directions of the enhancements for CO are similar to those for CO_2 except most notably for the lack of a reversed enhancement (Site 01 > Site 09) when the wind is from the south-southeast. In Indianapolis, the CO/CO_2 ratio of vehicular emissions range between 2.2 and 16.2 ppb/ppm with a large single polluter measuring 47.1 ppb/ppm (Vimont et al., 2016). The CO/CO_2 ratio emitted by power plants, however, tends to be either near zero or in the 5 – 7 ppb/ppm range, depending on plant operating conditions (Peischl et al., 2010, Table 3). Our measurements are consistent with the Eagle Valley Power Plant emitting low levels of CO but significant CO_2.

The CH_4 signal for Site 02 originates from a slightly shifted direction compared to CO_2 and CO, with the largest signals from the south-southwest, consistent with a large portion of the CH_4 emissions being from the landfill on the south side of the city (Cambaliza et al., 2015). The magnitude of the enhancement for wind directions between 150 and 300° is up to 21 ppb.

3.3 Time-averaged CO_2 mole fractions: Comparison to modeling results

Combining calculated tower footprints with Hestia emissions, we determine modeled CO_2 as described in Section 2.3 in order to compare with the observed CO_2. The atmospheric

inversion approach optimizes the spatially- and temporally-varying emissions by adjusting the emissions in order to minimize the difference between the observed and modeled CO_2 mole fraction. A first step in the process involves comparing the modeled ("forward") CO_2 with the observed. These results, averaged for afternoon hours (1700–2100 UTC) during a dormant period (January – April 2013), are shown in **Figure 13**. In both the modeled and observed results Sites 03 and 10 exhibit the largest enhancements in CO_2, and Sites 09, 04, and 05 show the smallest enhancements. The modeled CO_2 is highly correlated with the observed CO_2, with an r^2 of 0.94 (**Figure 13** inset). This correlation indicates the calculated footprints of the towers are qualitatively correctly sized; e.g., if the Site 09 footprint was actually an order of magnitude larger than modeled here, we would expect the urban signal for Site 09 to be more similar to that of Site 02. The overall magnitude of modeled enhancements is on average, however, only 53% of the observed enhancements (**Figure 13** inset). This result suggests that the either the meteorological model significantly overestimates vertical mixing, the actual emissions in Indianapolis are larger than reported in Hestia, or a combination of both. The mean errors in boundary layer depth, comparing INFLUX WRF modeling results to the Doppler lidar, are small: about 25 m (Deng et al., 2016), so the vertical mixing seems to be accurately modeled, on average. Shown in the **Figure 13** inset is the mean CO_2 mole fraction enhancement using the inverse fluxes (posterior) for 1 Jan-

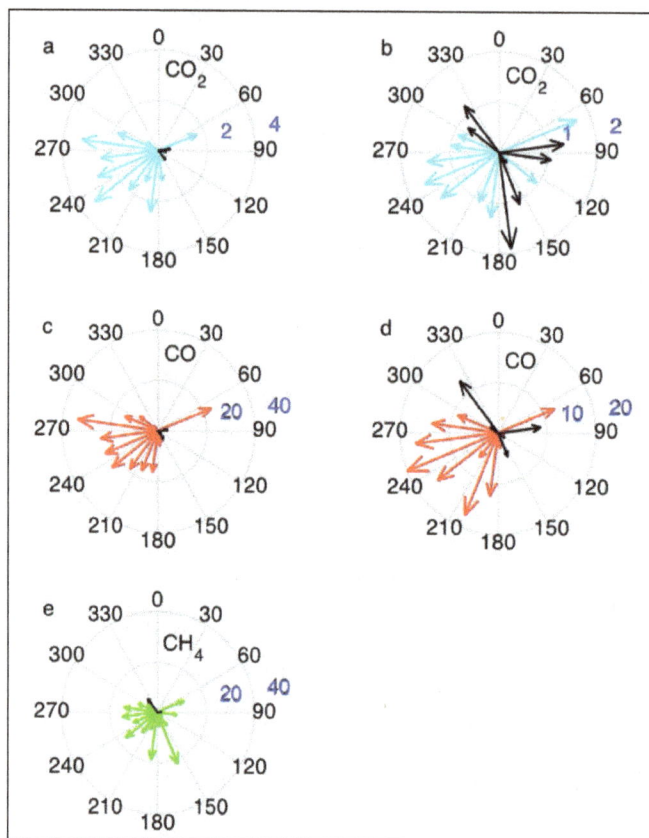

Figure 12: Urban enhancement of greenhouse gases as a function of near-surface wind direction. Here urban enhancement is defined as the mole fraction difference between predominantly downwind site and background Site 01. The averaging period is daytime hours on 1 January – 30 April 2013. Winds are measured at the Indianapolis International airport (http://cdo.ncdc.noaa.gov/qclcd/QCLCD). **a)** CO_2 at Site 02 – CO_2 at Site 01 (ppm), **b)** CO_2 at Site 09 – CO_2 at Site 01 (ppm), **c)** CO at Site 02 – CO at Site 01 (ppb), **d)** CO at Site 09 – CO at Site 01 (ppb), and **e)** CH_4 at Site 02 – CH_4 at Site 01 (ppb). Colored arrows indicate that the downwind site is larger than the background site from that direction, on average. Black arrows indicate that the downwind site is smaller than the background site from that direction, on average. Arrows point to the emission sources. Wind directions occurring on less than 3% of the days are excluded. The urban enhancement magnitude is denoted on the radial scale (in blue); note the differing scales. DOI: https://doi.org/10.1525/elementa.127.f12

uary – 30 April 2013. The posterior modeled enhancements agree well with the observations, as would be expected, and average 95% of the observed enhancements. When averaged over the entire time period of the inversion results in Lauvaux et al. (2016), the inversion results show an increase of about 20% in total emissions from the prior: 5.50 MtC for the period September 2012 – April 2013, compared to 4.56 MtC reported by Hestia. Emissions near the tower sites are increased more than the average pixel across the city, i.e., many pixels have near zero emissions both before and after the inversion. We also note that the inverse emission result also includes nighttime fluxes, which are not significantly modified by the daytime mole fractions at the towers, decreasing the overall change after inversion.

3.4 Vertical profiles of GHG dry mole fraction
3.4.1 CO₂ vertical profiles
Tower heights greater than 100 m AGL are generally considered desirable in order to measure within the well-mixed layer during the day (Bakwin et al., 1998),

in order to reduce interpretation problems induced by changing boundary layer depth and to mitigate sensitivity to nearby point sources. Because of the scarcity of tall towers within the city of Indianapolis, however, several towers in the 40–60 m AGL range are utilized in INFLUX. To investigate the ramifications of using shorter towers, composites of the difference between afternoon CO_2 measured at each level and that of the top level are shown in **Figure 14a** for Sites 01, 02, and 03. Averaging over a dormant period (1 November – 31 December 2013), the CO_2 profile at Site 01 is relatively constant compared to the others, with the 40-m level measuring 0.4 ppm higher CO_2 than the top level. The vertical gradient is more pronounced for Site 02, with the 40-m level being 1.2 ppm greater than the top level. Most of the INFLUX towers are between these two towers in terms of urban density. The vertical gradient is largest at Site 03, located downtown with large local emissions. The tower height at that location is 54 m AGL and the difference between the top level CO_2 and

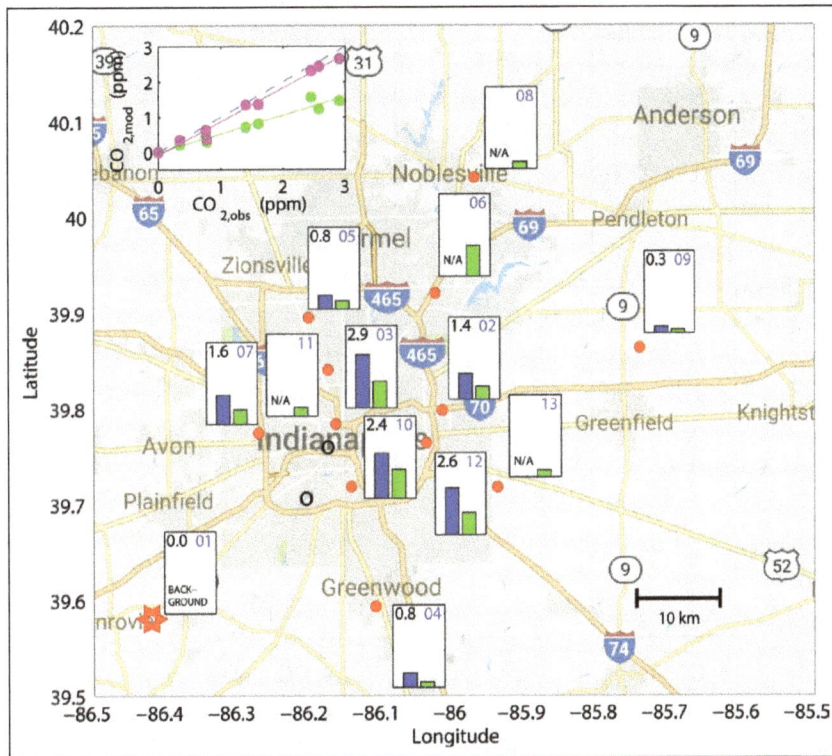

Figure 13: Observed and modeled CO_2 mole fraction above background. Observed (blue) time-averaged afternoon CO_2 mole fraction above background (Site 01) for INFLUX tower sites (1 January – 30 April 2013) at each of the INFLUX towers (represented by red dots), as in Figure 8. Corresponding model results (prior to inversion) using the 2014 release of Hestia describing emissions from 2013 are shown in green. The site number is shown in the upper right corner of each plot (in blue) and the average observed CO_2 above background (in ppm) in the upper left corner. The full y-axis scale for the plots is 4.5 ppm CO_2. Red filled circles indicate the locations of the sites, with the background Site 01 indicated by a red star. Note that not all the sites have CO_2 mole fraction data available during this period; these sites (Sites 06, 08, 11, and 13) are indicated by 'N/A'. The Harding Street Power Plant is indicated by the black circle south of the downtown area; the South Side Landfill is 2 km to the northwest. The smaller C.C. Perry Power Plant is indicated by the black circle near downtown. Shown in the inset is the modeled CO_2 prior to inversion (green) and using optimized fluxes from inversion (magenta) as a function of the observed CO_2. Green and magenta lines show the linear fit and blue dashed line is the 1: 1 line. DOI: https://doi.org/10.1525/elementa.127.f13

that at 40 m is 1.0 ppm. These results are, of course, specific to daytime in the dormant season. Changes in mole fraction as a function of height vary systematically with the sign and magnitude of the local surface fluxes.

The GHG results shown in this study are thus affected by the non-uniformity of tower heights. We now approximate how the results would differ if all the measurements had been at the same height above ground level. We note that the modeling results presented in this paper (Section 3.3) and in Lauvaux et al. (2016) take into account the differing tower heights. One possible approach to address the differing tower heights in the observed results is the virtual tall tower approximation (Davis et al., 2005; Haszpra et al., 2015), in which the CO_2 is normalized to a uniform height for tall towers (>100 m AGL). This approach utilizes a flux-gradient relationship (Wyngaard and Brost, 1984) that is a function of convective velocity scale, boundary layer depth, surface CO_2 flux and surface canopy structure (Wang et al., 2007; Patton et al., 2003). As these measurements are not available at the INFLUX towers, we instead use a simple approach based at CO_2 gradients measured

at the sites with multiple tower levels. Seven of the nine INFLUX towers have measurements available at 40–60 m AGL, including three towers with profile measurements. We therefore choose to approximate the CO_2 at 40 m AGL at all the towers for comparison, rather than correcting to >100 m AGL. The gradient between 40 m and the maximum height (>100 m AGL) at Site 01 is 0.5 ppm/100 m and that at Site 02 is 1.3 ppm/100 m. The landcover type at Site 09 is similar to Site 01. We thus use the Site 01 gradient to approximate the difference in CO_2 at 40 m AGL compared to 130 m AGL at Site 09. The urban density at Sites 04, 05 and 07 is between that of Site 01 and Site 02, and a range of values is therefore calculated. The towers at Sites 04 and 07 are 60 m AGL and 58 m AGL, so the adjustments are small (0.2 – 0.3 ppm). The adjustment in the enhancement at Site 05 (with a measurement height of 125 m AGL) to 40 m AGL is in the range of 0.1–0.6 ppm. The CO_2 mole fractions above background at maximum tower height and adjusted (if necessary) to 40 m AGL are both shown in **Table 2**. Note that the background value (Site 01) is 0.4 ppm higher at 40 m AGL compared to 121 m AGL, resulting in a shift in the reported enhancement

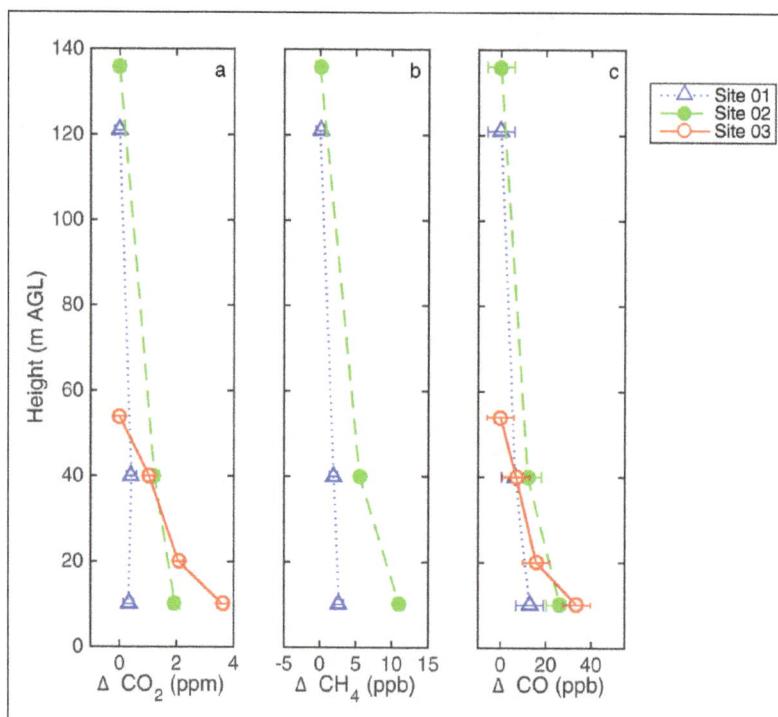

Figure 14: Composites of the vertical changes in greenhouse gas mole fraction. The vertical change is the difference between afternoon **a)** CO_2, **b)** CH_4, and **c)** CO measured at each level and that of the top level. Site 01 is shown as a dotted blue line, Site 02 as a green dashed line, and Site 03 as a solid red line. Shown profiles are using data from two months during the dormant season (1 November – 31 December 2013), as profiles from Sites 01 and 02 were not available for the period 1 January – 2 April 2013. The instrument at Site 03 does not measure CH_4. Here afternoon is defined as 1500–1600 LST. Bars indicate the compatibility of the mole fraction measurements (not visible for CO_2 and CH_4). DOI: https://doi.org/10.1525/elementa.127.f14

even for the sites at which there is a measurement at 40 m AGL (Sites 10 and 12). The spatial mean enhancement is nearly the same: 1.6 ppm compared to 1.7 ppm. While the overall spatial gradients are similar, there are differences for specific towers. For example, the Site 05 enhancement is 0.8 – 1.5 ppm rather than 0.8 ppm, and the Site 02 enhancement is 2.2 ppm rather than 1.4 ppm. At these sites in particular, the enhancement at 40 m AGL is increased by up to 50%, compared to the results using the maximum measurement height at each tower. Thus the height variability has an effect on the results presented here that is similar in magnitude to the effect of time of day for the averaging interval (**Figure 9**).

The 10-m AGL dormant-season mean CO_2 is also shown in **Figure 14a**. Near the source fluxes, the gradients for Sites 02 and 03 are larger: At Site 02 the 10-m CO_2 is 1.9 ppm larger than measured at the top level, and at Site 03 the 10-m CO_2 is 3.6 ppm larger than at the top level. Comparing the 10-m levels at the sites, the Site 02 enhancement at 10 m AGL is 3.2 ppm and that of Site 03 is 7.2 ppm. The measurements at this height are more affected by local signals, i.e., the footprints are smaller (Horst and Weil, 1992) and representative of a smaller area.

3.4.2 CH_4 and CO vertical profiles

The dormant-season averaged profiles of CH_4 mole fraction (**Figure 14b**) indicate a small difference (2.7 ppb) between the CH_4 measured at the lowest level (10 m AGL)

compared to the top level (121 m AGL) at the background Site 01. At Site 02, the gradient is larger, with the 10-m AGL CH_4 being 11.1 ppb higher than the highest level. For CO, the dormant-season average at Site 01 is 13 ppb higher at 10 m AGL, compared to the top level (**Figure 14c**). At Site 02 and 03, the differences between the 10 m AGL measurement and that at the top level are 26 and 34 ppb, respectively.

4 Discussion and Conclusions

In this paper, we present three years of CO_2, CH_4, and CO daytime dry mole fractions at towers between 39 and 136 m AGL, observed using cavity ring-down spectrometers at sites in and around Indianapolis, Indiana, in the U.S. Midwest. The differences among the smoothed CO_2 of the sites are small compared to the seasonal- and synoptic-scale variability, showing the importance of the synoptic-scale transport compared to the urban signal. The daily daytime urban signal is overwhelmed by the temporal variability unrelated to urban emissions. Typical synoptic, seasonal, and interannual cycles are apparent at all the sites. The seasonal amplitudes of CO_2 measured in and around Indianapolis average 36 ppm, nearly identical to the average of 35 ppm Miles et al. (2012) observed in the cornbelt region of the U.S. Upper Midwest. However, averaging over several months in the dormant season yields clear urban signals despite the temporal variability. The downtown Site 03 measures 2.9 ppm CO_2 than the background site, on average. CH_4 and CO are not measured at all of the sites, but for the subset for which these species are measured, the site with the largest magnitude CH_4 difference (21 ppb) from the background site is Site 10. And for CO, the downtown Site 03 measures 29 ppb above the background site. In terms of the range of signals, 90% of the afternoon enhancements at Site 02 are between −2.41 and 7.00 ppm CO_2, between −13.3 and 35.5 ppb CH_4 and between −13 and 53 ppb CO.

The GHG compatibility requirements for global studies (GAW Report No. 213, 2014) are not applicable in urban studies. Instead we define urban compatibility requirements based on this study. To achieve, for example, compatibility at 10% of the mean dormant-season urban signal shown in this study (2.9 ppm CO_2 at Site 03, 21 ppb CH_4 at Site 10 and 29 ppb CO at Site 03), the required compatibility is 0.29 ppm CO_2, 2.1 ppb CH_4 and 2.9 ppb CO. The signals are larger when the data is subset for downwind conditions. Another reasonable metric for the required compatibility is based on the distribution of afternoon-averaged dormant-season enhancements: 90% of the magnitudes of the urban enhancement are greater than 0.47 ppm CO_2, 2.2 ppb CH_4 and 2.3 ppb CO at Site 02. The compatibility for the measurements in this study: 0.18 CO_2, 0.6 ppb CH_4, and 6 ppb CO (Richardson et al., 2016), exceed the requirements for CO_2 and CH_4, but not for CO (based on the mean dormant-season urban signal). We also note that in other cities the compatibility requirements will differ, depending on the magnitude of urban fluxes and specific meteorology (e.g., presence of inversion layer).

The choice of background is crucial for urban GHG emission quantification. Turnbull et al. (2015) show that the total CO_2 is an appropriate proxy for fossil-fuel CO_2 for Indianapolis in the wintertime, but only when using a local background. The total CO_2 is only partially explained by fossil-fuel emissions when using a more distant background. In this paper, defining the urban enhancement as the difference between the mole fraction measured at each site and that measured at an appropriate background site (Site 01) eliminates the effects of synoptic-scale transport and allows interpretation of the urban signal. The location of the background site, however, is a factor in terms of quantification of site-to-site differences. Although Site 01 is the best overall choice of background given the current INFLUX network, it does shows evidence of a source(s) to the southeast, most likely the Eagle Valley Power Plant, 10 km to the south. In terms of the overall spatial patterns presented in this paper, choosing Site 01 as the only background site is not likely to largely affect the results, given that the mean difference between Site 01 and Site 09 is 0.3 ppm. Furthermore, inverse emission results indicate the difference in city emissions between using Site 01 as a background and using either Site 01 or Site 09 depending on the wind direction is low, about 4% (Lauvaux et al., 2016). However, when investigating time-variability in mole fraction enhancements or emissions, the effect may be larger.

Although the vertical gradients in CO_2 can be quite large, the effects of varying tower sampling heights were found to be significant but secondary to the observed spatial gradients for the various tower heights which are between 39 and 136 m AGL. For example, the enhancement of Site 03 compared to Site 01 considering the highest measurement level at both is 2.9 ppm, and for the 40-m measurement, the enhancement is 3.5 ppm. The atmospheric inversion results for INFLUX shown in Lauvaux et al. (2016) use transport resulting from releasing particles from the actual tower sampling heights, but the effect of the various tower heights on modeled emissions is dependent on the ability of the model to properly simulate vertical mixing. If, however, we consider the measurements at 10 m AGL, the Site 03 enhancement is 6.5 ppm, more than double the result for the highest tower measurements, and the results are representative of a smaller area.

Site 02 and Site 09 are both located predominantly downwind of the city. Therefore, the urban enhancements at these two sites can be used to characterize the relationship between the observed atmospheric signals and the distance to the metropolitan area. Subsampling for wind directions such that the sites are downwind of the city increases the urban signal, from 1.4 ppm to 3.3 ppm at Site 02 at the downwind edge of the city, for example. Similarly, the average signal at Site 09 (24 km east of the edge of the city) is 0.3 ppm, but is 1.6 ppm when the wind is aligned such that Site 09 is downwind of the city. Horizontal dispersion of the urban plume and entrainment decrease the signal by 51% over the 24 km between Site 02 and 09 when the wind is from the direction of the city. These results have potential implications for the satellite-based detection of mid- and small-sized cities (without topographical trapping of pollutants). Depending on the distance of satellite tracks from the urban center, this dramatic decrease could be a limiting factor. For example, the OCO-2 orbit tracks are separated by about 170 km (http:/oco.jpl.nasa.gov). Moreover, the results presented here are boundary layer measurements, and column-based measurements would be further diluted.

Similar measurements of urban greenhouse gas mole fractions are being performed in cities around the world. In Paris, Bréon et al. (2015) reported CO_2 mole fraction data for two 30-day periods using five instrumented tower sites. Two of the sites are located in mixed urban-rural areas and two sites used as background are part of the Integrated Carbon Observation System network 20 and 100 km from the urban center. An additional measurement, at the top of the Eiffel Tower, was determined to be poorly represented by the model for most wind speeds and directions. In Los Angeles, CO_2 and CH_4 are being measured at 14 tower and building roof-top sites within and near the Los Angeles basin (https://megacities.jpl.nasa.gov/; Verhulst et al., 2017). McKain et al. (2012) presented CO_2 data from Salt Lake City, Utah, and McKain et al. (2014) described the Boston, Massachusetts, network of five tower- and building-based observations.

The city of Indianapolis is readily detectable by the INFLUX network of in-situ tower-based greenhouse gas mole fraction measurements. The network represents one of the first urban deployments of multiple high-compatibility sensors. Spatial patterns in the observations are consistent with urban density and confirm the presence of high-resolution information for determination of spatial and temporal variability in emissions via an atmospheric inversion. The observed average dormant season CO_2 dry mole fraction and those predicted by a numerical modeling system are highly correlated. This paper represents an attempt to fully characterize and quantify urban GHG enhancements across space and time in a large metropolitan area.

Acknowledgements
D. Sarmiento (The Pennsylvania State University) provided building height analysis. B. Haupt (The Pennsylvania State University) contributed data ingest scripting and data pre-processing. Earth Networks, Inc. provided assistance in site maintenance. The authors would also like to thank J. Miles for help in maintaining the sites.

Funding information
This work is supported by the National Institute of Standards and Technology (Project # 70NANB10H245) and the National Oceanic and Atmospheric Administration (Award # NA13OAR4310076).

Competing interests
The authors have no competing interests to declare.

Contributions
- Contributed to conception and design: KJD, PBS, KG, TL, NLM, SJR, CS, JCT
- Contributed to acquisition of in-situ data: NLM, SJR
- Contributed to acquisition of flask data: JCT, MOC, CS
- Contributed to modeling results: TL, AD, KRG, RP, IR
- Contributed to analysis and interpretation of data: NLM, KJD, TL, SJR, MOC, PBS, CS, JCT, NVB
- Drafted and/or revised article: NLM, KJD, TL

References
Andres, RJ, Fielding, DJ, Marland, G, Boden, TA, Kumar, N and **Kearney, AT** 1999 Carbon dioxide emissions from fossil-fuel use, 1751–1950. *Tellus* **51B**: 759–765. DOI: https://doi.org/10.3402/tellusb.v51i4.16483

Asefi-Najafabady, S, Rayner, PJ, Gurney, KR, McRobert, A, Song, Y, et al. 2014 A multiyear, global gridded fossil fuel CO_2 emission data product: Evaluation and analysis of results. *J Geophys Res Atmos* **119**(17): 10213–10231. DOI: https://doi.org/10.1002/2013JD021296

Bakwin, PS, Tans, PP, Hurst, DF and **Zhao, C** 1998 Measurements of carbon dioxide on very tall towers: Results of the NOAA/CMDL program. *Tellus* **50B**: 401–415. DOI: https://doi.org/10.1034/j.1600-0889.1998.t01-4-00001.x

Bréon, FM, Broquet, G, Puygrenier, V, Chevallier, F, Xueref-Remy, I, et al. 2015 An attempt at estimating Paris area CO_2 emissions from atmospheric concentration measurements. *Atmos Chem Phys* **15**: 1707–1724. DOI: https://doi.org/10.5194/acp-15-1707-2015

Cambaliza, MOL, Shepson, PB, Bogner, J, Caulton, DR and **Stirm, B** 2015 Quantification and source apportionment of the methane emission flux from the city of Indianapolis. *Elem Sci Anth* **3**: 000037. DOI: https://doi.org/10.12952/journal.elementa.000037

Cambaliza, MO, Shepson, PB, Caulton, D, Stirm, B, Samarov, D, et al. 2014 Assessment of uncertainties of an aircraft-based mass-balance approach for quantifying urban greenhouse gas emissions. *Atmos Chem Phys* **14**: 9029–9050. www.atmos-chem-phys.net/14/9029/2014/, DOI: https://doi.org/10.5194/acp-14-9029-2014

Chen, F and **Dudhia, J** 2001 Coupling an advanced land surface–hydrology model with the Penn State–NCAR MM5 modeling system. Part I: Model implementation and sensitivity. *Mon Wea Rev* **129**: 569–585. DOI: https://doi.org/10.1175/1520-0493(2001)129<0569:CAALSH>2.0.CO;2

Ciais, P, Rayner, P, Chevallier, F, Bousquet, P, Logan, M, et al. 2010 Atmospheric inversions for estimating CO_2 fluxes: methods and perspectives. *Climatic Change* **103**: 69–92. DOI: https://doi.org/10.1007/s10584-010-9909-3

Davis, KJ 2005 Well-calibrated CO_2 mixing ratio measurements at flux towers: The virtual tall towers approach. 12th WMO/IAEA Meeting of Experts on Carbon Dioxide Concentration and Related Tracers Measurement Techniques, Toronto, Canada, 15–18 September 2003, WMO GAW Report no. 161, 101–108.

Deng, AJ, Lauvaux, T, Davis, KJ, Gaudet, BJ, Miles, N, et al. 2016 Toward reduced transport errors in a high resolution urban CO_2 inversion system. *Elem Sci Anth*, in press.

Deng, A, Seaman, NL, Hunter, GK and **Stauffer, DR** 2004 Evaluation of inter-regional transport using the MM5/SCIPUFF system. *J Appl Meteor*, **43**: 1864–1886, 2004. DOI: https://doi.org/10.1175/JAM2178.1

Deng, A, Stauffer, DR, Gaudet, BJ, Dudhia, J and **Hacker, J** 2009 Update on WRF-ARW end-to-end multi-scale FDDA system, 10th Annual WRF Users' Workshop, Boulder, CO, June 23, 14 pp. (Available at: http://www2.mmm.ucar.edu/wrf/users/workshops/WS2009/abstracts/1-09.pdf).

Durant, AJ, Le Quere, C, Hope, C and **Friend, AD** 2011 Economic value of improved quantification in global sources and sinks of carbon dioxide. *Philos Trans A* **369**: 1967–1979. DOI: https://doi.org/10.1098/rsta.2011.0002

European Commission, Joint Research Centre (JRC)/ Netherlands Environmental Assessment Agency (PBL) 2013 Emission Database for Global Atmospheric Research (EDGAR), release EDGARv4.2 FT2010. (Available at: http://edgar.jrc.ec.europa.eu).

GAW Report No. 229 2016 18th WMO/IAEA Meeting on Carbon Dioxide, Other Greenhouse Gases and Related Tracers Measurement Techniques (GGMT-2015), La Jolla, CA, USA, 13–17 September 2015.

Gurney, KR, Mendoza, D, Zhou, Y, Seib, B and **Fischer, M** 2009 The Vulcan Project: High resolution fossil fuel combustion CO_2 emissions fluxes for the United States. *Environ Sci Technol* **43**. DOI: https://doi.org/10.1021/es900806c

Gurney, KR, Razlivanov, I, Song, Y, Zhou, Y, Benes, B, et al. 2012 Quantification of fossil fuel CO_2 emissions on the building/street scale for a large U.S. city. *Environ Sci Technol* **46**: 12194–12202. DOI: https://doi.org/10.1021/es3011282

Haszpra, L, Barcza, Z, Haszpra, T, Pátkai, ZS and **Davis, KJ** 2015 How well do tall-tower measurements characterize the CO_2 mole fraction distribution in the planetary boundary layer? *Atmos Meas Tech* **8**: 1657–1671. DOI: https://doi.org/10.5194/amt-8-1657-2015

Horst, TW and **Weil, JC** 1992 Footprint estimation for scalar flux measurements in the atmospheric surface layer. *Bound Lay Meteorol* **59**: 279–296. DOI: https://doi.org/10.1007/BF00119817

IEA: World Energy Outlook 2008 Organization for Economic Co-operation and Development/International Energy Agency (OECD/IEA), 578 pp.

IPCC: Climate Change 2014 Synthesis Report. Contribution of Working Groups I, II and III to the Fifth Assessment Report of the Intergovernmental Panel on Climate Change [Core Writing Team, Pachauri, RK and Meyer, LA, (eds.)]. IPCC, Geneva, Switzerland, 151 pp.

Jin, S, Yang, L, Danielson, P, Homer, C, Fry, J, et al. 2013 A comprehensive change detection method for updating the National Land Cover Database to circa 2011. *Remote Sens Environ* **132**: 159–175. DOI: https://doi.org/10.1016/j.rse.2013.01.012

Jobson, BT, McKeen, SA, Parrish, DD, Fehsenfeld, FC, Blake, DR, et al. 1999 Trace gas mixing ratio variability versus lifetime in the troposphere and stratosphere: Observations. *J Geophys Res* **104**: 16,091–16,113. DOI: https://doi.org/10.1029/1999JD900126

Karion, A, Sweeney, C, Kort, EA, Shepson, PB, Brewer, A, et al. 2015 Aircraft-based estimate of total methane emissions from the Barnett Shale region. *Environ Sci Technol* **49**: 8124–8131. DOI: https://doi.org/10.1021/acs.est.5b00217

Lauvaux, T, Miles, N, Deng, A, Richardson, S, Cambaliza, MO, et al. 2016 High resolution atmospheric inversion of urban CO_2 emissions during the dormant season of the Indianapolis Flux Experiment (INFLUX). *J Geophys Res Atmos* **121**: 5213–5236. DOI: https://doi.org/10.1002/2015JD024473

Lauvaux, T, Miles, NL, Richardson, SJ, Deng, A, Stauffer, DR, et al. 2013 Urban emissions of CO_2 from Davos, Switzerland: The first real-time monitoring system using an atmospheric inversion technique. *J Appl Meteor Climatol* **52**: 2654–2668. DOI: https://doi.org/10.1175/JAMC-D-13-038.1

Lauvaux, T, Schuh, A, Uliasz, M, Richardson, S,

Miles, N, et al. 2012 Constraining the CO_2 budget of the corn belt: Exploring uncertainties from the assumptions in a mesoscale inverse system. *Atmos Chem Phys* **12**: 337–354. DOI: https://doi.org/10.5194/acp-12-337-2012

Mao, H and **Talbot, R** 2004 O_3 and CO in New England: Temporal variations and relationships. *J Geophys Res* **109**(D21): 304. DOI: https://doi.org/10.1029/2004JD004913

Marland, G and **Boden, T** 1993 The magnitude and distribution of fossil-fuel-related carbon releases. In NATO ASI Series, The Global Carbon Cycle, edited by Heimann, M, Spring Berlin Heidelberg, **15**: 117–138. DOI: https://doi.org/10.1007/978-3-642-84608-3_5

Marland, G, Rotty, RM and **Treat, NL** 1985 CO_2 from fossil fuel burning: global distribution of emissions. *Tellus* **37B**: 243–258. DOI: https://doi.org/10.1111/j.1600-0889.1985.tb00073.x

Mays, KL, Shepson, PB, Stirm, BH, Karion, A, Sweeney, C, et al. 2009 Aircraft-based measurements of the carbon footprint of Indianapolis. *Environ Sci Technol* **43**: 7316–7823. DOI: https://doi.org/10.1021/es901326b

McKain, K, Wofsy, SC, Nehrkorn, T, Eluszkiewicz, J, Ehleringer, JR, et al. 2012 Assessment of ground-based atmospheric observations for verification of greenhouse gas emissions from an urban region. *Proc Nat Acad Sci* **109**(22): 8423–8428. DOI: https://doi.org/10.1073/pnas.1116645109

Miles, NL, Richardson, SJ, Davis, KJ, Lauvaux, T, Andrews, AE, et al. 2012 Large amplitude spatial and temporal gradients in atmospheric boundary layer CO_2 mole fractions detected with a tower-based network in the U.S. Upper Midwest. *J Geophys Res B* **117**(G01): 019. DOI: https://doi.org/10.1029/2011JG001781

Nakanishi, M and **Niino, H** 2004 An improved Mellor-Yamada Level-3 model with condensation physics: Its design and verification. *Bound Layer Meteor* **112**: 1–31. DOI: https://doi.org/10.1023/B:BOUN.0000020164.04146.98

Nisbet, E and **Weiss, R** 2010 Top-down versus bottom-up. *Science* **328**(5983). DOI: https://doi.org/10.1126/science.1189936

Oda, T and **Maksyutov, S** 2011 A very high-resolution (1 km × 1 km) global fossil fuel CO_2 emission inventory derived using a point source database and satellite observations of nighttime lights. *Atmos Chem Phys* **11**: 543–556. www.atmos-chem-phys.net/11/543/2011/, DOI: https://doi.org/10.5194/acp-11-543-2011

Pacala, SW, Breidenich, C, Brewer, PG, Fung, IY, Gunson, MR, et al. 2010 Verifying greenhouse gas emissions: Methods to support international climate agreements. Committee on methods for estimating greenhouse gas emissions, National Research Council.

Patton, EG, Sullivan, PP and **Davis, KJ** 2003 The influence of a forest canopy on top-down and bottom-up diffusion in the planetary boundary layer. *Quart J*

Roy Meteor **129**(590): 1415–1434. DOI: https://doi.org/10.1256/qj.01.175

Peischl, J, Ryerson, TB, Holloway, JS, Parrish, DD, Trainer, M, et al. 2010 A top-down analysis of emissions from selected Texas power plants during TexAQS 2000 and 2006. *J Geophys Res* **115**(D16): 303. DOI: https://doi.org/10.1029/2009JD013527

Rayner, PJ, Raupach, MR, Paget, M, Peylin, P and **Koff, E** 2010 A new global gridded data set of CO_2 emissions from fossil fuel combustion: Methodology and evaluation, *J Geophys Res* **115**(D19): 306. DOI: https://doi.org/10.1029/2009JD013439

Richardson, SJ, Miles, NL, Davis, KJ, Lauvaux, T, Martins, D, et al. 2016 CO_2, CO, and CH_4 surface in situ measurement network in support of the Indianapolis FLUX (INFLUX) Experiment. *Elem Sci Anth*, under review.

Rogers, RE, Deng, A, Stauffer, DR, Gaudet, BJ, Jia, Y, et al. 2013 Application of the Weather Research and Forecasting Model for air quality modeling in the San Francisco Bay Area. *J Appl Meteor* **52**: 1953–1973. DOI: https://doi.org/10.1175/JAMC-D-12-0280.1

Schuh, AE, Lauvaux, T, West, TO, Denning, AS, Davis, KJ, et al. 2013 Evaluating atmospheric CO_2 inversions at multiple scales over a highly inventories agricultural landscape. *Global Change Biol* **19**(5): 1424–1439. DOI: https://doi.org/10.1111/gcb.12141

Swanson, AL, Blake, NJ, Atlas, E, Flocke, F, Blake, DR, et al. 2003 Seasonal variations of C_2–C_4 nonmethane hydrocarbons and C_1–C_4 alkyl nitrates at the Summit research station in Greenland. *J Geophys Res* **108**(D2): 4065. DOI: https://doi.org/10.1029/2001JD001445

Turnbull, J, Guenther, D, Karion, A, Sweeney, C, Anderson, E, et al. 2012 An integrated flask sample collection system for greenhouse gas measurements. *Atmos Meas Tech* **5**: 2321–2327. DOI: https://doi.org/10.5194/amt-5-2321-2012

Turnbull, J, Sweeney, C, Karion, A, Newberger, T,

Tans, P, et al. 2015 Towards quantification and source sector identification of fossil fuel CO_2 emissions from an urban area: Results from the INFLUX experiment. *J Geophys Res Atmos* **120**. DOI: https://doi.org/10.1002/2014JD022555

Uliasz, M 1994 Lagrangian particle modeling in mesoscale applications. In Environmental Modelling II, ed. Zanetti, P, Computational Mechanics Publications, 71–102.

U.S. Energy Information Administration 2016 Available at: http://www.eia.gov/electricity/data/browser/, last accessed 2 October 2016.

Verhulst, KR, Karion, A, Kim, J, Salameh, PK, Keeling, RF, et al. 2017 Carbon dioxide and methane measurements from the Los Angeles Megacity Carbon Project: 1. Calibration, urban enhancements, and uncertainty estimates. *Atmos Chem Phys Discuss*, under review. DOI: https://doi.org/10.5194/acp-2016-850

Vimont, IJ, Turnbull, JC, Petrenko, VV, Place, PP, Karion, A, et al. 2016 Carbon monoxide isotopic measurements in a US urban center confirm traffic emissions as the dominant wintertime carbon monoxide source. *Elem Sci Anth*, under review.

Wang, W, Davis, KJ, Yi, C, Patton, EG, Butler, MP, et al. 2007 Estimating daytime CO_2 fluxes over a mixed forest from tall tower mixing ratio measurements. *Bound Layer Meteor* **124**: 305–314. DOI: https://doi.org/10.1007/s10546-007-9162-0

Wyngaard, JC and **Brost, RA** 1984 Top-down and bottom-up diffusion of a scalar in the convective boundary layer. *J Atmos Sci* **41**: 102–112. DOI: https://doi.org/10.1175/1520-0469(1984)041<0102:TDABUD>2.0.CO;2

Zhou, Y and **Gurney, K** 2010 A new methodology for quantifying on-site residential and commercial fossil fuel CO_2 emissions at the building spatial scale and hourly time scale. *Carbon Manage* **1**(1): 45–56. DOI: https://doi.org/10.4155/cmt.10.7

3

Toward reduced transport errors in a high resolution urban CO_2 inversion system

Aijun Deng*, Thomas Lauvaux§, Kenneth J. Davis§, Brian J. Gaudet§, Natasha Miles§, Scott J. Richardson§, Kai Wu§, Daniel P. Sarmiento§, R. Michael Hardesty[†], Timothy A. Bonin[†], W. Alan Brewer[†] and Kevin R. Gurney[‡]

We present a high-resolution atmospheric inversion system combining a Lagrangian Particle Dispersion Model (LPDM) and the Weather Research and Forecasting model (WRF), and test the impact of assimilating meteorological observation on transport accuracy. A Four Dimensional Data Assimilation (FDDA) technique continuously assimilates meteorological observations from various observing systems into the transport modeling system, and is coupled to the high resolution CO_2 emission product Hestia to simulate the atmospheric mole fractions of CO_2. For the Indianapolis Flux Experiment (INFLUX) project, we evaluated the impact of assimilating different meteorological observation systems on the linearized adjoint solutions and the CO_2 inverse fluxes estimated using observed CO_2 mole fractions from 11 out of 12 communications towers over Indianapolis for the Sep.-Nov. 2013 period. While assimilating WMO surface measurements improved the simulated wind speed and direction, their impact on the planetary boundary layer (PBL) was limited. Simulated PBL wind statistics improved significantly when assimilating upper-air observations from the commercial airline program Aircraft Communications Addressing and Reporting System (ACARS) and continuous ground-based Doppler lidar wind observations. Wind direction mean absolute error (MAE) decreased from 26 to 14 degrees and the wind speed MAE decreased from 2.0 to 1.2 m s[-1], while the bias remains small in all configurations (< 6 degrees and 0.2 m s[-1]). Wind speed MAE and ME are larger in daytime than in nighttime. PBL depth MAE is reduced by ~10%, with little bias reduction. The inverse results indicate that the spatial distribution of CO_2 inverse fluxes were affected by the model performance while the overall flux estimates changed little across WRF simulations when aggregated over the entire domain. Our results show that PBL wind observations are a potent tool for increasing the precision of urban meteorological reanalyses, but that the impact on inverse flux estimates is dependent on the specific urban environment.

Keywords: Greenhouse gas; Transport; Weather and Research Forecasting Model (WRF); Inversion

1. Introduction

Inversion of atmospheric tracers (Tarantola 2005) is a widely used method to determine the surface fluxes of greenhouse gases (GHGs) such as carbon dioxide (CO_2) (Tans et al. 1990; Enting and Mansbridge 1989) using observations from various observing platforms such as towers (Richardson et al. 2012, Andrews et al. 2014, Miles et al. 2016, Richardson et al. 2016), aircraft (e.g. Gerbig et al. 2003), ground-based remote sensing (e.g. Wunch et al. 2010) and more recently satellite measurements (Crisp et al. 2004). Inverse methods have the potential to support both global- and local-scale GHG emissions monitoring

(Pacala et al., 2010), pending adequate accuracy and precision of the inverse flux estimates. Inverse GHG flux estimate methods require an accurate representation of the physical relationship between the GHG mole fractions measured by these various observing platforms and the original surface fluxes of GHGs (Enting 2002). If atmospheric transport uncertainty is large, the observational data have limited impact in determining the true GHG surface fluxes. However, atmospheric transport errors are not typically rigorously quantified for inverse flux estimates, but research suggests that transport errors have a large impact on atmospheric GHG mole fraction estimates (Lin and Gerbig 2005, Díaz-Isaac et al. 2014) and flux estimates (Baker et al 2006, Lauvaux et al. 2009, Lauvaux and Davis 2014). Few studies have attempted to assimilate jointly meteorological observations and atmospheric greehouse gas measurements (e.g. Kang et al., 2012; Agustí-Panareda et al., 2016). Therefore, improving atmospheric transport model performance should

* Utopus Insights, Inc., New York, US

§ The Pennsylvania State University, Pennsylvania, US

† Cooperative Institute for Research in Environmental Sciences, University of Colorado/NOAA, Chemical Sciences Division, Colorado, US

‡ Arizona State University, Arizona, US

Corresponding author: Aijun Deng (axd157@psu.edu)

substantially improve the accuracy and precision of inverse GHG flux estimates.

An objective of the Indianapolis Flux Experiment (INFLUX) is to quantify and improve the precision and accuracy of atmospheric inverse methods for determining urban CO_2 emissions (Davis et al. 2016, this issue). Improving the accuracy and precision of atmospheric transport model solutions is thus a critical element of this study. Methods for improving transport model solutions can be divided into four broad classifications, increasing model resolution, improving model physical parameterizations, improving input data used to drive the simulation, and assimilation of meteorological observations.

Law et al. (2003) attempted to reduce transport model errors by increasing model resolution, potentially reducing the representation errors affecting global scale models (Ahmadov et al. 2007, Carouge et al. 2010). Higher resolutions such as regional applications of inverse modeling (e.g. Schuh et al., 2010; 2013; Lauvaux et al. 2016), demand greater atmospheric transport fidelity due to highly dynamic continental meteorology (Law et al. 2008), and the presence of observations obtained largely within the planetary boundary layer (PBL) and in close proximity to strong sources and sinks (e.g. Peters et al. 2007, Miles et al. 2016, this issue). Simply increasing resolution is thus unlikely to solve the issue of transport model accuracy since resolution and complexity increase jointly. In addition, increased resolution does not ensure accuracy in model physical parameterizations or input data. Urban inversions are thus likely to suffer from significant transport errors similar to regional and global inversions (Feng et al. 2016).

The Weather Research and Forecasting (WRF) model is a state-of-the-science community-supported numerical weather prediction (NWP) and atmospheric simulation system. WRF has been used worldwide for both research and operational applications (Skamarock et al. 2008). Its ability to simulate atmospheric processes relevant to atmospheric transport and dispersion has been tested widely (Cintineo et al. 2014, Clark et al. 2015, Coniglio et al. 2013, Hariprasad et al. 2014, Rogers et al. 2013, Lauvaux et al. 2013, Karion et al. 2015). In addition to its advanced numerical scheme and continuously upgraded array of model physics parameterizations, WRF has a four dimensional data assimilation (FDDA) capability implemented by Penn State University (Deng et al. 2009) that allows meteorological observations to be continuously assimilated, allowing WRF to produce dynamic analyses at user-desired resolution. Rogers et al. (2013) investigated the effect of various FDDA strategies on the accuracy of the WRF-simulated mesoscale features over the Central Valley of California. The optimal FDDA configuration identified in Rogers et al. (2013) has been used in many recent modeling studies involving studying GHGs (Lauvaux et al. 2013, Karion et al. 2015).

Towards that end, we test the impact of assimilating various meteorological data on the transport model accuracy for the INFLUX region, and the effect of assimilating these meteorological data sources on the inverse CO_2 fluxes derived for the city. Since wind speed, wind direction, and the depth of mixing in the PBL are expected to be critical to accurate and precise GHG flux estimation, the regional operational weather observation network was enhanced with a commercial compact Halo Streamline Doppler lidar (Pearson et al. 2009) capable of continuous observations of those key atmospheric variables. The lidar data complement the standard WMO rawinsondes (only available at 00 and 12 UTC), and aircraft measurements from the commercial airline program Aircraft Communications Addressing and Reporting System (ACARS, Anderson 2010). We evaluate the performance of the inversion system over 2 months (September-October 2013) using WRF atmospheric simulations that assimilate different combinations of data sources from WMO surface stations, the lidar, and in-situ data from ACARS, and assess the overall impact on WRF performance for each of the meteorological instrumentation used in our assimilation system. Finally, we conduct several sensitivity studies to assess the impact of these different meteorological data assimilation choices on the inverse CO_2 flux estimates obtained over the two-month period. Section 2 briefly describes the WRF model used in this study. Model configurations and experimental design are discussed in Section 3. Model results and discussions are given in Section 4, and a summary and conclusions are given in Section 5. This study complements a companion study by Sarmiento et al (2016, this issue) that explores the dependence of model performance on both the choice of land surface and PBL parameterizations, and the input data used to describe the urban surface.

2. Model description

The modeling system used in this research consists of a mesoscale atmospheric modeling component that handles forward transport and dispersion, a Lagrangian Particle Dispersion Model (LPDM, Uliasz 1994) that is run backward in time and driven by the flow fields simulated by the atmospheric transport model, WRF, and a Bayesian the inversion modeling system to compute the posterior fluxes given the prior fluxes, the transport adjoint solutions, and the estimates of prior emissions errors, transport model errors and measurement errors. The mesoscale atmospheric model used here is the Weather Research and Forecasting (WRF) model, a mesoscale atmospheric modeling system that has been used worldwide for both research and operational applications (Skamarock et al. 2008). WRF's development is supported by the broad scientific community, along with very active participation of university scientists worldwide. WRF has a flexible, portable code that runs efficiently in computing environments ranging from massively parallel supercomputers and clusters to laptops.

WRF is a non-hydrostatic, fully compressible three dimensional (3D) primitive equation model with a terrain-following, hydrostatic pressure vertical coordinate, and is designed for simulating atmospheric phenomena across scales ranging from large eddies (~100 m) to mesoscale circulations and waves (~1 km to 100 km) to synoptic-scale weather systems (~1000 km). These applications include real-time NWP, model physics research, regional climate simulation, hazard prediction modeling and, with the addition of a chemistry module (WRF-Chem, Grell et al. 2005), air-quality studies.

The WRF model includes a complete suite of atmospheric physical processes that interact with the model's dynamics and thermodynamics core. These physical processes include cloud microphysics (MP), cumulus parameterization needed on coarser grids ($\Delta x > O(10 \text{ km})$) for representing the un-resolved atmospheric convection, atmospheric radiation, PBL/turbulence physics, and land surface models (LSMs). Selection of the model physics suite in this research is based on previous research (e.g., Lauvaux et al. 2013, Rogers et al. 2013). For cloud microphysics, this study uses the WRF single-moment five-class (WSM5) simple ice scheme (Hong et al. 2004) that assumes no mixed-phase conditions. For cumulus parameterization, the Kain–Fritsch scheme (Kain and Fritsch 1990, 1993; Kain 2004) is used on the 9-km grid (see next section for grid configuration). For atmospheric radiation, the Rapid Radiative Transfer Model (RRTM; Mlawer et al. 1997) longwave (LW)/Dudhia shortwave (SW; Dudhia 1989) scheme is used. For PBL turbulent processes, the turbulent kinetic energy (TKE)-predicting Mellor-Yamada Nakanishi Niino (MYNN) Level 2.5 turbulent closure scheme (Nakanishi and Niino 2006) is used, along with the MYNN surface layer scheme to preserve consistency. The decision of selecting the MYNN PBL scheme is based on our experiences and previous studies where MYNN appeared to produce the most accurate PBL temperature and moisture profiles (Cintineo et al. 2014, Clark et al. 2015) as well as the most accurate PBL depth (Coniglio et al. 2013, Hariprasad et al. 2014) in simulations of the PBL over land, all of which are highly important to simulating transport and dispersion of surface emissions. For land surface processes, the Noah LSM (Chen and Dudhia 2001, Tewari et al. 2004) is used. The Noah LSM is a four-layer soil temperature and moisture scheme and includes plant root zone, evapotranspiration, soil drainage, and runoff, taking into account vegetation categories, monthly vegetation fraction, soil texture, and snow and ice cover.

The WRF modeling system has four dimensional data assimilation (FDDA) capabilities that allow continuous assimilation of meteorological observations during the model simulation, unlike intermittent approaches used in variational data assimilation techniques. For retrospective applications, FDDA can be used in numerical models to produce accurate dynamic analyses at the desired temporal and spatial resolution. FDDA has been widely used in studying atmospheric transport and dispersion processes (Deng et al. 2004, 2006, Rogers et al. 2013, Lauvaux et al. 2013, Karion et al. 2015). The version of FDDA used in this research was originally developed for MM5 and was enhanced and implemented into WRF (Deng et al. 2009, Rogers et al. 2013). Further enhancements to the observation nudging technique in WRF have brought more flexibility to control how surface observations influence meteorology aloft. WRF users have freedom to choose different vertical weighting functions for the surface observations (Rogers et al. 2013). Unlike Rogers et al. (2013) in which various FDDA strategies were evaluated to identify

the optimal FDDA settings to produce the most accurate model solutions to represent the meteorological conditions, this research uses the findings from Rogers et al. (2013) to focus on exploring the effect of assimilating various meteorological observation types on the model solution as well as on the inverse flux solutions driven by the WRF-simulated meteorological solutions. The FDDA strategy used in this research includes using analysis nudging and observation nudging on the 9-km coarse grid, and only observation nudging on the 3- and 1-km fine grids, with similar multiscale configurations and parameter settings to those used in Rogers et al. (2013), Lauvaux et al. (2013), and Karion et al. (2015).

To solve the inverse problem, surface fluxes are related to the observed atmospheric mole fraction via a representation of atmospheric dynamics referred to as the observation operator (Lauvaux et al. 2012). This linear operator is then inverted to allow for the optimization of fluxes given prior flux estimates, atmospheric observations and the associated uncertainties. Whereas the WRF model represents the projection of the surface fluxes into atmospheric mole fraction, the inverse of the operator is still required in order to solve for the inverse problem. To represent the inverse operator, also called influence function, we use the Lagrangian Particle Dispersion Model (Uliasz 1994) driven by the hourly mean wind fields, potential temperature, and the turbulence simulated by the WRF model as discussed above. The mean wind fields and the TKE will drive the LPDM backward in time so that areas of influence at the surface for any given tower observation can be represented by the model solution. LPDM has been used extensively in the past for various applications (e.g. Pielke and Uliasz, 1993, Schuh et al. 2010, Lauvaux et al. 2012), and more recently to perform urban inversions of CO_2 over Indianapolis (Lauvaux et al. 2016). Lauvaux et al. (2016) provides a complete description of the coupling between WRF and LPDM used here.

The inversion system is solving for the exact solution of the Bayesian inverse problem, i.e. producing the analytical flux solution with its associated error covariances (i.e. Kalman method) (Tarantola 2005). Prior CO_2 emissions estimates within the inner domain are drawn from the Hestia 2012 emissions product (Gurney et al. 2012). Hestia is a high-resolution (i.e., building-level) inventory-based data product created to estimate anthropogenic CO_2 emissions for the Indianapolis area. It combines several datasets such as energy consumption, traffic data, industrial productivity, and electricity generation from the power plant, with models such as a building energy model. Forwards-in-time atmospheric CO_2 simulations based on emissions from the Hestia dataset within WRF-Chem are only used to compare to the mole fractions estimated with WRF-LPDM. Time periods with large discrepancies between forwards and adjoint atmospheric CO_2 mole fractions reveal unusually difficult transport conditions, and these times are excluded from the inverse analyses.

The inverse system for Indianapolis has been described in Lauvaux et al. (2016) for the period September 2012 to May 2013. Here, we perform five-day emission inversions for September and October 2013 using the reference configuration as described in Lauvaux et al. (2016). The minimzation is performed using a Kalman filter producing the analytical solution to the inverse problem (Tarantola et al., 2005). Similar approaches have been used at global (e.g. Law et al., 2003), continental (Carouge et al., 2010), or regional (Lauvaux et al., 2008). The prior error structures prescribed in the inversion follow the urban land mask and a correlation length scale of 4 km, both convolved to generate the prior emission covariance matrix similar to Carouge et al. (2010) and Lauvaux et al. (2012). The control vector to be optimized in the system includes only the fossil fuel emissions, ignoring biogenic fluxes for this time of year. Turnbull et al. (2015) demonstrated the limited influence of biogenic signals during October to April (less than 5%) relatively too small to be optimized over that time of year. The background conditions were determined using the optimal tower location depending on wind speed, as detailed in Lauvaux et al. (2016). Of most importance, we use the wind direction to define the optimal background sites for each observation time and perform multiple inversions using identical assumptions only swapping the influence functions from the different

WRF-FDDA simulations. This comparison allows us to isolate and quantify the sensitivity of the inverse estimate of CO_2 emissions to the different transport solutions.

3. Model configuration, meteorological observations, and experiment design

3.1. Model configuration

The WRF modeling system used in this research is based on WRF V3.5.1, released in September 2013. The INFLUX WRF configuration consists of three nested grids with 9-/3-/1-km horizontal resolutions (**Figure 1a**), with the focus on the 1-km grid. The topographic and landuse database needed to initialize the WRF model is based on the U.S. Geological Survey (USGS) 30-second terrain and 24-category landuse. As indicated by the landuse distributions (**Figure 1b, c, d**), a significant fraction of the 1-km grid is characterized as urban (**Figure 1d**). In the vertical, fifty nine (59) vertical terrain-following layers are used, with the center point of the lowest model layer located ~7 m above ground level (AGL). The thickness of the layers increases gradually with height, with 24 layers below 850 hPa (~1550 m above sea level). The top of the model is set at 100 hPa.

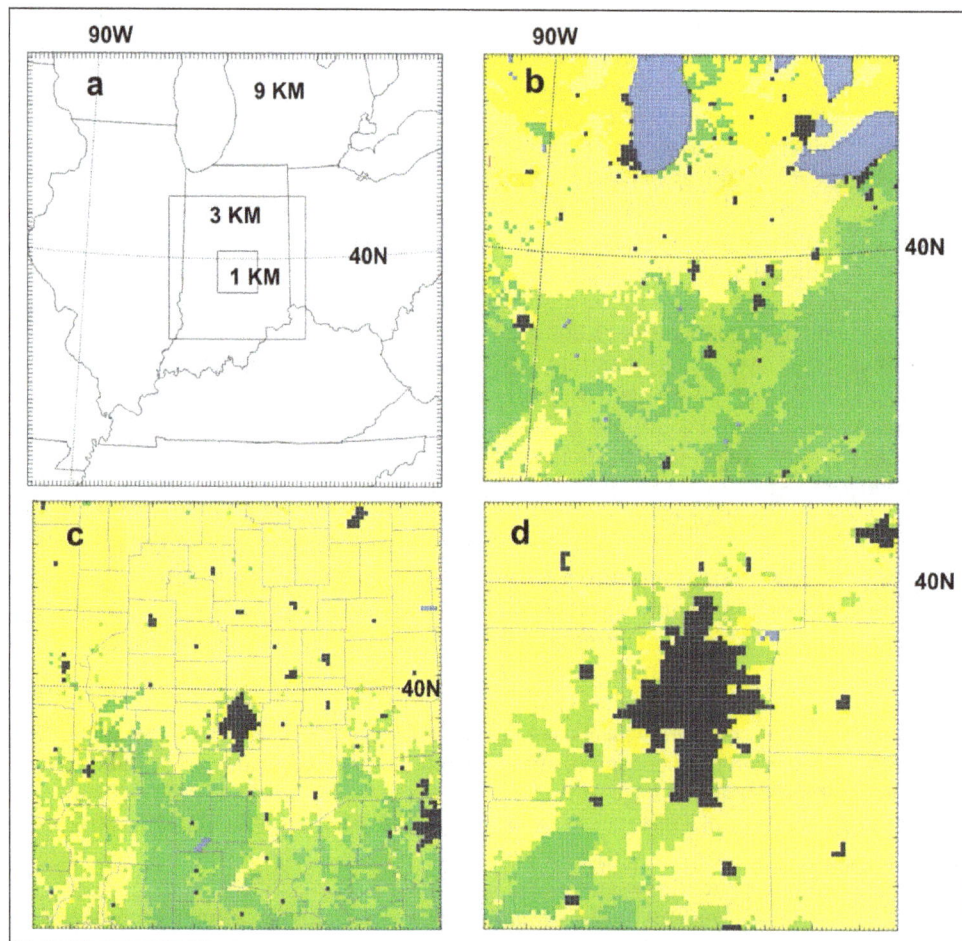

Figure 1: WRF grid configuration. a) WRF 9-/3-/1-km resolution grids, **b)** 9-km grid landuse, **c)** 3-km grid landuse, and **d)** 1-km grid landuse. Graph b corresponds to the full areas shown in graph a, and graphs c and d are enlargements corresponding to the areas indicated by the black boxes in graph a. Color notation for the landuse panels: Black: urban; Light Yellow: dryland cropland and pasture; Yellow: mixed dryland-irrigated cropland and pasture; Light Green: irrigated cropland and pasture; Green: forest; Blue: water bodies. DOI: https://doi.org/10.1525/elementa.133.f1

The WRF-Chem system was configured to run for a two-month period (Sept.–Oct. 2013), in five-day segments with a 12-hour overlapping time-window. The WRF model solutions are then used to drive a LPDM that calculates the CO_2 footprints for each of the CO_2 tower observations (Lauvaux et al., 2016). The footprints or influence functions are used in the inversion system to compute the updated posterior CO_2 fluxes.

3.2. Meteorological observations

The meteorological observations assimilated in WRF-FDDA include the standard measurements of 10-m wind, 2-m temperature and moisture fields from World Meteorological Organization (WMO) surface stations at hourly intervals and radiosondes at 12-houly intervals, as shown in **Figure 2**. There are five upper-air observations spread across the 9-km grid. However, none of them are located on the 3- and 1-km grids and only three WMO stations are available on the 1-km grid.

The 20-minute wind profiles measured by the lidar are used in this study. The lidar is deployed in Lawrence, IN, about 15 km to the northeast of downtown Indianapolis.

The lidar has operated and provided data continually since April 2013, with a 6-month gap at the end of 2015 when the system was temporarily removed and upgraded. Additional information about the deployment and data are available online (http://esrl.noaa.gov/csd/projects/influx/). Pearson et al. (2009) provides a complete description of the system and its operating principles. For data presented and used here, range gates (i.e., spatial resolution) were set to be equally spaced at 38.4 m apart. Thus, scales of motion and turbulence larger than the range gate size were explicitly resolved by the lidar.

The lidar directly measures range resolved line-of-sight velocities and backscatter intensity from aerosols and other scatterers. The lidar performs a sequence of velocity azimuth display (VAD) scans at multiple elevations, range height indicator (RHI) scans, and stares. This suite of scans that repeats every 20 min is used to make profiles of the wind speed, wind direction, velocity variances, and backscatter intensity. These profiles are used together to estimate the PBL depth. The 20-minute wind profiles measured by the lidar are also available for assimilation into WRF.

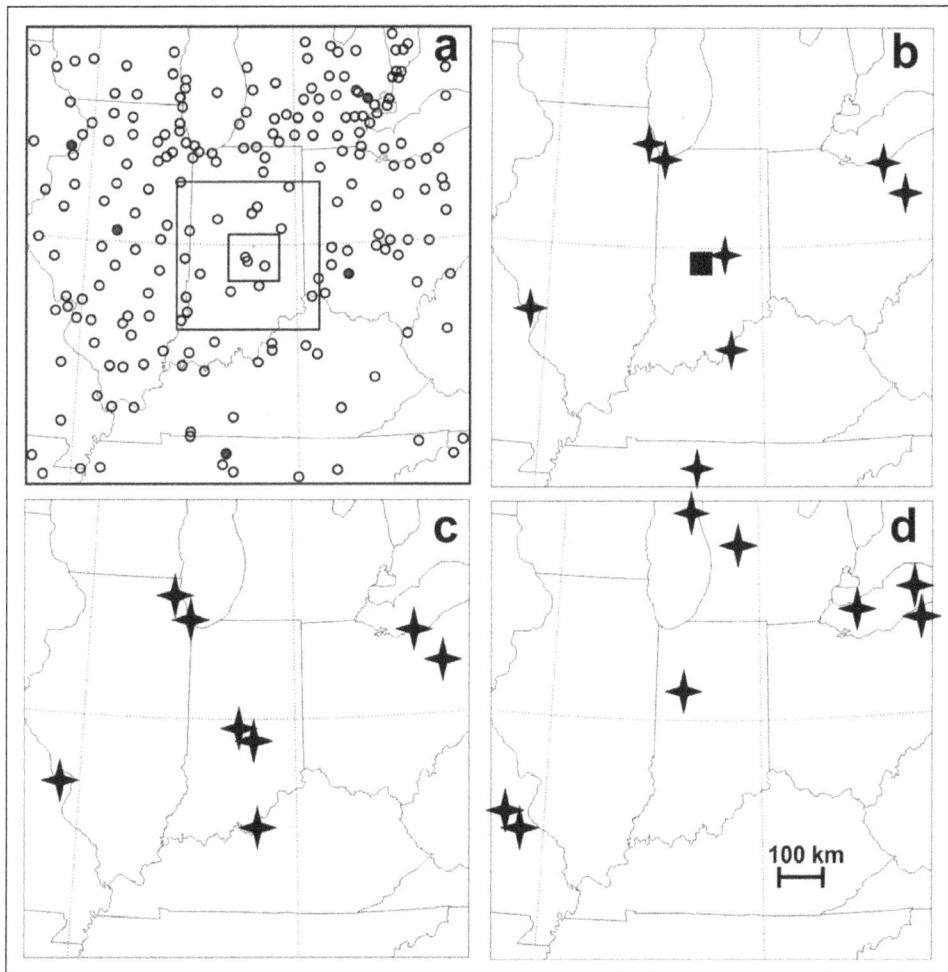

Figure 2: Distribution of the assimilated observations, for 21 UTC 15 October 2013. a) National Weather Service surface (open circles) and radiosonde (solid circles) observations, **b)** LIDAR and ACARS at 935 hPa level, **c)** LIDAR and ACARS at 885 hPa level, and **d)** ACARS at 250 hPa level. Note that star symbols denote the ACARS observations and solid square denotes the Halo lidar observations. DOI: https://doi.org/10.1525/elementa.133.f2

In addition to the WMO and lidar observations, the winds, temperatures and moisture fields observed from the ACARS are also assimilated. The ACARS observations are available at times when observations are taken from commercial aircraft, and are distributed in the vertical from near-surface to cruising altitudes above the tropopause, and in the horizontal along the flight tracks. PBL data density is highest near major airports as aircraft ascend or descend.

3.3. Experimental design

In order to evaluate the effect of assimilating various observations, as shown in **Table 1**, four different WRF configurations (or experiments) are conducted and results of both meteorological fields and posterior CO_2 fluxes are compared among the four experiments: 1) NOFDDA – No data assimilation of any form is applied, and WRF is purely driven by the North America Regional Reanalysis (NARR), an analysis product that is a combination of model background and observations but on a coarse spatial and temporal scale (i.e., 32 km); 2) FDDA_WMO – Only standard WMO hourly surface (winds) and 12-hourly upper-air observations (winds, temperature and water vapor) are assimilated; 3) FDDA_WMO_Lidar – In addition to WMO observations, wind profiles from the local INFLUX lidar are also assimilated; and 4) FDDA_WMO_Lidar_ACARS – In addition to the WMO and lidar observations, the ACARS observations are also assimilated. Since all five WMO sondes are outside the 3- and 1-km grids, the effect of assimilating WMO sondes can only propagate into the 3- and 1-km through the grid boundaries between the 3- and 9-km grids. Since in FDDA the impact of surface observations is limited to the lowest portion of atmosphere up to the model-simulated PBL top (Rogers et al. 2013), it is anticipated that assimilating additional upper air observations such as Lidar and ACARS observations (**Figure 2b, 2c, 2d**) can cause additional improvements in the WRF model solutions so that the transport error in the inversion system is further reduced. During model integration, observations are continuously assimilated using the WRF FDDA technique that is described in Deng et al. (2009) and Rogers et al. (2013). Similar to previous studies (i.e., Deng et al. 2004, 2006, Rogers et al. 2013, Lauvaux et al. 2013, Karion et al. 2015), assimilation of temperature and moisture observations are only allowed above the model-predicted PBL top so that the internal model PBL physics may operate without interference from the data assimilation, while winds are assimilated through the entire atmosphere.

All experiments use identical model physics for all grids except the cumulus parameterization scheme (that is: 4-layer Noah LSM, the 2.5-level MYNN PBL scheme, and the RRTM longwave/Dudhia shortwave scheme are used on all grids, while the KF cumulus scheme is used only on the 9-km grid). For the preliminary evaluations conducted in this paper, each WRF simulation segment of the four experiments is five days long, and was initialized with 3-hourly NARR at 32-km × 32-km resolution for the initial conditions and lateral boundary conditions (ICs/LBCs). The NARR analyses were downloaded from the Research Data Archive (RDA) maintained by the Computational and Information Systems Laboratory (CISL) at the National Center for Atmospheric Research (NCAR).

In addition to assimilating observations during the model integration, the initial condition fields are further enhanced by sonde and surface data through the WRF objective analysis process, Obsgrid, using a modified Cressman analysis method (Deng et al. 2009, Rogers et al. 2013). The three-dimensional (3D) analyses and the surface analysis fields used for analysis FDDA are also enhanced by the objective analysis process and are defined at three-hour intervals (Deng et al. 2009), which means that WRF solutions are nudged towards more accurate analyses than NARR on the WRF grid by including observations.

3.4. Model evaluation methods

The WRF-simulated meteorological fields are evaluated quantitatively by comparing the error statistics of the model-simulated wind speed, wind direction, and temperature. Evaluation is performed on the 1-km grid only since the high-resolution grid is our primary interest. Mean absolute error (MAE) and root mean squared error (RMSE) are calculated to measure how close the model values are to the observed values. Mean error (ME) is calculated to measure the model bias for a given variable. MAE and ME are computed for both the surface and upper air observation locations. For the surface, the WRF (2-m temperature and 10-m wind) values derived from the lowest model layer using similarity theory are compared with the surface observations. For the upper air, the model values are interpolated onto the observation locations in both horizontal and vertical pressure space, and are then compared with the observations. A calm wind threshold was used in this study to remove situations with very light winds (less than or equal to 1 m s⁻¹) for the wind direction statistics calculation because the wind direction for near-calm wind is undefined. Note that due to the limited size of the 1-km grid, there are only three WMO surface stations and no

Table 1: FDDA configuration for WRF simulations. DOI: https://doi.org/10.1525/elementa.133.t1

Exp. Name	Data assimilated
NOFDDA	No meteorological observations
FDDA_WMO	WMO surface and upper-air sonde observations of winds, temperature and water vapor
FDDA_WMO_Lidar	Same as above with addition of INFLUX Halo Doppler lidar winds
FDDA_WMO_Lidar_ACARS	Same as above with addition of ACARS wind, temperature and water vapor

sondes available within the grid, thus reserving a separate set of WMO observations for independent verification (as done in Rogers et al. 2013) is not possible. Since the WRF FDDA technique uses a relaxation term in the model tendency equations rather than conform the model solution to an observation, one should not expect data assimilation to produce an exact match between the modeled and observed values, and spatial correlations in the nudging coefficients limit the extent of the observation influence. As shown in Rogers et al. (2013), the accuracy of a simulation using FDDA scheme highly depends on the accuracy of the model dynamics. Therefore, the model statistics (i.e. MAE and ME) should remain similar with either assimilated or independent data. In addition, as shown in Rogers et al. (2013), meteorological observations within a distance (e.g., ~60 km) are correlated. For these two reasons, the complete set of meteorological observations including WMO, lidar and ACARS was used in model verification. However, the ACARS data could be considered to be independent observations for the first three numerical experiments since they are not assimilated in them; similarly, lidar is not assimilated in the first two experiments and so is independent of those simulations.

In addition to wind speed and wind direction, we present the evaluation of PBL depth which impacts directly the transport and dispersion of trace gases near the surface, and therefore the estimation of surface fluxes in the inversion. Note that PBL depth is not assimilated, thus it is an independent variable for validation. WRF model can produce a diagnosed PBL depth from its PBL sub-model or PBL scheme. Several methods have been used to diagnose the PBL depth within the different PBL schemes that have been proposed in the literature, usually based on either the vertical thermal profile or the vertical profile of TKE as predicted by the specific PBL scheme. Here, we used the 2.5-level MYNN scheme (Nakanishi and Niino 2006) which is a TKE-predictive scheme. We compare the model-predicted PBL-depth with the observed PBL depth to measure how well the model represents PBL processes. Nighttime comparisons of PBL depth were not performed due to the fact that modeled PBL depth in stable conditions is often less reliable and harder to define, and thus nighttime tower data are not currently utilized in our inverse CO_2 flux estimates.

To evaluate the impact of the different model configurations on the CO_2 emissions from the inversion,

we compute the ratios of the Bayes Factors (BF) which represent the ratios of the marginal likelihoods for the different model configurations. BFs relate to the relative values of posterior conditional probabilities for a given model and can be expressed as follows:

$$-2log(BF)=log(|HBH^T+R|)+(y-Hx_b)^T\times(HPH^T+R)^{-1}\times(y-Hx_b)$$

where H is the Jacobian, B and R the prior and data covariances, y the observations and x_b the prior estimate. Larger values of BF correspond to more probable models. By calculating the ratios of BF across our model configurations, we can evaluate the improvement relative to one another. Ratios of BF greater than 10 suggest a strong evidence for improvement, moderate evidence between 3 and 10, and anecdotal evidence between 1 and 3.

4. Results
4.1. Meteorological evaluation
Table 2 shows the ME and MAE of the WRF-predicted 10-m wind direction, wind speed and 2-m temperature over the 1-km grid verified hourly (both day and night) against three WMO surface measurements, averaged over the period between 00 UTC 27 August and 00 UTC 3 November 2013. Comparing the model surface MAE and ME scores among all four numerical experiments, we notice that the greatest error reductions occur between experiments NOFDDA and FDDA_WMO. Surface wind direction MAE (ME) is reduced from 30 to 19 (6 to 2) degrees, and surface wind speed MAE (ME) is reduced from 1.0 to 0.8 (0.2 to 0.1) m s⁻¹. Since lidar and ACARS observations are all above the surface, assimilating lidar and ACARS does not directly improve the surface MAE and ME scores for experiments FDDA_WMO_Lidar and FDDA_WMO_Lidar_ACARS. The MAE and ME scores for both experiments remain similar to the FDDA_WMO experiment (e.g., 19-degree wind direction MAE, 0.8 m s⁻¹ -wind speed MAE for both experiments). Although some slight degradation in wind speed and temperature ME scores is seen in the FDDA_WMO_Lidar_ACARS experiment as compared to the FDDA_WMO experiment, the FDDA_WMO_Lidar_ACARS experiment has the overall smallest MAE scores out of all four experiments. Surface temperature improvements are minimal since temperature assimilation is only allowed above the model-predicted PBL.

Table 2: Mean error and mean absolute error of the WRF-predicted 10-m wind direction, wind speed and 2-m temperature on the 1-km grid, averaged for the three WMO surface measurements and for all times (hourly) over the period between 00 UTC 27 August and 00 UTC 3 November 2013. DOI: https://doi.org/10.1525/elementa.133.t2

		NOFDDA	FDDA_WMO	FDDA_WMO_Lidar	FDDA_WMO_Lidar_ACARS
Wind Direction (Degree)	ME	6	2	2	1
	MAE	30	19	19	19
Wind Speed (m s⁻¹)	ME	0.2	0.1	0	−0.2
	MAE	1.0	0.8	0.8	0.8
Temperature (K)	ME	−1.0	−0.8	−0.9	−1.4
	MAE	2.3	2.3	2.4	2.2

Table 3: Same as Table 2, but verified against the upper-air observations, INFLUX lidar measurements (winds only) and ACARS measurements (winds and temperatures) below 2 km AGL. DOI: https://doi.org/10.1525/elementa.133.t3

		NOFDDA	FDDA_WMO	FDDA_WMO_Lidar	FDDA_WMO_Lidar_ACARS
Wind Direction (Degree)	ME	4	2	−1	0
	MAE	26	24	15	14
Wind Speed (m s⁻¹)	ME	0.2	−0.2	−0.2	−0.2
	MAE	2.0	2.0	1.3	1.2
Temperature (K)	ME	0.8	1.0	1.0	0.5
	MAE	1.3	1.4	1.4	0.8

Table 3 shows ME and MAE of the WRF-predicted wind direction, wind speed, and temperature over the 1-km grid verified hourly (both day and night) against the lower tropospheric (below 2 km AGL) INFLUX lidar measurements (winds only) and ACARS measurements (winds and temperatures), averaged over the period between 00 UTC 27 August and 00 UTC 3 November 2013. The information shown in **Table 3** is the same as **Table 2** except that validations are now performed against all lower-tropospheric (below 2 km AGL) measurements excluding the three surface stations. Now we clearly see error magnitudes that gradually decrease as additional observations are assimilated into the WRF model from NOFDDA to the best model performance with FDDA_WMO_Lidar_ACARS.

As shown in **Table 3**, clear wind error reductions, both MAE and ME, are achieved by assimilating the INFLUX lidar wind measurements. For example, there is a 9-degree reduction in wind direction MAE and 0.7 m s⁻¹ reduction in wind speed MAE. There are no temperature improvements since no temperature observations are available from the lidar. Assimilation of ACARS observations further reduces model MAE and ME error consistently in both wind and temperature fields except the wind speed ME which is already very small. For example, there is a 0.5 °C ME reduction and 0.6 °C MAE reduction in temperature error when ACARS observations are assimilated. Similar to the surface layer error statistics, the FDDA_WMO_Lidar_ACARS case has the smallest overall errors.

To demonstrate and summarize the effect of data assimilation more directly, **Figure 3** and **4** show the scatterplots comparing the simulation without assimilating observations (i.e., NOFDDA) and the best simulation with assimilating all the available observations (i.e., FDDA_WMO_Lidar_ACARS) for the INFLUX project. The improvement due to assimilating observations is evident and the error reduction is significant, especially for wind direction (e.g., nearly 40% reduction in the surface wind direction MAE). Assimilation of upper air wind observations from the INFLUX lidar and ACARS, and temperatures from ACARS clearly improves model performance, especially for wind speed and wind direction (e.g., nearly 50% reduction in upper-air wind direction MAE).

Model errors averaged in the vertical and over the entire two-month period do not represent the temporal and spatial error distributions, both of which are important for the inverse flux corrections at 1-km, five-day resolution.

Figure 5 shows the model error diurnal variation of the WRF-simulated wind direction and wind speed, averaged over the two-month period. It is shown that the wind direction MAEs for all experiments do not show apparent diurnal variations (**Figure 5a**), while wind speed MAEs are larger in daytime than in nighttime (**Figure 5b**), likely due to larger wind speed in the daytime. It is also possible that the larger daytime errors are due to the effect of PBL large eddies that is not well represented in the model, while during the nighttime winds are more influenced by large-scale weather patterns that are controlled by the large-scale dynamics. FDDA reduces the model MAE quite significantly due to assimilating the three WMO surface observations (e.g., nearly 40% error reduction in the surface wind direction). Note that assimilating lidar in FDDA_WMO_Lidar and ACARS in FDDA_WMO_Lidar_ACARS is not expected to further reduce the model surface errors since they are upper-air observations.

Similar to the MAEs, the wind direction MEs (**Figure 5c**) do not show diurnal variations. Although the wind direction MEs are quite small (< 10 degrees), error reductions are noticeable in the three FDDA simulations. For wind speed (**Figure 5d**), like its MAEs, the MEs are larger and positive (model is faster than the observations) in daytime, and FDDA tends to reduce the daytime biases, although the overall MEs are quite small (< 0.6 m s⁻¹). Note that assimilating upper-air observations from INFLUX lidar and ACARS pushed the model bias slightly to the negative side. Since the lidar and ACARS observations do not directly influence the surface errors and the magnitudes of the MEs are quite small (**Figure 5d** and **Table 2**), these small negative biases are likely the artifact of imbalances introduced by insertion of lidar and ACARS observations, or possibly due to the behavior of the model vertical mixing.

Figure 6 shows vertical MAE and ME distributions for WRF-predicted wind direction (**Figure 6a** and **c**) and wind speed (**Figure 6b** and **d**) within the lowest 2.5 km AGL, averaged over the two-month period, comparing among four numerical experiments listed in **Table 1**. For both wind speed and wind direction, NOFDDA has the largest error through the entire atmosphere below ~2 km AGL (except for the wind speed near the 1.5-km level where FDDA_WMO appears to have slightly larger error), with a ~30-degree wind direction error and 2–2.5 m s⁻¹ wind speed error at the surface. Model errors generally decrease

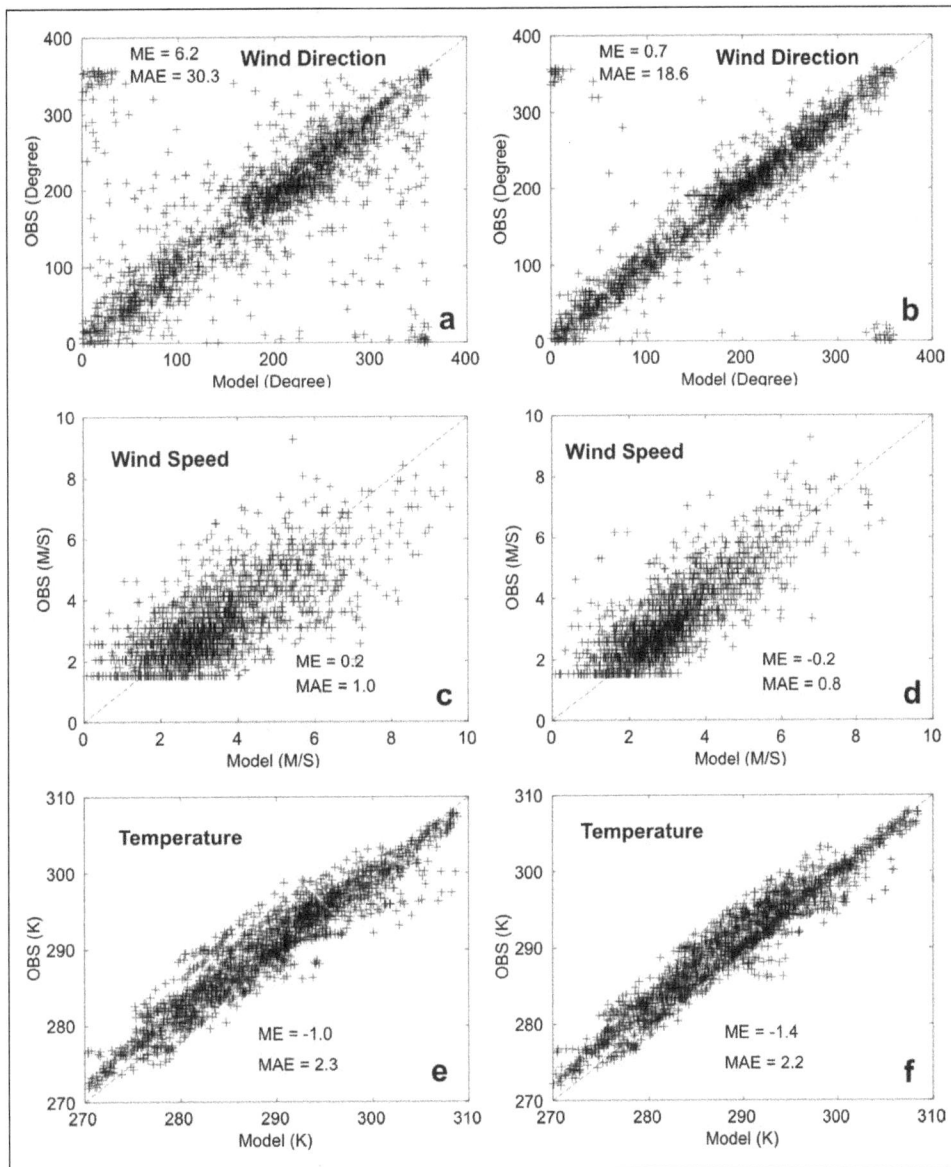

Figure 3: Surface layer model performance. WRF-predicted versus observed wind direction (**a**: NOFDDA, **b**: FDDA_WMO_Lidar_ACARS), wind speed (**c**: NOFDDA, **d**: FDDA_WMO_Lidar_ACARS) and temperature (**e**: NOFDDA, **f**: FDDA_WMO_Lidar_ACARS) on the 1-km resolution grid, between 00 UTC 27 August and 00 UTC 3 November 2013, comparing between NOFDDA (left) and FDDA_WMO_Lidar_ACARS (right) experiments. Note that x-axis represents the model value and y-axis represents the observed value, and each data point represents a model-observation pair averaged for three National Weather Service surface stations inside the 1-km resolution grid at a given hour. The dashed line in the figure represents the y = x line that indicate a perfect model-observation match; thus the points above y = x line represents the situation where model underpredicts and points below represents the situation where model overpredicts. Note that ME represents mean error and MAE represents mean absolute error. DOI: https://doi.org/10.1525/elementa.133.f3

with height. Comparing the NOFDDA and FDDA_WMO experiments, we see that assimilating only WMO surface observations improves the model-predicted winds. Although not significant, the improvement over a vertical layer is expected since the FDDA sub-model in WRF allows the influence of the surface observations to spread to the entire depth of the PBL depending on the stability regime (Deng et al. 2009, Roger et al. 2013); however, the magnitudes of improvements, especially for wind direction, gradually decreases with height. The addition of the INFLUX

lidar winds to the data assimilation further reduces model error. Consistent with **Tables 2** and **3**, the large gap between the FDDA_WMO and FDDA_WMO_Lidar shows that addition of upper air observations creates substantially improved model solutions beyond what can be obtained from assimilating surface observations alone. It is clear that addition of ACARS observations further reduces model errors although the further improvement is not large (likely because after assimilating the lidar data the model errors are already quite small).

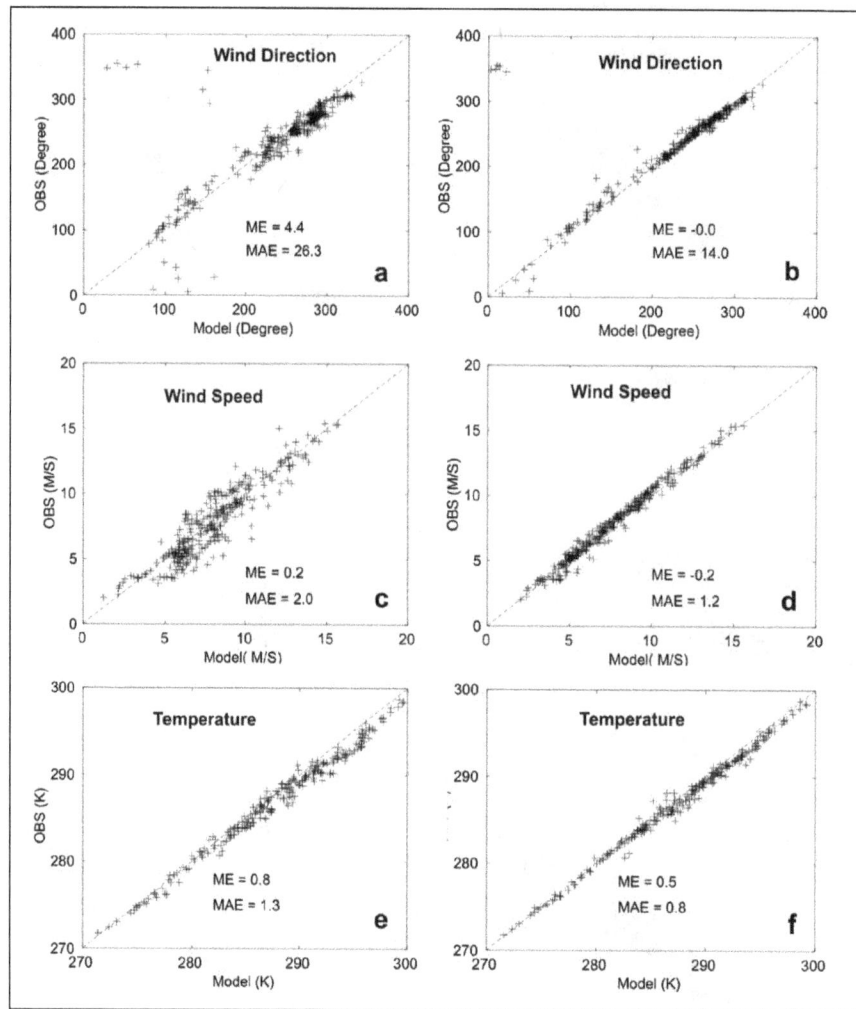

Figure 4: Model performances for the lower troposphere within 2 km AGL. Same as Fig. 3 but for the lower troposphere. Each data point represents a model-observation pair averaged for all the available observations (i.e., Halo LIDAR and ACARS) at a given hour. DOI: https://doi.org/10.1525/elementa.133.f4

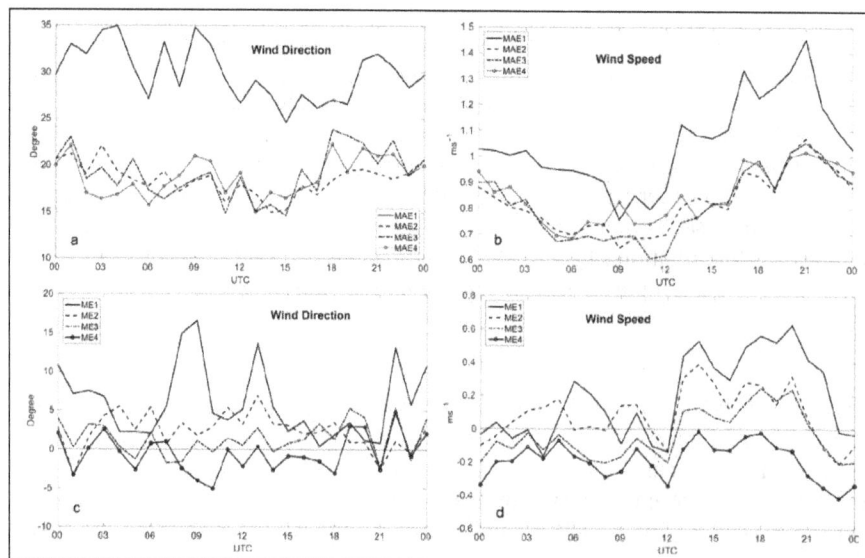

Figure 5: Model surface errors vs UTC time of a day. Daily time series of model errors averaged over the two-month period, for the 1-km grid, for **a)** mean absolute error for wind direction, **b)** mean absolute error for wind speed, **c)** mean error for wind direction, and **d)** mean error for wind speed, for all four experiments: NOFDDA (MAE1/ME1), FDDA_WMO (MAE2/ME2), FDDA_WMO_Lidar (MAE3/ME3) and FDDA_WMO_Lidar_ACARS (MAE4/ME4). Note that ME represents mean error and MAE represents mean absolute error. DOI: https://doi.org/10.1525/elementa.133.f5

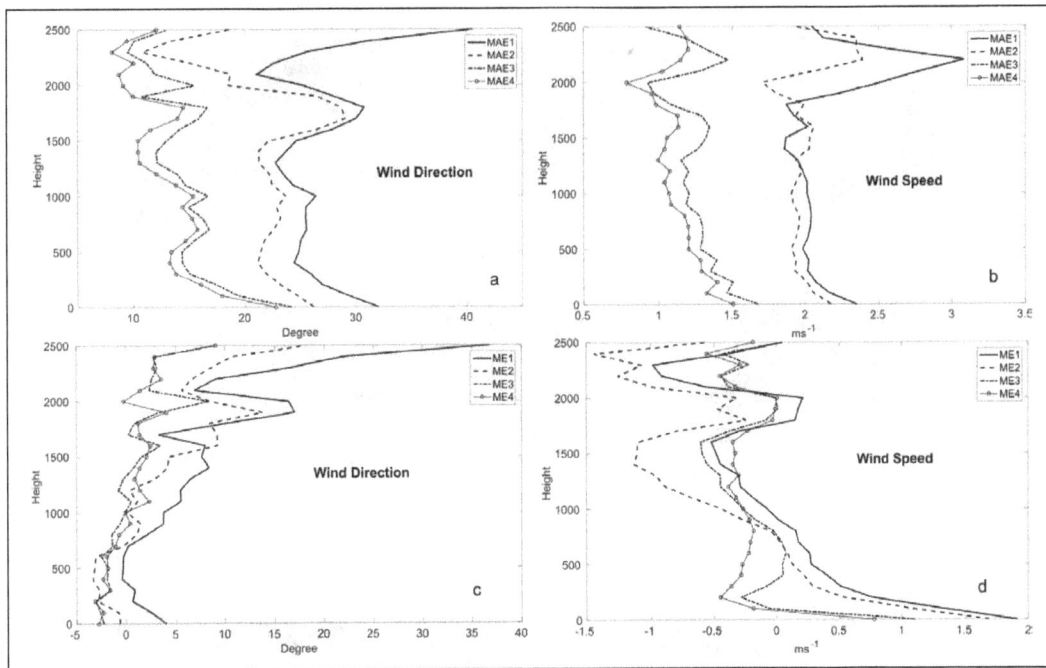

Figure 6: Vertical distribution of model errors. Vertical profile of model errors averaged over the two-month period, for the 1-km grid, for **a)** mean absolute error for wind direction, **b)** mean absolute error for wind speed, **c)** mean error for wind direction, and **d)** mean error for wind speed, for all four experiments: NOFDDA (MAE1/ME1), FDDA_WMO (MAE2/ME2), FDDA_WMO_Lidar (MAE3/ME3) and FDDA_WMO_Lidar_ACARS (MAE4/ME4). Note that ME represents mean error and MAE represents mean absolute error. DOI: https://doi.org/10.1525/elementa.133.f6

The wind direction biases (**Figure 6c**) are quite small for the low levels (< 5 degrees), and biases increases with height. It is apparent that FDDA reduces the biases. Wind speed biases (**Figure 6d**) in NOFDDA demonstrate large positive biases within the lowest 1 km although the magnitudes of biases decrease with height, and becomes somewhat negative in mid- and upper- PBL. Assimilating surface observation slightly reduces the low-level biases, and the error reduction appears to be limited to below 1 km, while assimilating lidar and ACARS observations more significantly reduces model biases.

4.2. Evaluation of model-predicted PBL depth using the Halo Doppler Lidar data

The PBL depth for this study is manually estimated from the lidar observations for each 20-min time period, identified by large gradients in Signal-to-Noise Ratio (SNR) and the height where the vertical velocity variance becomes small (less than ~0.1 $m^2 s^{-2}$). NOAA is currently implementing and testing an algorithm to automate the PBL depth estimation process for the INFLUX data set. As an example, a comparison of PBL structures between WRF and the INFLUX lidar observations at Indianapolis for 29 and 30 August 2013 is shown in **Figure 7**. Generally, the model-predicted TKE structures are highly correlated with the lidar-observed vertical velocity variances and large gradients in SNR. The diurnal variation of the PBL structure can be clearly seen within both model output and observations. However, differences in the vertical extent of the PBL depth between the model output and observations are apparent.

Table 4 compares the MAE and ME of the WRF-predicted PBL depth on the 1-km grid verified hourly against the lidar

estimates of PBL depth at the lidar site in Indianapolis. The evaluation of PBL depth is conducted only for the daytime period between 17 and 22 UTC when a well-mixed PBL is fully developed and quasi-stationary, for the 2-month period between 00 UTC 27 August and 00 UTC 3 November 2013. Our results show that for all experiments the MAE of model-predicted PBL depth is quite similar, in a range between 223 and 272 m.

4.3. Evaluation of inverse emissions

To evaluate the impact of the different WRF simulations on the posterior fluxes estimated from the inversion system, we coupled the WRF-FDDA modeled variables (mean winds and turbulence) to generate the corresponding LPDM tower footprints, or influence functions. Using the different WRF simulations discussed above and their corresponding LPDM tower footprints, the five-day inverse emissions were computed for whole-city emissions using a Bayesian inversion system at 1-km resolution over the urban area of Indianapolis, and the CO_2 mole fraction from the 11 of the 12 towers from the INFLUX tower network (**Figure 8**), for the entire two-month period (Sept–Oct 2013). Inverse CO_2 emissions were computed over five-day periods and the configuration was similar to Lauvaux et al. (2016) and kept identical across the four inversions. Therefore, the differences in the inverse emissions correspond to the impact of different transport model realizations, and more precisely the impact of the FDDA systems, assimilating various data sources.

Figure 9 illustrates the different influence functions (or tower footprints) for the 12 tower locations for one single observation time. The extent and the main axes of

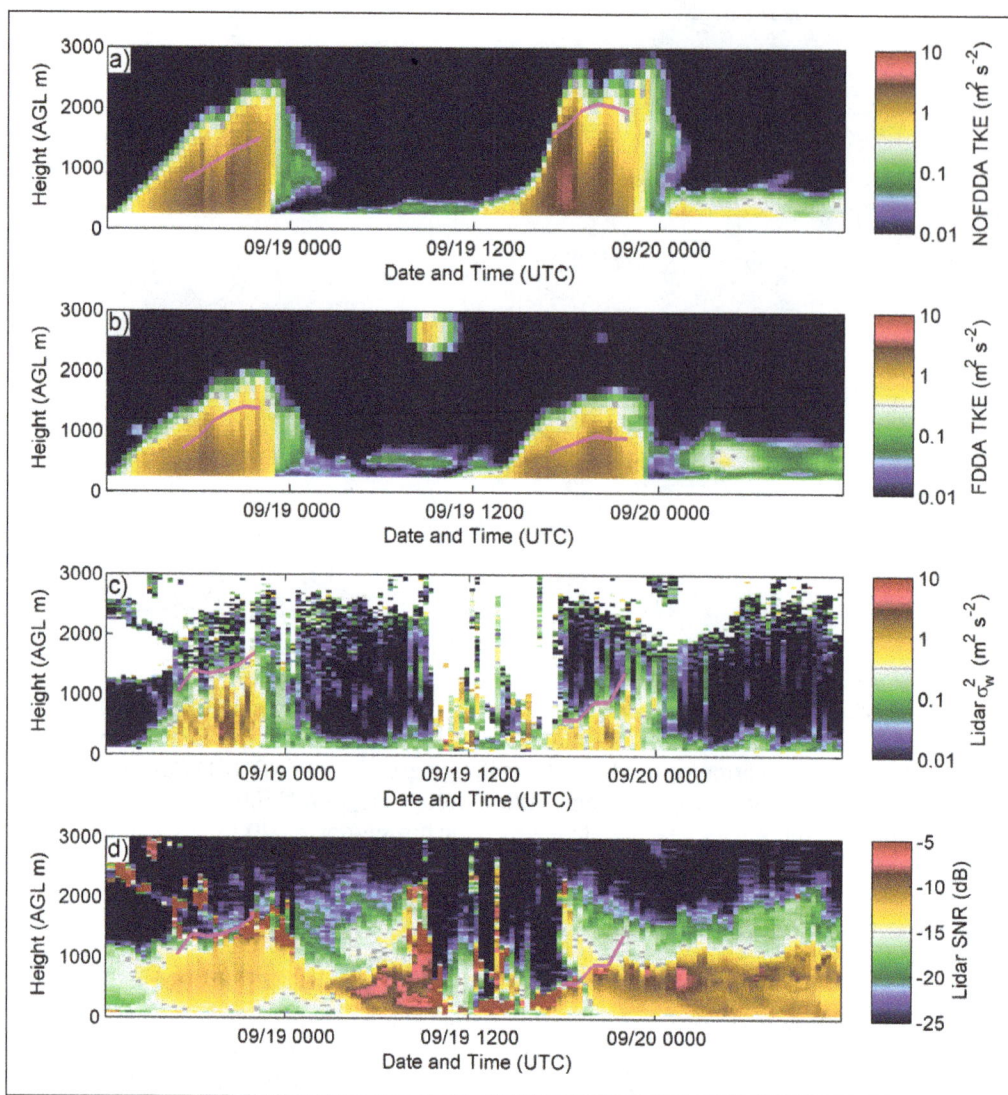

Figure 7: Planetary Boundary Layer verification. Comparison of Planetary Boundary Layer structures between WRF and the INFLUX lidar observations at Indianapolis for 28 and 30 August 2013: **a)** WRF-predicted turbulent kinetic energy from NOFDDA, **b)** WRF-predicted turbulent kinetic energy from FDDA_WMO_Lidar_ACARS, **c)** Lidar-observed vertical velocity variances, and **d)** Lidar-observed Signal-to-Noise Ratio (SNR). DOI: https://doi.org/10.1525/elementa.133.f7

Table 4: Mean error and mean absolute error of the WRF-predicted PBL depth on the 1-km grid verified against the Indianapolis INFLUX lidar measurements, averaged for all times between 17 and 22 UTC each day for the period between 00 UTC 27 August and 00 UTC 3 November 2013. DOI: https://doi.org/10.1525/elementa.133.t4

	NOFDDA	FDDA_WMO	FDDA_WMO_Lidar	FDDA_WMO_Lidar_ACARS
ME (m)	25	103	83	−23
MAE (m)	259	272	254	223

the tower footprints vary significantly depending on the WRF-FDDA model results, which translates into varying spatial attributions of the observed atmospheric CO_2 mole fractions. Overall, four atmospheric inversions were performed for the period September-October 2013 producing five-day emissions estimates at 1 km resolution over the domain. **Figure 10** shows the relative differences among the different transport solutions (influence functions) in space over the simulation domain (equivalent to the 1-km grid of WRF) represented by average pairwise differences. The maximum differences, up to 50%, are located at some of the tower locations emphasizing the importance of the near-field contribution and the high sensitivity of the footprints to wind direction changes. Differences average around 15% across the domain. Once combined with prior information, the impact of different transport fields is

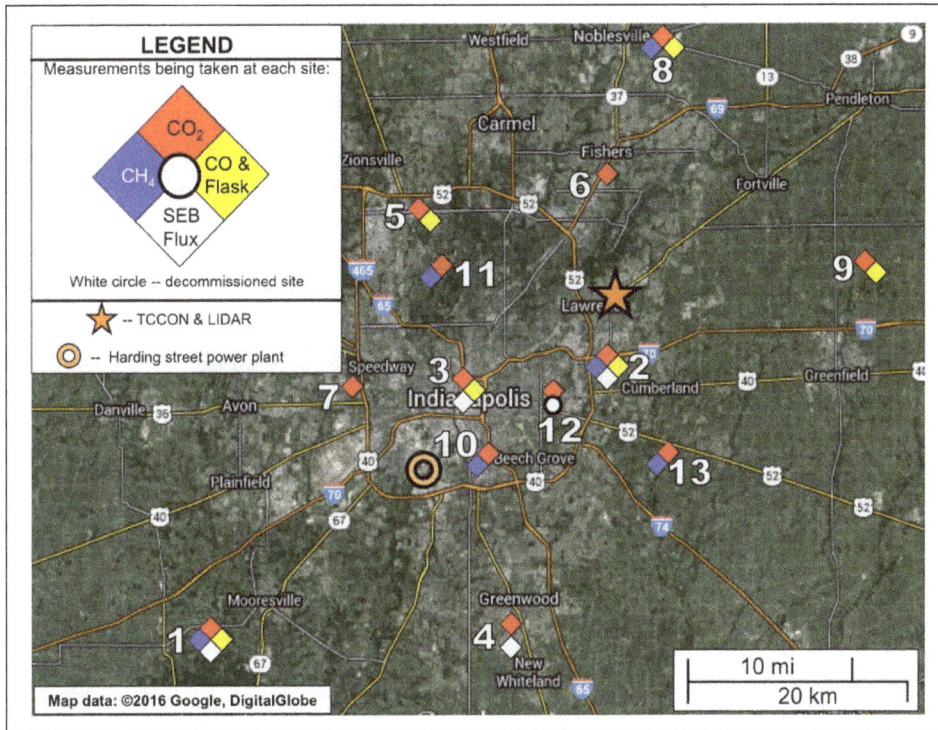

Figure 8: Observation network showing 12 INFLUX towers. Observation network showing location of 12 INFLUX towers. Particular site details and coordinates are given in Miles et al. (2016) and Richardson et al. (2016). DOI: https://doi.org/10.1525/elementa.133.f8

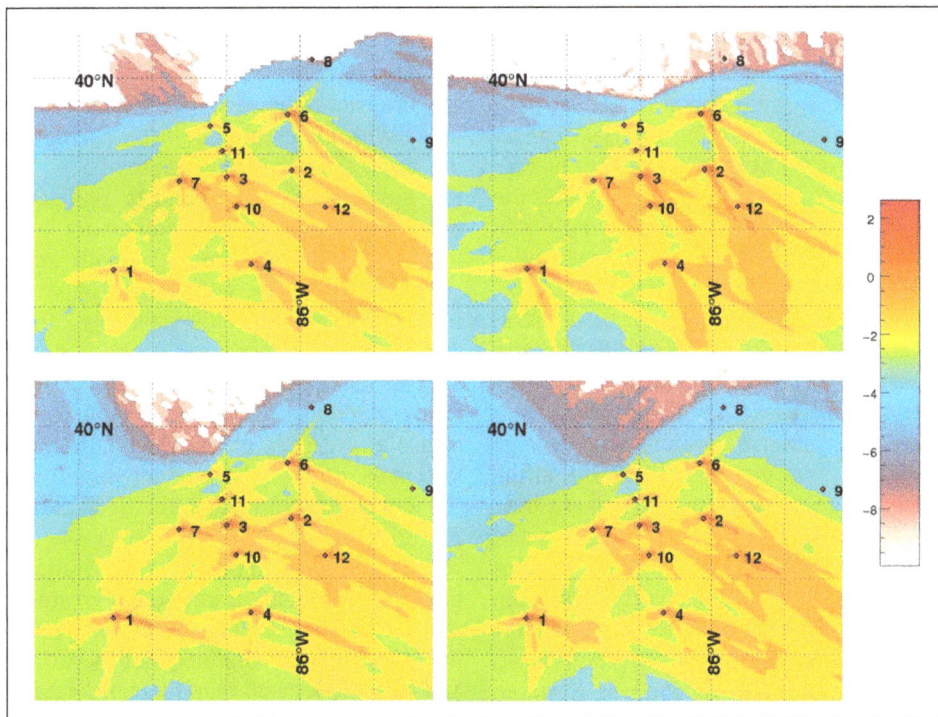

Figure 9: Influence functions. Influence functions over the INFLUX 1-km resolution domain for 10 of the 12 tower locations of the INFLUX network using the Lagrangian Particle Dispersion Model, averaged for 26–30 October 2013 (corresponding to the observation time 17–22UTC) driven by the meteorological variables from the four different WRF configurations: NOFDDA (Upper left), FDDA_WMO (Upper right), FDDA_WMO_Lidar (Lower left), and FDDA_WMO_Lidar_ACARS (Lower right), in log scale (ppm hour m^2 g^{-1}). Note that numbers 1–12 indicate the tower locations as detailed in Figure 8 and two towers were not operational during the time period Oct 26–30. DOI: https://doi.org/10.1525/elementa.133.f9

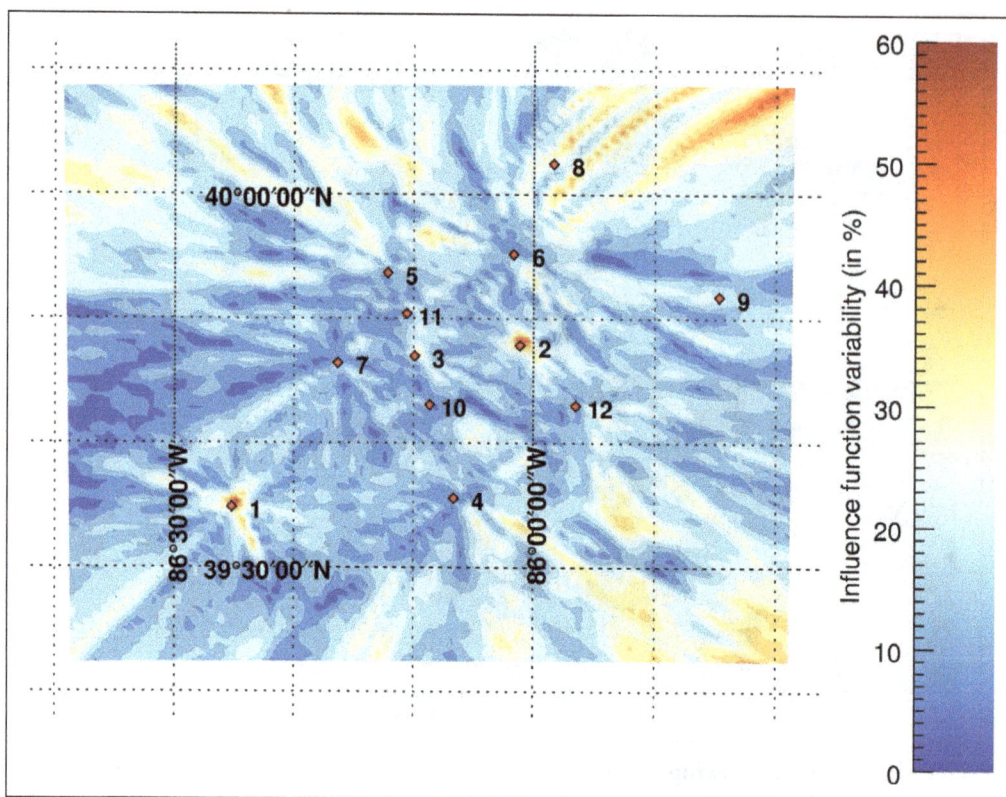

Figure 10: Spread of influence functions. Spatial distribution of influence function (averaged pairwise differences) over the two months (September–October 2013, 17–22UTC) representing the variability in the tower footprints across the different WRF-FDDA experiments. Note that numbers 1–12 indicate the tower locations as detailed in Figure 8 and two towers were not operational during the time period Oct 26–30. DOI: https://doi.org/10.1525/elementa.133.f10

decreased as inverse emissions are also constrained by prior fluxes and their associated error covariances. Here, similar to Lauvaux et al. (2016), we assume that the error variances scale with the prior emissions, which amplifies the beltway, for example, as a likely locations for corrections in the optimization procedure. **Figure 11** shows the maps of the differences across the four different CO_2 inverse flux estimates. As expected, differences are distributed following the error variances and not the spatial differences in the influence functions. As shown in Lauvaux et al. (2016), prior emission errors impact significantly the spatial distribution and the whole-city inverse emissions. Because large spatial gradients are present in the emissions and in the emissions error variances, the minimzation will attribute differences to high-variance areas, lessening the impact of the transport. Here, varying influence functions remains less influential on our inverse solutions than the constraints imposed by structures in the prior emission error covariances. The magnitudes of the flux corrections are as high as 15%, similar to the influence functions. However, the local maxima observed previously at the tower locations are not visible in the CO_2 flux differences.

Finally, we show the aggregated impact over the domain for each five-day periods of the two-month inversion. **Figure 12** shows that the total impact of meteorological data assimilation on each five-day segment is relatively small compared to the correction by

the inversion (shown by the distance to the prior in blue). With the exception of few five-day periods showing larger differences (e.g. September 6–11, 2013), most inversion periods correspond to similar inverse fluxes; this is explained by the compensation of higher and lower emissions distributed spatially across the inversion domain. Differences remain small at the five-day time scale (less than 10%) and over the two months (less than 5% change of the total 2-month emissions). We conclude based on the inversion results that all the transport configurations do not introduce any significant bias in the solution, which confirms that the low systematic errors in the meteorological variables across the four simulations match the small differences in total posterior emissions. Random errors in meteorological variables propagates into the flux solution as additional posterior uncertainties but do not create any systematic errors in the fluxes. While this ensemble of simulations is not calibrated to meteorological observations (e.g. Grimit and Mass, 2007), and therefore does not necessarily represent the true transport errors in the simulation, we show here that for this experiment, the different observations used in the WRF-FDDA simulations do affect the spatial attribution of flux corrections but have a limited impact on total inverse emissions over the entire domain. Furthermore, we computed the Bayes Factors for each model configurations. Results are presented in **Table 5**.

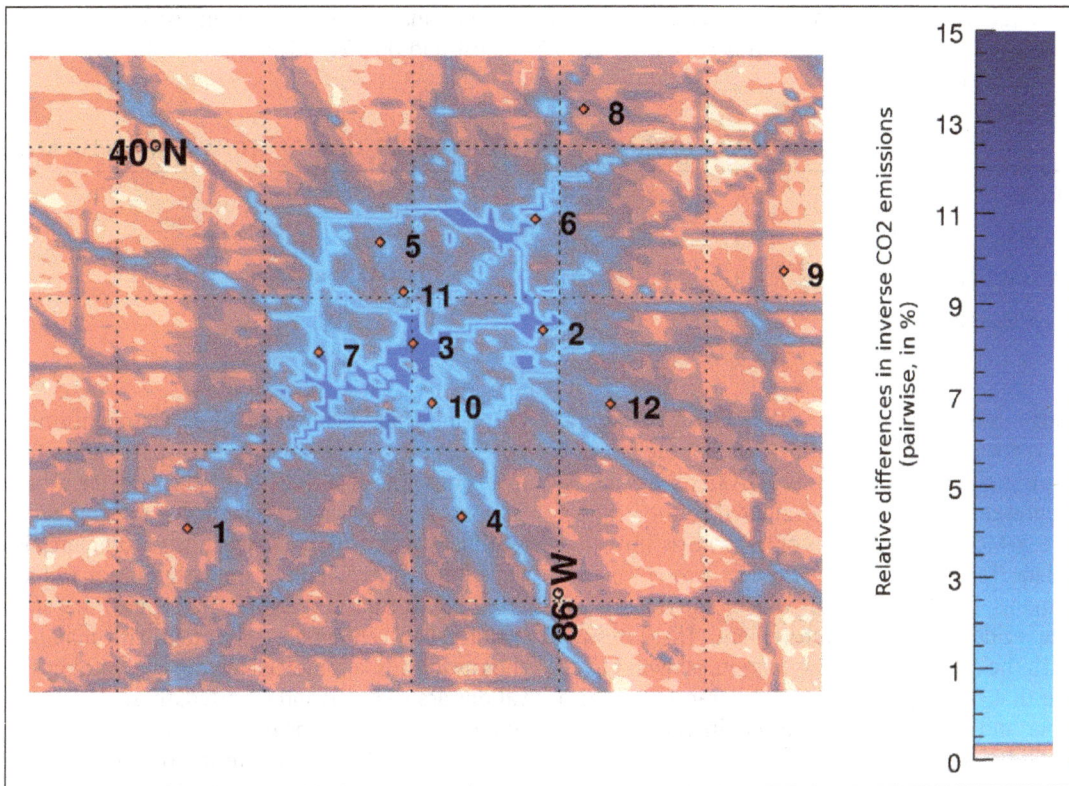

Figure 11: Spread of the inverse emission. Relative differences among the four inversion configurations (in % – averaged pairwise differences) representing the variability in the inverse emissions due to the transport fields, averaged over the two-month period. The assimilation of different meteorological data types impacts the inverse CO_2 emissions by up to 15% in the downtown area and around the beltway. Tower indices are similar to the site locations in Figure 8, except for Site 13 which was not operational in 2013. Note that numbers 1–12 indicate the tower locations as detailed in Figure 8 and two towers were not operational during the time period Oct 26–30. DOI: https://doi.org/10.1525/elementa.133.f11

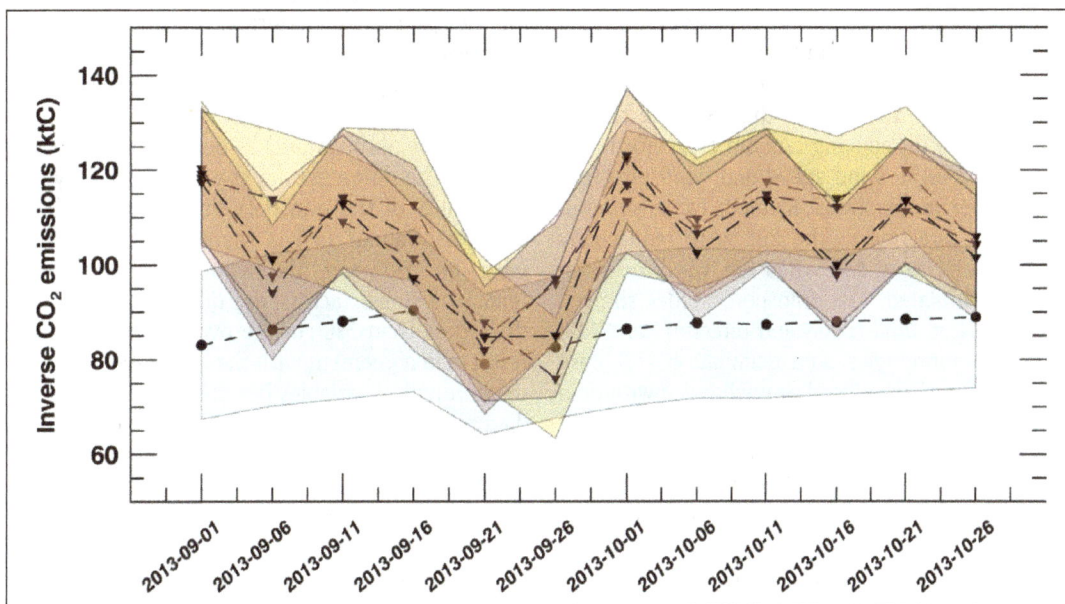

Figure 12: Inverse emissions. Inverse emissions in ktC per 5-day periods for the entire 1-km resolution INFLUX domain (87 km × 87 km) from the four WRF configurations, respectively WRF (no FDDA), WRF-FDDA with WMO data, WRF-FDDA with WMO and Lidar data, and WRF-FDDA with WMO, Lidar, and ACARS data. The prior emissions (from Hestia) are indicated in blue. DOI: https://doi.org/10.1525/elementa.133.f12

Table 5: Median ratio of the Bayes Factors derived from the Kalman matrix inversion when comparing the different model configurations for the period between 00 UTC 27 August and 00 UTC 3 November 2013. DOI: https://doi.org/10.1525/elementa.133.t5

	NOFDDA	FDDA_WMO	FDDA_WMO_Lidar	FDDA_WMO_Lidar_ACARS
NOFDDA	N/A	2.14	0.12	0.03
FDDA_WMO	0.5	N/A	0.31	0.09
FDDA_WMO_Lidar	8.3	3.2	N/A	0.5
FDDA_WMO_Lidar_ACARS	30.3	11.5	2.6	N/A

The ratios between the different WRF-FDDA configurations show strong evidences (ratio > 10) that the last two configurations (i.e. with lidar and ACARS observations) better represent the transport compared to the original FDDA (i.e. WMO surface stations only) or no assimilation of any data. The ratio between the last two configurations remains small (about 2.6) which suggests only a moderate evidence of an improvement. These results agree with the meteorological evaluation, allowing us to conclude that the improvement of the transport is also noticeable and statistically significant in the CO_2 flux inverse solutions.

5. Conclusions

Atmospheric transport model errors limit the accuracy and precision of inverse flux estimates. Assimilation of meteorological observations is a well-established method for reducing transport model errors. This paper represents a first quantification of the impact of meteorological data assimilation on urban-scale CO_2 inverse flux estimates.

As expected, meteorological data assimilation significantly reduces transport model error. For example, ~40% error reduction in the surface wind direction and ~50% error reduction in upper-air wind direction are achieved due to data assimilation. This paper demonstrates that observations of winds throughout the PBL have significantly greater value in reducing random errors (or mean absolute error, MAE) in wind speed and wind direction than surface layer observations. Random errors in wind speed and wind direction in the PBL were reduced by approximately a factor of two when PBL wind measurements were assimilated. The transport model showed very small biases in wind speed and direction, and PBL depth prior to meteorological data assimilation, and these small biases were either reduced or unchanged with data assimilation, with the exception that the assimilation of only surface layer observations did cause a noticeable increase in the PBL depth bias. We expect that if our initial meteorological simulation had been biased, that meteorological data assimilation would have reduced these errors as well. Sarmiento et al., (2016, this issue), for example, show that for Indianapolis model-data biases can be significant at different times of year for this configuration, and that biases vary across choices of model configurations and land surface data.

Dedicated observational systems such as the lidar deployed for INFLUX, capable of continuous profiling of

PBL winds, are recommended as a straightforward and potent means of minimizing transport errors for urban inversions. The relatively small domains for urban inversions make direct measurements of regional wind fields feasible. We also show, however, that commercial aircraft data from ACARS have a similarly potent influence on atmospheric transport errors. This is promising since most major urban centers are collocated with major airports, thus as long as commercial aircraft meteorological observations are recorded and reported, these data can be used to improve atmospheric transport model performance.

We were unable to explore the quantitative impact of data assimilation on the performance of a biased meteorological model configuration. Further, the data assimilated here did not significantly improve the random errors in PBL depth, and might not have a significant impact on a model configuration with a PBL depth bias. Lidar and ACARS observations, however, could clearly identify such biases, and additional data assimilation approaches could be adopted to address this issue. Improvements to model physics and input data (e.g. Sarmiento et al. 2016, this issue) represent another important approach to improving transport model performance.

Inverse emissions from the four simulations show a significant improvement when using the transport configurations assimilating vertical observations (with lidar and aircraft observations) as demonstrated by the Bayes Factors. The differences in space were directly related to the quality of the transport simulations, with local differences of about 15% in the emission corrections after inversion. However, the whole-city inverse emissions remained similar across the different model configurations (less than 5% change over the two months). This result is in agreement with the low biases in the different meteorological variables before and after meteorological data assimilation. It is reasonable to expect that if model meteorology was initially biased, meteorological data assimilation could have a substantial impact on the time-integrated, whole-city emissions as well.

In summary, this work shows the benefit of meteorological data assimilation on urban transport model performance, especially in reducing PBL wind speed and direction MAE when assimilating PBL wind profile observations. This reduction in the transport errors for wind speed, direction, and PBL height will provide more robust CO_2 inverse emissions at the city-scale, by improving the spatial attribution of the emission corrections. We expect that similar results would be achieved for biases given transport simulations with initially biased wind speed and wind direction.

Acknowledgements
We thank Arkayan Samaddar at Penn State University for assisting with some of the figures and Art Person at Penn State University for helping process WMO observations.

Funding information
This research is supported by the National Institute of Standards and Technology (Project # 70 NANB10H245).

Competing interests
The authors have no competing interests to declare.

Author contribution
- Contributed to conception and design: AD, TL, KJD, BJG
- Contributed to acquisition of data: AD, NLM, SJR, RMH, TAB, WAB, KRG
- Contributed to modeling results: AD, TL, BJG, DPS
- Contributed to analysis and interpretation of data: AD, TL, KJD, KW
- Drafted and/or revised article: AD, KJD, TL, BJG, NLM

References

Agustí-Panareda, A, Massart, S, Chevallier, F, Balsamo, G, Boussetta, S, Dutra, E and Beljaars, A 2016 A biogenic CO_2 flux adjustment scheme for the mitigation of large-scale biases in global atmospheric CO_2 analyses and forecasts, *Atmos. Chem. Phys.*, **16**: 10399–10418. DOI: https://doi.org/10.5194/acp-16-10399-2016

Ahmadov, R, Gerbig, C, Kretschmer, R, Koerner, S, Neininger, B, Dolman, AJ and Sarrat, C 2007 Mesoscale covariance of transport and CO_2 fluxes: Evidence from observations and simulations using the WRF-VPRM coupled atmosphere-biosphere model, *J. Geophys. Res.*, **112**(D22): 107. DOI: https://doi.org/10.1029/2007JD008552

Anderson, LK 2010 ACARS – A Users Guide. Las Atalayas. p. 5. ISBN 978-1-4457-8847-0 (https://books.google.com/books?id=sS3pAQAAQBAJ).

Andrews, AE, Kofler, JD, Trudeau, ME, Williams, JC, Neff, DH, Masarie, KA, Chao, DY, Kitzis, DR, Novelli, PC, Zhao, CL, Dlugokencky, EJ, Lang, PM, Crotwell, MJ, Fischer, ML, Parker, MJ, Lee, JT, Baumann, DD, Desai, AR, Stanier, CO, De Wekker, SFJ, Wolfe, DE, Munger, JW and Tans, PP 2014 CO_2, CO, and CH_4 measurements from tall towers in the NOAA Earth System Research Laboratory's Global Greenhouse Gas Reference Network: instrumentation, uncertainty analysis, and recommendations for future high-accuracy greenhouse gas monitoring efforts. *Atmos. Meas. Tech.*, **7**: 647–687, www.atmos-meas-tech.net/7/647/2014/. DOI: https://doi.org/10.5194/amt-7-647-2014

Baker, DF, Law, RM, Gurney, KR, Rayner, P, Peylin, P, Denning, AS, Bousquet, P, Bruhwiler, L, Chen, YH, Ciais, P, Fung, IY, Heimann, M, John, J, Maki, T, Maksyutov, S Masarie, K, Prather, M, Pak, B, Taguchi, S and Zhu, Z 2006 TransCom 3 inversion intercomparison: Impact of transport model errors on the interannual variability of regional CO_2 fluxes, 1988–2003, *Global Biogeochemical Cycles*, **20**(GB1): 002. DOI: https://doi.org/10.1029/2004GB002439

Carouge, C, Bousquet, P, Peylin, P, Rayner, PJ and Ciais, P 2010 What can we learn from European continuous atmospheric CO_2 measurements to quantify regional fluxes – Part 1: Potential of the 2001 network, *Atmos. Chem. Phys.*, **10**: 3107–3117. DOI: https://doi.org/10.5194/acp-10-3107-2010

Chen, F and Dudhia, J 2001 Coupling an advanced land surface-hydrology model with the Penn State–NCAR MM5 modeling system. Part I: Model implementation and sensitivity. *Mon. Wea. Rev.*, **129**: 569–585. DOI: https://doi.org/10.1175/1520-0493(2001)129<0569:CAALSH>2.0.CO;2

Cintineo, R, Otkin, JA, Xue, M and Kong, F 2014 Evaluating the Performance of Planetary Boundary Layer and Cloud Microphysical Parameterization Schemes in Convection-Permitting Ensemble Forecasts Using Synthetic GOES-13 Satellite Observations. *Mon. Wea. Rev.*, **142**: 163–182. DOI: https://doi.org/10.1175/MWR-D-13-00143.1

Clark, AJ, Michael, C, Coniglio, B, Coffer, E, Greg, T, Ming, X and Fanyou, K 2015: Sensitivity of 24-h Forecast Dryline Position and Structure to Boundary Layer Parameterizations in Convection-Allowing WRF Model Simulations. *Wea. Forecasting*, **30**: 613–638. DOI: https://doi.org/10.1175/WAF-D-14-00078.1

Coniglio, MC, James, C, Jr., Patrick, TM and Fanyou, K 2013 Verification of Convection-Allowing WRF Model Forecasts of the Planetary Boundary Layer Using Sounding Observations. *Wea. Forecasting*, **28**: 842–862. DOI: https://doi.org/10.1175/WAF-D-12-00103.1

Crisp, D, Atlas, RM, Breon, F-M, Brown, LR, Burrows, JP, Ciais, P, Connor, BJ, Doney, SC, Fung, I, Jacob, D, Miller, CE, O'Brien, D, Pawson, S, Randerson, JT, Rayner, P, Salawitch, R, Sander, SP, Sen, B, Stephens, GL, Tans, PP, Toon, G, Wennberg, P, Wofsy, SC, Yung, YL, Kuang, Z, Chudasama, B, Sprague, G, Weiss, B, Pollock, R, Kenyon, D and Schroll, S 2004 The Orbiting Carbon Observatory (OCO) mission, *Advances in Space Research*, **34**: 700–709. DOI: https://doi.org/10.1016/j.asr.2003.08.062

Davis, KJ, Lauvaux, T, Miles, NL, Richardson, SJ, Gurney, KR, Hardesty, RM, Brewer, A, Shepson, PB, Cambaliza, MO, Sweeney, C, Turbull, J, Karion, A and Whetstone, J 2016 The Indianapolis Flux Experiment (INFLUX): A test-bed for developing anthropogenic greenhouse gas emission measurements, *Elem Sci Anth.* In press for the INFLUX Special Feature.

Deng, A, Seaman, NL, Hunter, GK and **Stauffer, DR** 2004 Evaluation of interregional transport using the MM5-SCIPUFF system. *J. Appl. Meteor.,* **43**: 1864–1886. DOI: https://doi.org/10.1175/JAM2178.1

Deng, A and **Stauffer, DR** 2006 On improving 4-km mesoscale model simulations. *J. Appl. Meteor.,* **45**: 361–381. DOI: https://doi.org/10.1175/JAM2341.1

Deng, A, Stauffer, DR, Gaudet, BJ, Dudhia, J, Hacker, J, Bruyere, C, Wu, W, Vandenberghe, F, Liu, Y and **Bourgeois, A** 2009 Update on WRF-ARW end-to-end multi-scale FDDA system, *10th Annual WRF Users' Workshop,* Boulder, CO, June 23, 14 pp.

Díaz Isaac, LI, Lauvaux, T, Davis, KJ, Miles, NL, Richardson, SJ, Jacobson, AR and **Andrews, AE** 2014 Model-data comparison of MCI field campaign atmospheric CO_2 mole fractions, *J. Geophys. Res. Atmos.,* **119**: 10,536–10,551. DOI: https://doi.org/10.1002/2014JD021593

Dudhia, J 1989 Numerical study of convection observed during the Winter Monsoon Experiment using a mesoscale two–dimensional model. *J. Atmos. Sci.,* **46**: 3077–3107. DOI: https://doi.org/10.1175/1520-0469(1989)046<3077:NSOCOD>2.0.CO;2

Enting 2002 Inverse Problems in Atmospheric Constituent Transport. Cambridge Atmospheric and Space Science Series.

Enting, IG and **Mansbridge, JV** 1989 Seasonal sources and sinks of atmospheric CO_2 Direct inversion of filtered data. *Tellus B,* **41B**: 111–126. DOI: https://doi.org/10.1111/j.1600-0889.1989.tb00129.x

Feng, S, Lauvaux, T, Newman, S, Rao, P, Ahmadov, R, Deng, A, Díaz-Isaac, LI, Duren, RM, Fischer, ML, Gerbig, C, Gurney, KR, Huang, J, Jeong, S, Li, Z, Miller, CE, O'Keeffe, D, Patarasuk, R, Sander, SP, Song, Y, Wong, KW and **Yung, YL** 2016 Los Angeles megacity: a high-resolution land–atmosphere modelling system for urban CO_2 emissions, *Atmos. Chem. Phys.,* **16**: 9019–9045. DOI: https://doi.org/10.5194/acp-16-9019-2016

Gerbig, C, Lin, JC, Wofsy, SC, Daube, BC, Andrews, AE, Stephens, BB, Bakwin, PS and **Grainger, CA** 2003 Toward constraining regional-scale fluxes of CO_2 with atmospheric observations over a continent: 1. Observed spatial variability from airborne platforms, *J. Geophys. Res.,* **108**: 4756. DOI: https://doi.org/10.1029/2002JD003018

Grell, GA, Peckhama, SE, Schmitzc, R McKeenb, SA, Frostb, G, Skamarockd, WC and **Edere, B** 2005 Fully coupled "online" chemistry within the WRF model, *Atmos. Environ.* **39**: 6957–6975. DOI: https://doi.org/10.1016/j.atmosenv.2005.04.027

Grimit, EP and **Mass, CF** 2007 Measuring the ensemble spread-error relationship with a probabilistic approach: Stochastic ensemble results. *Mon. Wea. Rev.,* **135**: 203–221. DOI: https://doi.org/10.1175/MWR3262.1

Gurney, KR, Razlivanov, I, Song, Y, Zhou, Y, Benes, B and **Abdul-Massih, M** 2012 Quantification of fossil fuel CO_2 emissions at the building/street scale for a large US city. *Environmental Science & Technology,* 120815073657007. DOI: https://doi.org/10.1021/es3011282

Hariprasad, KBRR, Srinivas, CV, Bagavath, SA, Vijaya Bhaskara Rao, S, Baskaran, R and **Venkatraman, B** 2014 Numerical simulation and intercomparison of boundary layer structure with different PBL schemes in WRF using experimental observations at a tropical site. *Atmos. Res.* **145**: 27e44.

Hong, S-Y, Jimy, D and **Chen, S-H** 2004 A revised approach to ice microphysical processes for the bulk parameterization of clouds and precipitation. *Mon. Wea. Rev.,* **132**: 103–120. DOI: https://doi.org/10.1175/1520-0493(2004)132<0103:ARATIM>2.0.CO;2

Kain, JS 2004 The Kain-Fritsch convective parameterization: An update, *J. Appl. Meteor,* **43**: 170–181. DOI: https://doi.org/10.1175/1520-0450(2004)043<0170:TKCPAU>2.0.CO;2

Kain, JS and **Fritsch, JM** 1990 A one-dimensional entraining/detraining plume model and its application in convective parameterization. *J. Atmos. Sci.,* **47**: 2784–2802. DOI: https://doi.org/10.1175/1520-0469(1990)047<2784:AODEPM>2.0.CO;2

Kain, JS and **Fritsch, JM** 1993 Convective parameterization in mesoscale models: the Kain-Fritsch scheme. *The representation of cumulus convection in numerical models, AMS. Monograph,* Emanuel, KA, and Raymond, DJ, (eds.), 165–170.

Kang, J-S, Kalnay, E, Miyoshi, T, Liu, J and **Fung, I** 2012 Estimation of surface carbon fluxes with an advanced data assimilation methodology, *J. Geophys. Res.,* **117**(D24): 101. DOI: https://doi.org/10.1029/2012JD018259

Karion, A, Sweeney, C, Kort, EA, Shepson, PB, Brewer, A, Cambaliza, M, Conley, SA, Davis, K, Deng, A, Hardesty, M, Herndon, SC, Lauvaux, T, Lavoie, T, Lyon, D, Newberger, T, Pétron, G, Rella, C, Smith, M, Wolter, S, Yacovitch, TI and **Tans, P** 2015 Aircraft-based estimate of total methane emissions from the Barnett Shale region, *Environ. Sci. Technol.,* **49**: 8124–8131. DOI: https://doi.org/10.1021/acs.est.5b00217

Lauvaux, T and **Davis, KJ** 2014 Planetary boundary layer errors in mesoscale inversions of column-integrated CO_2 measurements, *J. Geophys. Res. Atmos.,* **119**: 490–508. DOI: https://doi.org/10.1002/2013JD020175

Lauvaux, T, Miles, NL, Deng, A, Richardson, SJ, Cambaliza, MO, Davis, KJ, Gaudet, B, Gurney, KR, Huang, J, O'Keefe, D et al. 2016 High resolution atmospheric inversion of urban CO_2 emissions during the dormant season of the Indianapolis Flux Experiment (INFLUX), *J. Geophys. Res. Atmos.,* **121**: DOI: https://doi.org/10.1002/2015JD024473

Lauvaux T, Miles, NL, Richardson, SJ, Deng, A, Staufer, D, Davis, KJ, Jacobson, G, Rella, C,

Calonder, G-P and DeCola, PL 2013 Urban emissions of CO_2 from Davos, Switzerland: the first real-time monitoring system using an atmospheric inversion technique. *J. Appl Meteor. and Climatol.*, **52**: 2654–2668. DOI: https://doi.org/10.1175/JAMC-D-13-038.1

Lauvaux, T, Pannekoucke, O, Sarrat, C, Chevallier, F, Ciais, P, Noilhan, J and Rayner, PJ 2009 Structure of the transport uncertainty in mesoscale inversions of CO_2 sources and sinks using ensemble model simulations, *Biogeosciences*, **6**: 1089–1102. DOI: https://doi.org/10.5194/bg-6-1089-2009

Lauvaux, T, Schuh, AE, Uliasz, M, Richardson, S, Miles, N, Andrews, AE, Sweeney, C, Diaz, LI, Martins, D, Shepson, PB and Davis, KJ 2012 Constraining the CO_2 budget of the corn belt: exploring uncertainties from the assumptions in a mesoscale inverse system, *Atmos. Chem. Phys.*, **12**: 337–354. DOI: https://doi.org/10.5194/acp-12-337-2012

Law, RM, et al. 2008 TransCom model simulations of hourly atmospheric CO_2: Experimental overview and diurnal cycle results for 2002, Global Biogeochem. *Cycles*, **22**(GB3): 009. DOI: https://doi.org/10.1029/2007GB003050

Law, RM, Rayner, PJ, Steele, LP and Enting, IG 2003 Data and modelling requirements for CO_2 inversions using high-frequency data. *Tellus B*, **55**: 512–521. DOI: https://doi.org/10.1034/j.1600-0889.2003.00029.x

Lin, JC and Gerbig, C 2005 Accounting for the effect of transport errors on tracer inversions. *Geophysical Research Letters*, **32**(L01): 802. DOI: https://doi.org/10.1029/2004GL021127

Miles, LN, Richardson, SJ, Lauvaux, T, Davis, KJ, Deng, A, Turnbull, J, Karion, A, Sweeney, C, Gurney, KR, Patarasuk, R, Razlivanov, I, Cambaliza, MO and Shepson, PB 2016 Quantification of urban atmospheric boundary layer greenhouse gas dry mole fraction enhancements: Results from the Indianapolis Flux Experiment (INFLUX), *Elem Sci Anth*. In press for the INFLUX Special Feature.

Mlawer, EJ, Steven, J, Taubman, P, Brown, D, Iacono, MJ and Clough, SA (1997), Radiative transfer for inhomogeneous atmospheres: RRTM, a validated correlated–k model for the longwave. *J. Geophys. Res.*, **102**: 16663–16682. DOI: https://doi.org/10.1029/97jd00237

Nakanishi, M and Niino, H 2006 An improved Mellor-Yamada level 3 model: its numerical stability and application to a regional prediction of advecting fog. *Bound. Layer Meteor.* **119**: 397–407. DOI: https://doi.org/10.1007/s10546-005-9030-8

Pacala, SW and Coauthors 2010 Verifying Greenhouse Gas Emissions: Methods to Support International Climate Agreements, *National Research Council Draft Report*, The National Academies Press, Washington, DC, USA.

Pearson, G, Davies, F and Collier, C 2009 An analysis of the performance of the UFAM pulsed Doppler lidar for observing the boundary layer. *J. Atmos. Oceanic Technol.*, **26**: 240–250. DOI: https://doi.org/10.1175/2008JTECHA1128.1

Peters, W, Jacobson, AR, Sweeney, C, Andrews, AE, Conway, TJ, Masarie, K, Miller, JB, Bruhwiler, LMP, Pétron, G, Hirsch, AI, Worthy, DEJ, vanderWerf, GR, Randerson, JT, Wennberg, PO, Krol, MC and Tans, PP 2007 An atmospheric perspective on North American carbon dioxide exchange: CarbonTracker, *PNAS* 2007 **104**(48): 18925–18930. DOI: https://doi.org/10.1073/pnas.0708986104

Pielke, RA and Uliasz, M 1993 Influence of landscape variability on atmospheric dispersion. *J. Air Waste Mgt.*, **43**: 989–994. DOI: https://doi.org/10.1080/1073161X.1993.10467181

Richardson, SJ, Miles, NL, Davis, KJ, Crosson, ER, Rella, C and Andrews, AE 2012 Field testing of cavity ring-down spectroscopic analyzers measuring carbon dioxide and water vapor. *J. Atmos. Oceanic Tech.*, **29**: 397 – 406. DOI: https://doi.org/10.1175/JTECH-D-11-00063.1

Richardson, SJ, Miles, NL, Davis, KJ, Lauvaux, T, Martins, D, et al. 2016 CO_2, CO, and CH_4 surface in situ measurement network in support of the Indianapolis FLUX (INFLUX) Experiment. To be submitted to *Elementa*.

Rogers, RE, Deng, A, Stauffer, DR, Gaudet, BJ, Jia, Y, Soong, S, Tanrikulu, S 2013 Application of the Weather Research and Forecasting Model for Air Quality Modeling in the San Francisco Bay Area. *J. Appl. Meteor.*, **52**: 1953–1973. DOI: https://doi.org/10.1175/JAMC-D-12-0280.1

Sarmiento, DP, Davis, KJ, Deng, A, Lauvaux, T, Brewer, A and Hardesty, M 2016 A comprehensive assessment of land surface-atmosphere interactions in a WRF/Urban modeling system for Indianapolis, IN, *Elem Sci Anth*. In press for the INFLUX Special Feature.

Schuh, AE, Denning, AS, Corbin, KD, Baker, IT, Uliasz, M, Parazoo, N, Andrews, AE and Worthy, DEJ 2010 A regional high-resolution carbon flux inversion of North America for 2004, *Biogeosciences*, **7**: 1625–1644. DOI: https://doi.org/10.5194/bg-7-1625-2010

Schuh, AE, Lauvaux, T, West, TO, Denning, AS, Davis, KJ, Miles, N, Richardson, S, Uliasz, M, Lokupitiya, E, Cooley, D, Andrews, A and Ogle, S 2013 Evaluating atmospheric CO_2 inversions at multiple scales over a highly inventoried agricultural landscape. *Glob Change Biol*, **19**: 1424–1439. DOI: https://doi.org/10.1111/gcb.12141

Skamarock, WC, Klemp, JB, Dudhia, J, Gill, DO, Barker, DM, Duda, MG, Huang, X-Y, Wang, W and Powers, JG 2008 A description of the Advanced Research WRF Version 3. *NCAR Technical Note NCAR/TN-475+STR*. 113 pp.

Tans, PP, Fung, IY and Takahashi, T 1990 Observational constraints on the global atmospheric CO2 budget. *Science*, **247**: 1431–1439. DOI: https://doi.org/10.1126/science.247.4949.1431

Tarantola, A 2005 Inverse Problem Theory and Model Parameter Estimation. *SIAM*. DOI: https://doi.org/10.1137/1.9780898717921

Tewari, M, Chen, F, Wang, W, Dudhia, J, LeMone, MA, Mitchell, K, Ek, M, Gayno, G,

Wegiel, J and **Cuenca, RH** 2004 Implementation and verification of the unified NOAH land surface model in the WRF model. 20th conference on weather analysis and forecasting/16th conference on numerical weather prediction, pp. 11–15.

Uliasz, M 1994 Lagrangian particle modeling in mesoscale applications, Environmental Modelling II, ed. Zanetti, P, *Computational Mechanics Publications*, 71–102.

Wunch, D, Toon, GC, Wennberg, PO, Wofsy, SC, Stephens, BB, Fischer, ML, Uchino, O, Abshire, JB, Bernath, P, Biraud, SC, Blavier, J-FL, Boone, C, Bowman, KP, Browell, EV, Campos, T, Connor, BJ, Daube, BC, Deutscher, NM, Diao, M, Elkins, JW, Gerbig, C, Gottlieb, E, Griffith, DWT, Hurst, DF, Jiménez, R, Keppel-Aleks, G, Kort, EA, Macatangay, R, Machida, T, Matsueda, H, Moore, F, Morino, I, Park, S, Robinson, J, Roehl, CM, Sawa, Y, Sherlock, V, Sweeney, C, Tanaka, T and **Zondlo, MA** 2010 Calibration of the Total Carbon Column Observing Network using aircraft profile data, *Atmos. Meas. Tech.*, **3**: 1351–1362. DOI: https://doi.org/10.5194/amt-3-1351-2010

4

Assessing the optimized precision of the aircraft mass balance method for measurement of urban greenhouse gas emission rates through averaging

Alexie M. F. Heimburger[*], Rebecca M. Harvey[*], Paul B. Shepson[*,†], Brian H. Stirm[‡], Chloe Gore[*], Jocelyn Turnbull[§], Maria O. L. Cambaliza[‖], Olivia E. Salmon[*], Anna-Elodie M. Kerlo[*], Tegan N. Lavoie[*], Kenneth J. Davis[¶], Thomas Lauvaux[¶], Anna Karion[**,††,§§], Colm Sweeney[**,††], W. Allen Brewer[**], R. Michael Hardesty[††] and Kevin R. Gurney[‡‡]

To effectively address climate change, aggressive mitigation policies need to be implemented to reduce greenhouse gas emissions. Anthropogenic carbon emissions are mostly generated from urban environments, where human activities are spatially concentrated. Improvements in uncertainty determinations and precision of measurement techniques are critical to permit accurate and precise tracking of emissions changes relative to the reduction targets. As part of the INFLUX project, we quantified carbon dioxide (CO_2), carbon monoxide (CO) and methane (CH_4) emission rates for the city of Indianapolis by averaging results from nine aircraft-based mass balance experiments performed in November-December 2014. Our goal was to assess the achievable precision of the aircraft-based mass balance method through averaging, assuming constant CO_2, CH_4 and CO emissions during a three-week field campaign in late fall. The averaging method leads to an emission rate of 14,600 mol/s for CO_2, assumed to be largely fossil-derived for this period of the year, and 108 mol/s for CO. The relative standard error of the mean is 17% and 16%, for CO_2 and CO, respectively, at the 95% confidence level (CL), i.e. a more than 2-fold improvement from the previous estimate of ~40% for single-flight measurements for Indianapolis. For CH_4, the averaged emission rate is 67 mol/s, while the standard error of the mean at 95% CL is large, i.e. ±60%. Given the results for CO_2 and CO for the same flight data, we conclude that this much larger scatter in the observed CH_4 emission rate is most likely due to variability of CH_4 emissions, suggesting that the assumption of constant daily emissions is not correct for CH_4 sources. This work shows that repeated measurements using aircraft-based mass balance methods can yield sufficient precision of the mean to inform emissions reduction efforts by detecting changes over time in urban emissions.

Keywords: greenhouse gas; emission rates; precision; urban; quantification

1. Introduction

Urban greenhouse gas emissions, as well as city expansion, are expected to continue to grow in the coming years (IEA, 2008; Seto et al., 2012). Currently, fossil fuel-related carbon dioxide (CO_2) emissions from cities represent more than 70% of the global total, mainly due to energy production and use (IEA, 2008; IPCC 5th Assessment Report, 2014), which also represents the second major anthropogenic source of carbon monoxide (CO) after biomass burning (Olivier et al., 1996). Recent studies have shown that

[*] Department of Chemistry, Purdue University, West Lafayette, Indiana, US
[†] Department of Earth, Atmospheric and Planetary Science and Purdue Climate Change Research Center, Purdue University, West Lafayette, Indiana, US
[‡] Department of Aviation and Transportation Technology, Purdue University, West Lafayette, Indiana, US
[§] National Isotope Center, GNS Science, Lower Hutt, NZ
[‖] Department of Physics, Ateneo de Manila University, Loyola Heights, Quezon City, PH
[¶] The Pennsylvania State University, Department of Meteorology, University Park, PA
[**] NOAA/ESRL, Colorado, US
[††] CIRES, University of Colorado at Boulder, Boulder, Colorado, US
[‡‡] School of Life Sciences, Arizona State University, Tempe, Arizona, US
[§§] NIST, Gaithersburg, Maryland, US
Corresponding author: Alexie M. F. Heimburger (alexie.heimburger@gmail.com)

urban environments also emit large amounts of methane (CH_4) (Wunch et al., 2009, Wennberg et al., 2012, McKain et al., 2015), a potent greenhouse gas (GHG) with a global warming potential 28–34 times greater than that of CO_2 over a 100-year time period (Myhre et al., 2013). Despite ocean and land sinks, ~55% of CO_2 emissions accumulate in the atmosphere and can thus impact the global climate (Kirschke et al., 2013; Le Quéré et al., 2014). Quantification of both the magnitude and uncertainty of GHG emissions in urban environments is therefore critical for implementing coherent and effective policies to mitigate such emissions, and to reduce their effects on climate change (IEA, 2008; Hutyra et al., 2014).

Since the Copenhagen Accord in 2009, several countries have confirmed their commitment to reduce their GHG emissions (UNFCCC, 2010; President's Climate Action Plan, 2013, 2014a, 2014b; European Commission, 2014). During the recent United Nations Framework Convention on Climate Change negotiating session held in Paris 2015 (COP21/CMP11), the United States announced a goal of 26–28% reduction in emissions by 2025 compared to 2005 levels, while China committed to a carbon dioxide emissions reduction of 60–65% per unit of gross domestic product by 2030 (China's INDC, June 2015). Europe adopted a reduction target of at least 40% below 1990 levels by 2030 (30–40% below 2005 depending on activity sectors; http://ec.europa.eu/clima/policies/strategies/2030/index_en.htm, accessed on 12/07/2015). Such goals are achievable, but reduction efforts need to be "measurable", "reportable", and "verifiable" (Vine and Sathaye, 1999; Schakenbach et al., 2006; NRC, 2010; UNFCCC, 2015). Measurements of GHG emissions often have significant uncertainties, ranging from less than 10% to 100% (NRC, 2010), depending on the methods used (e.g., bottom-up inventories, which assess and integrate emissions from specific sources and activities; top-down techniques, which rely on greenhouse gas concentration measurements conducted within the atmosphere), the type of sources and target gases, and the spatial scale considered (national, regional, and local) (Marland, 2008, 2012; NRC, 2010; Peylin et al., 2011; Kirschke et al., 2013; Cambaliza et al., 2014; Bergamaschi et al., 2015). Uncertainties for GHG emissions at sub-national scales (e.g. city/county/state/province) are usually significantly greater (~50% to >100%, Gurney et al., 2009; Mays et al., 2009; NRC, 2010; Cambaliza et al., 2014; 2015) than those at national or continental scales (< 25% at national levels, Marland, 2008, 2012; Gurney et al., 2009; NRC, 2010), and are in many cases larger than the emission reduction targets themselves, making it difficult to assess the efficacy of local scale efforts. These large uncertainties can be partially explained by the dearth of local-scale measurements and method development, coupled with the general lack of local-scale mitigation efforts (Rosenzweig et al., 2010; Hutyra et al., 2014). However, local policy plans initiated by proactive cities and networks of local political decision-makers (i.e. World Mayors Council on Climate Change, Local Governments for Sustainability) have recently emerged and highlighted the urgent need of new and timely information on urban GHG emissions from the

scientific community (Rosenzweig et al., 2010; Hutyra et al., 2014).

Urban emissions of CO_2, CH_4, and CO (a proxy for combustion) are usually derived from several natural and anthropogenic sources that are difficult to separate (Hutyra et al., 2014). Both fossil fuel- and biogenic-related CO_2 concentrations in urban environments are affected by many sources and sinks. Fossil fuel-related CO_2 is largely emitted by electricity generation, mobile source combustion, and point sources, such as large industrial facilities. Biogenic CO_2 sources and sinks include biosphere respiration, biofuel use, and biomass burning. CO has been used as a tracer for anthropogenic sources (Parrish et al., 1993; Turnbull et al., 2011), and is emitted during incomplete combustion of fossil- and bio-fuels from vehicles, agricultural waste (not at Indianapolis), and industrial processes. CO is also produced by oxidation of biogenic volatile carbon compounds, especially during the summer time, and is consumed in the atmosphere by reaction with the OH radical (Parrish et al., 1993). CH_4 can be emitted from natural wetlands, or from anthropogenic sources (Bousquet et al., 2006; Kirschke et al., 2013), such as rice production (not at Indianapolis), ruminant animals (not at Indianapolis), natural gas infrastructure, biomass burning, landfills, and wastewater treatment plants. CH_4 emission estimates are usually less certain than CO_2 estimates because many CH_4 sources are more diffuse, as a result of unintended release or more complex and temporally and spatially variable biological decay processes (Forster et al., 2007; Kirschke et al., 2013; President's Climate Action Plan, 2014b).

Improvement in the quality of GHG measurements represents an urgent challenge. The development of low cost, precise emissions measurement techniques that can be applied to a range of urban environments is needed to provide scientifically sound information for future GHG mitigation strategies, whether local, regional or national. The multi-institution collaborative Indianapolis Flux Experiment project (INFLUX, http://sites.psu.edu/influx/) was designed to evaluate and minimize GHG emissions uncertainties at the city-scale, by developing, assessing, combining, and improving top-down and bottom-up approaches to quantify urban GHG emissions. Indianapolis is an advantageous test area, due to its physical separation from adjacent cities by surrounding agricultural lands, effectively isolating the city from other major sources of anthropogenic pollution (Cambaliza et al., 2014, 2015). The INFLUX project involves various top-down approaches to estimate CO_2, CH_4, and CO emissions from the city using continuous measurements and flask sampling from 12 towers situated within the city and in the surrounding suburbs (Miles et al., 2015; Turnbull et al., 2012, 2015), periodic aircraft measurements (Mays et al., 2009; Cambaliza et al., 2014, 2015), inverse modeling (Lauvaux et al., 2016), and a high resolution model-data fusion product, Hestia, providing a bottom-up estimate of fossil fuel-related CO_2 emissions for Indianapolis (Gurney et al., 2012). Integration of multiple top-down and bottom-up approaches are needed to converge on the most accurate, policy-relevant, and mechanistically representative emission estimate (Nisbet and Weiss, 2010).

As part of INFLUX, we have been quantifying the emission rates for CO_2 and CH_4 for the city of Indianapolis since 2008, and for CO since 2014, using a top-down aircraft-based mass balance approach. Top-down aircraft-based approaches are particularly effective at quantifying urban emissions because of their capability to sample the entire plume of a large emission source, with a short deployment time. Limitations of this method of measurement include initial costs, sporadic frequency of measurements, reliance on weather conditions (e.g. constant wind direction), and challenges in the definition of background conditions (e.g. White et al., 1976; Trainer et al., 1995; Lind and Kok, 1999; Kalthoff et al., 2002; Carras et al., 2002; Mays et al., 2009; Turnbull et al., 2011; O'Shea et al., 2014a, 2014b; Gioli et al., 2014; Cambaliza et al., 2014, 2015; Karion et al., 2015). However, aircraft-based measurement approaches have been successfully compared to bottom-up inventories (Cambaliza et al., 2014; Karion et al., 2015; Lamb et al., 2016). While accurate average results are achievable, individual flight realizations of emission rates have an uncertainty of ~ ±40% (Cambaliza et al., 2014).

In this paper, we describe our efforts to evaluate the potential of the mass balance experiment (MBE) approach to achieve useful precision through averaging of assumed random error. Karion et al. (2015) showed that, by averaging eight flight experiments, the precision of the mean for the MBE approach can be significantly improved when data are collected in a relatively short period of time, enabling the assumption of constant GHG emissions during the sampling period. They demonstrated that the averaged CH_4 emission rate in the Barnett Shale region, which includes eight different counties with dense natural gas production and urban areas, can be known with a precision of ~28% using the 1-sigma standard deviation of the mean, and of 17% at the 95% confidence level (CL) using a statistical bootstrapping method, while the relative uncertainties they found for a single-flight MBE can be as high as ~40%. In this paper, we investigate the efficacy of this method at the city scale through averaging of nine MBEs conducted from November 13th to December 3rd 2014 (3 weeks). We present and discuss the results for CO_2, CO, and CH_4 emission rates for the nine MBEs, and their average and standard error of the mean at 95% CL. Finally, we discuss the variability of the emission rates, and areas for improvement of this method.

2. Methods
2.1 Study area
Indianapolis, Indiana (39.79°N, 86.15°W, ~240 m above sea level) is the 12th largest city in the United States with a population of 848,800 (2014 United States Census Bureau) (**Figure 1**). The metropolitan area, combining Indianapolis, Carmel and Anderson, has (for the study period) a population of 1,971,000. Indianapolis is located on a flat plain, at least 120 km away from other metropolitan areas and is surrounded in all directions by rural land-use, primarily cropland. This isolated urban geography, often referred to as an "island city", is also accompanied by a relatively uniform inflow of boundary layer GHG concentrations, since upwind anthropogenic sources are relatively well-mixed when they reach Indianapolis (Cambaliza et al., 2014). These features make atmospheric measurements easier to interpret and simulate.

2.2 Aircraft instrumentation
CH_4, CO, and CO_2 emissions were quantified using an aircraft-based platform, combined with a mass balance approach. Flight experiments were performed using Purdue University's Airborne Laboratory for Atmospheric Research (ALAR, http://science.purdue.edu/shepson/research/bai/alar.html), a light twin-engine Beechcraft Duchess aircraft with an instrumentation compartment space of ~1 m³. This aircraft is equipped with i) a global positioning and inertial navigation system (GPS/INS), ii) a Best Air Turbulence (BAT) probe for wind measurements (Crawford and Dobosy, 1992; Garman et al. 2006, 2008), iii) a Picarro cavity ring-down spectrometer (CRDS, model G2401-m) for in-situ, real-time CO_2, CO, CH_4, and H_2O measurements (Crosson, 2008; see Karion et al. (2013a) for more details on the instrument performance), iv) an in-flight CO_2/CH_4 calibration system, and v) a programmable flask package (PFP) system for discrete ambient air sampling (Karion et al., 2013a; Sweeney et al., 2015).

Wind speeds and wind directions were obtained at 50 Hz using data from the nine-port differential pressure BAT probe that extends from the nose of the aircraft. Atmospheric temperature was measured using a microbead thermistor located at the center of the probe. A fast response thermocouple was also attached to the probe for comparison with the microbead observations. The measured pressure variations across the hemisphere of the probe were combined with 50 Hz spatial data from both GPS/INS and temperature sensors to obtain the three-dimensional wind vectors (Garman et al, 2006).

The CRDS measures gas concentrations at 0.5 Hz. Ambient air is pulled from the nose of the aircraft through a 5 cm diameter PFA Teflon tube at a flow rate of 1840 l/min (residence time: ~0.1 s) using a high-capacity blower located at the rear of the aircraft. The spectrometer is connected to the Teflon tubing using a tee and a 0.64 cm diameter Teflon inlet line, allowing ambient air to be continuously pumped through the analyzer at a flow rate of ~300 sccm (residence time: ~10 s). Just before the beginning and at the end of each MBE, both CO_2 and CH_4 were calibrated in-flight using three NOAA/ESRL reference cylinders. The certified mole fractions in each tank were: i) low concentrations: 368.02 ppm and 1781.05 ppb for CO_2 and CH_4, respectively, ii) medium concentrations: 410.73 ppm and 2222.42 ppb, iii) high concentrations: 447.11 ppm and 3261.49 ppb. The NOAA reproducibility (1 s) on the cylinder measurements are 0.03 ppm for CO_2 and 0.35 ppb for CH_4. The CRDS exhibits consistent reproducibility and linearity over time (Supplementary information: Figure S1). CO was calibrated at NOAA/ESLR before and after the field campaign, i.e. in Sept. 2014 and July 2015. The two calibrations exhibited the same calibration coefficients to within the uncertainty of the measurement, which was determined using in-flight calibrations, as described in Section S1.

2.3 Flight design and boundary layer height determination

Nine MBEs were performed during the late fall, on weekdays, from Nov. 13th to Dec. 3rd 2014 (Table S1). They were conducted when the convective boundary layer (CBL) was most likely fully developed and relatively constant in height throughout the duration of the experiment (about 4 hours on average), i.e. between 12:00 and 16:00 local time (Cambaliza et al., 2014). Prior to each flight, we ensured that morning wind conditions were sufficiently strong (wind speed at the surface and aloft) and consistent (wind direction) to avoid significant GHG accumulation in the boundary layer prior to our measurements (wind conditions monitored via Forecast (http://www.wunderground.com/, Terminal Aerodrome Forecast/TAF, Model Output Statistics/MOS) and real time weather at Indianapolis Automatic Weather Observation System/ASOS, METeorological Airport Report/METAR). For a typical MBE, the aircraft groundtrack was oriented perpendicular to the wind direction to intercept the polluted plume from the city (e.g. Carras et al., 2002). Three to five horizontal transects were performed at one downwind distance from the city and at different constant altitudes up to close to the top of the CBL (z_i) (**Figure 1**, Figure S2). Downwind transects were flown approximately 30 km from the city center, a distance far enough from the city that the plume was well-mixed but close enough to measure the urban plume above the background and to minimize mixing to and contact with the top of the boundary layer. The downwind transects allow for the construction of a two-dimensional plane onto which measurements are projected (e.g. Carras

et al. 2002; Kalthoff et al., 2002; Cambaliza et al., 2014). Indianapolis is approximately 70 km in width and therefore transect lengths are extended to 80–100 km to allow the capture of downwind air from beyond the city limits for regional background concentration determination (Cambaliza et al., 2014). One upwind horizontal transect (30 km upwind of the city center) was flown at a constant altitude (approximately 350 m above ground, Figure S2) prior to the downwind transects to identify possible significant CO, CO_2, and CH_4 sources upwind of Indianapolis, which would contribute to the observed city plume. In this case and since only one upwind transect is performed by MBE, the emission rate of the upwind point source was calculated by a single transect approach (Turnbull et al., 2011; Karion et al., 2013b). The emission rate is then subtracted from the final downwind emission rate exiting the city. A significant upwind source of CO_2 from the Eagle Valley power plant was transported to the city on Nov. 19th, the only day where upwind CO_2 emissions were observed (Figure S2). No identifiable upwind CO emissions were observed during the field campaign. Depending on the wind direction, CH_4 emissions from two different landfills located outside the city are evident in the sampled air and thus mixed into the city plume (TwinBridges if westerly winds, Caldwell if easterly winds, see **Figures 1** and S2). We subtracted the averaged emission rate of the respective landfill reported in Cambaliza et al. (2017), when necessary. Specifically, we subtracted an emission rate of 12.2 mol/s estimated for the TwinBridges landfill (Cambaliza et al., 2017) from the city-wide CH_4 emission rate obtained on Nov. 13th, 14th, 17th, 19th, 20th, 25th and

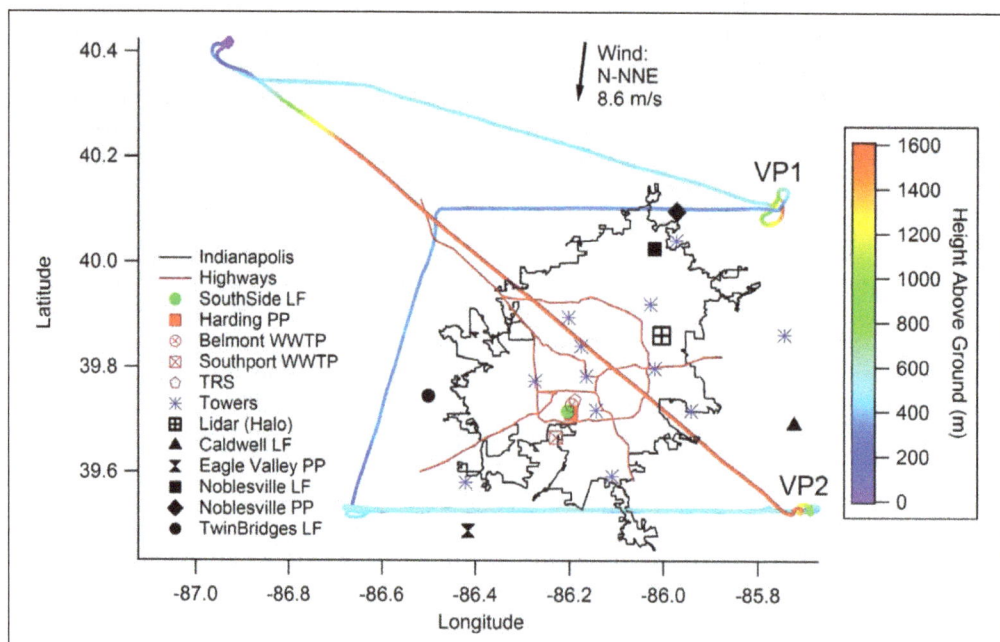

Figure 1: Flight path on December 01, 2014. Example of flight path (Dec 01 2014) performed over Indianapolis for the November-December field campaign, comprising two vertical profiles at the beginning and end of the experiment (VP1 and VP2, respectively), one upwind transect and several downwind transects flown at one downwind distance and different altitudes. DOI: https://doi.org/10.1525/elementa.134.f1

Abbreviations used in figure legend: PP: power plant, LF: landfill, WWTP: wastewater treatment plant, Towers: location of INFLUX towers, TRS: natural gas transmission regulating station.

Dec. 3rd (westerly winds, Figure S2), and 8.5 mol/s estimated for the Caldwell landfill (Cambaliza et al., 2015) on Nov. 21st (easterly winds, Figure S2). Emissions from the two landfills were not advected into the city plume on Dec. 1st (southerly winds).

The CBL depth was determined using measured vertical profiles (VPs) of H_2O, CO_2 and CH_4 concentrations and potential temperature (θ) (Figure S3). When conditions allowed, VPs were performed from as close to the ground as is safe (usually 150 m above ground level (agl)), to above the top of the CBL and into the free troposphere (~1500 m agl). Typically, two vertical profiles are conducted for each experiment (e.g., before starting the upwind transect and after finishing the last downwind transect) to assess the change in CBL height over the course of the experiment (**Figures 1** and S2, Table S1). The average of the upwind and downwind z_i from the two VPs is used as the upper bound of the vertical integration for the emission rate calculation (Eq. 2, see section 2.4). For days of thick cloud cover, only one VP (Nov. 17th and 19th) or no VP (Nov. 13th and 25th) was achieved (Table S1). To determine the full range of z_i for these particulars days, we used observations from a Doppler LIDAR (Light Detection and Ranging), Halo, located on the roof of Ivy Tech Community College northeast of Indianapolis (GPS coordinates: 39.8615°N, 86.0038°W, http://www.esrl.noaa.gov/csd/groups/csd3/measurements/influx/) (**Figures 1** and S4). The LIDAR started to measure boundary layer parameters beginning in April, 2013. A fixed scan pattern, repeated every 20–30 minutes, provides vertical profiles of horizontal wind speed and direction, vertical velocity variance and aerosol backscatter intensity. z_i values from the LIDAR were determined using the vertical velocity variance data available on the NOAA/ESRL website (see link above), after we ensured that both the vertical velocity variance and the aerosol backscatter signal strength provided similar results. From August 2013 to November 2014, nine VPs were flown above the LIDAR, allowing the comparison of z_i from both aircraft and LIDAR observations (**Figure 2**). The linear fit (**Figure 2**), forced through zero, exhibits a slope equal to 1.00 ± 0.02 (1σ) with a R^2 = 0.90 (Pearson coefficient is equal to 0.950, when the critical value is 0.798 at 99% CL, n−2 = 7), highlighting a good correlation between the two methods, with no significant bias. We thus considered that the full range of z_i for flights when only one or no VP was performed can be determined using the LIDAR observations. We averaged the z_i observations from the LIDAR corresponding to the duration of a single MBE, and used this average in Eq. 2 for emission rate integration.

2.4 Aircraft emission rate calculation and background selection

CH_4, CO_2, and CO concentrations (Figure S5), temperature, pressure and perpendicular wind speed recorded on the downwind transects were used to calculate the emission rates (in mol/s) from the city. First, fluxes (F_{ijk}, in mol/m²/s) were calculated at each data point (Figure S6) using the following formula:

$$F_{ijk} = \left(C_{M,ijk} - C_{bg,ijk}\right) . U_{ijk} \qquad (1)$$

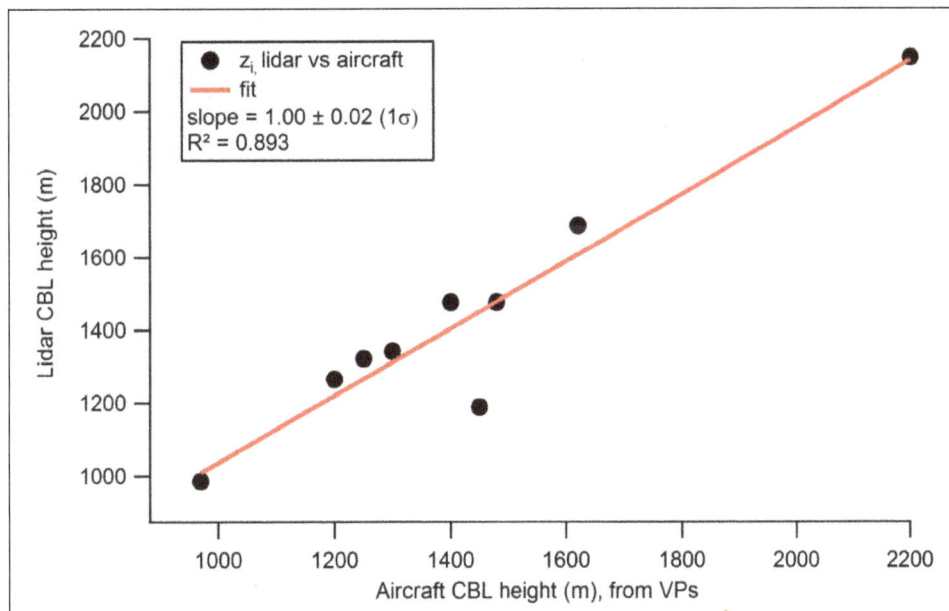

Figure 2: Comparison between the observed CBL height from aircraft vertical profile measurements and LIDAR observation. Comparison of the top of the convective boundary layer (CBL, z_i), determined by aircraft (x-axis) and LIDAR (y-axis) observations. Aircraft vertical profiles were flown above the LIDAR, inside the city boundaries. The nine flights were performed on (mm/dd/yy): 08/14/13 (VP from 17:01 pm to 17:23 pm GMT), 07/11/13 (from 21:31 to 21:46), 02/24/14 (18:49 to 19:05), 06/12/14 (20:14 to 8:39), 06/15/14 (17:44 to 17:54), 06/17/14 (19:56 to 20:19), 07/25/14 (19:25 to 19:34), 11/10/14 (20:28 to 20:33) and 11/14/14 (20:04 to 20:18). DOI: https://doi.org/10.1525/elementa.134.f2

where $C_{M,ijk}$ is the concentration measured for a specific latitude (i), longitude (j) and height (k) and converted to mol/m³ using the ideal gas law (pressure and temperature are used to calculate molar density). $C_{bg,ijk}$ represents background concentration (mol/m³, see below for more details), U_{ijk} is the 10 s averaged perpendicular wind speed (m/s, Cambaliza et al., 2014). The term ($C_{M,ijk} - C_{bg,ijk}$), also referred to as E_{ijk} in the supplementary information, represents the enhancement from the city. All the single-point horizontal fluxes are then projected onto a 2-D plane, and interpolated from the ground to the top of the boundary layer (z_i, in m), to the edges of the downwind distance flown by the airplane using a multi-transect kriging approach, at 10 m (z-axis) × 100 m (x-axis) resolution (Matlab-based EasyKrig 3.0; Chu 2004; Mays et al., 2009; Cambaliza et al., 2014) (Figure S7), for integration (Eq. 2).

$$ER = \int_0^{z_i} \int_{-x}^{+x} F_{ijk} dx\ dz \qquad (2)$$

Here, ER (mol/s) is the integrated emission rate for one MBE, $-x$ and $+x$ are the effective horizontal boundaries of the city, and dx and dz are the discrete horizontal and vertical distances (in m), over which the emission rate values are calculated, respectively. Whereas in Eq. 1 we calculate the flux through a vertical plane downwind of the city, in Eq. 2 we sum the flux in each interpolated pixel of that plane to calculate the total integrated emission rate from the city for a MBE. This produces a result which is consistent with what we observe, which is the total signal from all emission sources in the city. This distinguishes it from a surface flux, which can be calculated as a spatial average, e.g. as done by Mays et al. (2009). However, the surface fluxes are spatially highly heterogeneous (the power plant is a point source that represents roughly a third of the total, and mobile source emissions follow the major roads), and so calculated fluxes are not readily comparable between observations, while total emissions are much more so. And, the area used by Mays et al. (2009) represents an arbitrary choice of the definition of city boundaries. Thus we report only the total urban emissions for Indianapolis, in moles/s.

While entrainment/detrainment is possible, this has not been observed to date in our downwind vertical profiles. Typically, the boundary layer height does not increase by more than 15% and often less, during the course of the experiment. Because of this and the relatively short distance of the downwind transects from city sources (~30 km), chosen to minimize time for full mixing to the top of the boundary layer, our method assumes that entrainment/detrainment effects are minimal. We should also note that since we designed these flights for the fully developed boundary layer afternoon period, our measurements reflect emission rates for that time of day, as well as the Fall season. We also calculated enhancements, and then fluxes, after interpolation of the data (kriging), i.e. at each gridded cell using interpolated GHG concentrations, pressure, temperature and perpendicular wind speed (Mays et al., 2009; Cambaliza et al., 2014, 2015), to assess any changes in emission rates due to the choice of the kriging approach. Results from the both methods are presented in the "Results and Discussion".

Concentrations from both edges of the downwind transects are considered to be equivalent to the inflow of background air on the upwind side of the city, as previously defined by Mays et al. (2009) and Cambaliza et al. (2014, 2015) for aircraft measurements at Indianapolis. This approach was also used by Karion et al. (2015) for the Barnett Shale region. The location and distance of the transect edges were determined using a similar technique proposed by Cambaliza et al. (2014), i.e. by *i*) projecting the city boundaries onto the downwind transects using the observed mean wind direction, *ii*) observing the Gaussian shape of the CO_2, CH_4 and CO plumes recorded on these transects, and *iii*) defining the "edge" as the area of the transect where mixing ratios decrease to a constant concentration, and are likely not influenced by urban emissions. For all the flights, plumes from the city are well defined and concentrations decrease before reaching a constant concentration on both edges of the downwind transects. These edge concentrations are used for background determination (**Figures 3**, S5).

In the interest of consistency, we defined backgrounds in two ways for all the flights. First, background concentrations were defined as a single horizontally constant value, but varying with altitude (Cambaliza et al., 2014). This choice was motivated by the fact that significant background vertical gradients are observed for several flights, demonstrated on Dec. 3rd for CO_2, CH_4 and CO (Figure S5, see also vertical profiles on Figure S3), suggesting that altitude-dependent background concentrations, which might also vary with time and the growth of the CBL, should be determined for each individual downwind transect. Additionally, for some downwind transects, we observed that background concentrations on one edge can be significantly different from background concentrations observed on the opposite edge, suggesting a horizontal gradient in background concentration along the transect (see Nov. 25th, **Figures 3** and S5). Therefore, backgrounds were also defined using a linear function passing through median concentrations of the two sets of edge concentrations for each downwind transect (**Figures 3**, S5). The slope and the y-intercept of the linear function are used to calculate background concentrations at each data point where a measurement was recorded ($C_{bg,ijk}$ in Eq. 1). Consequences of background determination on emission rate results are discussed in the "Results and Discussion".

For the Nov–Dec 2014 field campaign, we assumed constant emissions from CO_2, CH_4 and CO sources at Indianapolis such that each of the nine MBEs can be considered as statistically repeated independent sampling of carbon emissions from the city. This is a reasonable assumption for CO and CO_2 at a given climate condition for weekdays, and for CH_4, given the diversity of sources in the city, either biogenic, or from the natural gas distribution system (Lamb et al., 2016). Emission rates from the nine MBEs were then averaged and the standard error of the mean at 95% CL (also referred SEM95 hereafter) calculated as $\frac{t*s}{\sqrt{n}}$, with t-student = 2.306, s is the sample standard deviation from the nine MBEs, and n the number of experiments.

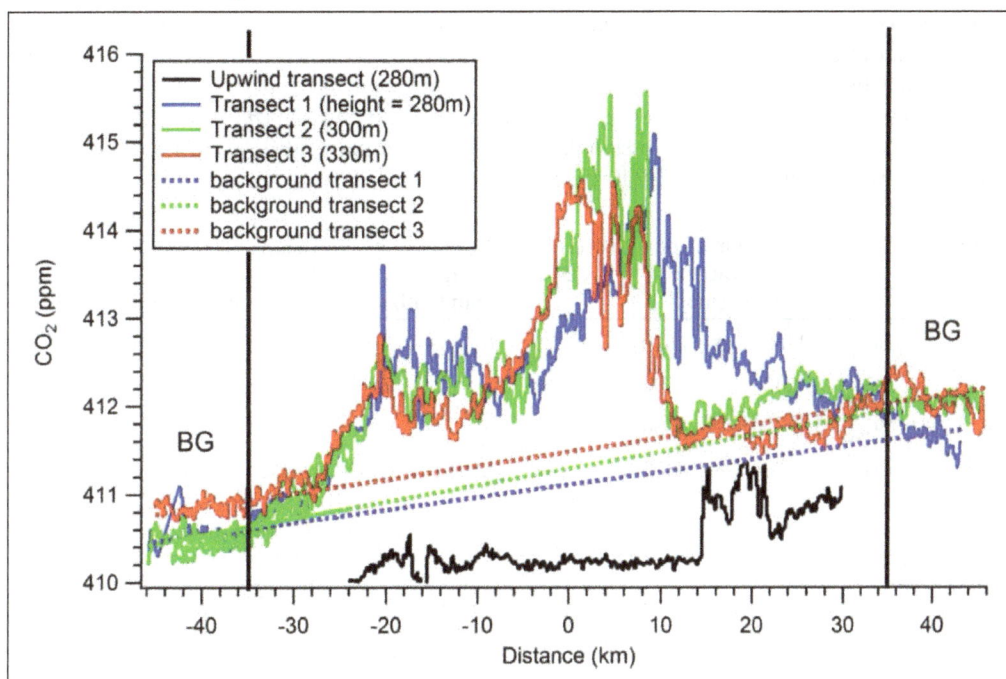

Figure 3: Downwind and upwind CO_2 concentrations, and background as a function of distance. CO_2 mole fraction in the downwind plume at different altitudes (blue, red and green full lines) and upwind the city (black full line), observed on November 25th, 2014 (wind direction = west, see Table S1). Dashed lines represent the background (as a linear regression) for each downwind transect. Higher background concentrations on the southern part of transects (positive distances) might be attributed to Eagle Valley power plant plume (see Figure 1). Vertical back lines represent the limits between background (BG, edges) and the city plume. DOI: https://doi.org/10.1525/elementa.134.f3

3. Results and discussion

3.1 Method comparison

To optimize our analysis method, data were processed using two different kriging approaches and background determinations. The background was defined *i*) as a single horizontal constant background concentration varying with altitude (BG_{avg}) (Cambaliza et al., 2014) and *ii*) as a linear function to account for any horizontal gradient in background concentrations along the length of the downwind transects (BG_{linReg}). We also interpolated *i*) concentrations, temperature, pressure and winds before calculating fluxes at each gridded cell (kriging_CTPU) (Mays et al., 2009; Cambaliza et al., 2014, 2015) or *ii*) individual point determinations of the calculated fluxes (kriging_F_{ijk}).

Calculated average emission rates for CO_2, CO, and CH_4 are presented in **Table 1**, as a function of background selection and interpolation approach. When averaged, CO_2, CO, and CH_4 emission rates are statistically indistinguishable using either BG_{avg} or BG_{linReg} and kriging_F_{ijk} or kriging_CTPU, as shown in **Table 1**. When BG_{avg} is used with both the kriging_F_{ijk} and kriging_CTPU approaches, CO_2 SEM95 is equal to 45% and 39%, respectively, i.e. poorer than the 17% SEM95 (kriging_F_{ijk}) and 25% SEM95 (kriging_CTPU) found when BG_{linReg} is applied. This result suggests that linear functions used to determine background concentrations greatly reduce variability of CO_2 emission rates calculated from replicate measurements. However, the SEM95 is similar for CO and CH_4 regardless of the method of background determination. The use of kriging_CTPU or kriging_F_{ijk} does not significantly change

the SEM95 of the method for CO_2 and CO. For CH_4, the SEM95 is improved by ~20% using kriging_CPTU instead of kriging_F_{ijk}. However, the CH_4 SEM95 remains large (41–60%, **Table 1**) compared to those for CO_2 and CO for the same set of flights, suggesting that the CH_4 emission rate variability is driven by factors other than the data analysis approach, and that for CH_4 the assumption of constant emission rate is not robust. In the following, we use the kriging_F_{ijk} approach with BG_{linReg}, since these two choices appear to yield the lowest SEM95 for the averaged CO_2 and CO emission rates.

Cambaliza et al. (2014) demonstrated that the choice in background concentration determination is one of the most sensitive parameters impacting uncertainties and variability in the emission rate results. Of course, uncertainties in the calculated emission rates due to background depend significantly on the magnitude of the enhancement, i.e. that the uncertainties are larger for smaller enhancements. For the Nov.–Dec. 2014 flights, we observed that change in background determination can lead to significant change in the emission rate of a single MBE. When the method kriging_F_{ijk} is applied, the average difference between emission rates of a MBE calculated using BG_{linReg} and using BG_{avg} is 50% for CO_2 (median = 9%), 32% for CH_4 (median = 18%) and 13% for CO (median = 9%), and range from 1% – 198%, 7% – 150% and 1% – 35%, respectively (**Figure 4**, Table S2). The most significant differences are observed on Nov. 17th and 19th for CO_2, and Dec. 3rd for CH_4 (**Figure 4**, Table S2). However, given the presence of spatial gradients in

Table 1: Comparison method: background selection and application of the kriging approach. Averaged CO_2, CH_4 and CO emission rates from the nine mass balance experiments (MBEs) *i*) when the emission rates were calculated by kriging the point fluxes (kriging_F_{ijk}) and *ii*) when they were calculated and integrated after interpolation of concentrations, pressure, temperature and perpendicular wind speed (kriging_CTPU). For both interpolation methods, backgrounds were defined *iii*) as linear regression (BG_{linReg}) and *iv*) as averages (BG_{avg}). Precision is the standard error of the mean at 95% CL. DOI: https://doi.org/10.1525/elementa.134.t1

Method	BG	CO_2 emission rate mean (mol/s)			Precision CO_2	CH_4 emission rate mean (mol/s)			Precision CH_4	CO emission rate mean (mol/s)			Precision CO
kriging_F_{ijk}	BG_{linReg}	14600	±	2500	17%	67	±	40	60%	108	±	17	16%
	BG_{avg}	10400	±	4600	45%	81	±	47	58%	98	±	19	19%
kriging_CTPU	BG_{linReg}	12400	±	3200	25%	56	±	23	42%	92	±	13	14%
	BG_{avg}	11024	±	4300	39%	65	±	27	41%	91	±	16	17%

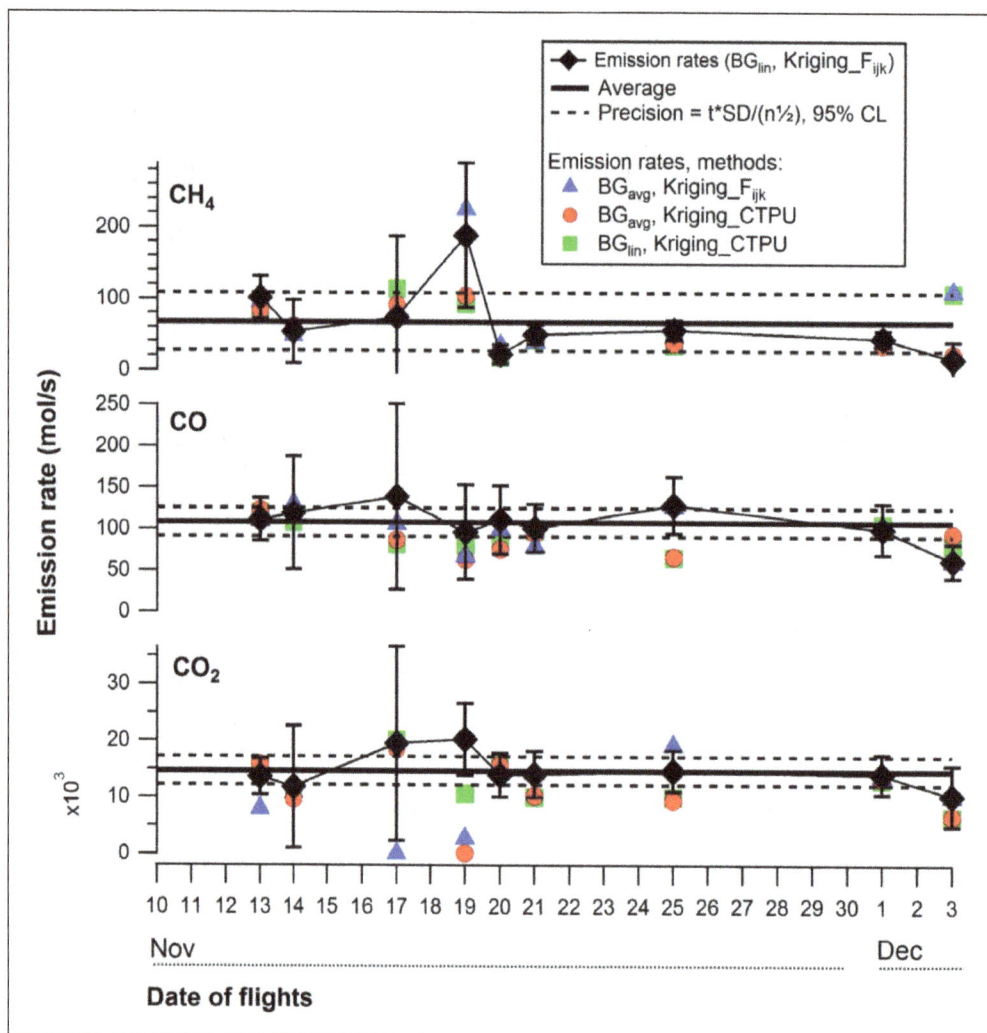

Figure 4: Overview of emission rates results. Calculated emission rates for each flight experiment using the four approaches described in Section 3.2. DOI: https://doi.org/10.1525/elementa.134.f4

background, we regard it best to use the linear regression approach. These results do indicate that the CH_4 emission rate determination is not more sensitive to background than is CO_2, supporting our conclusions that the actual CH_4 emission rates are variable day-to-day.

We also observed differences in calculated emission rates for a single MBE depending on the choice of the kriging method (and for a same background determination) (**Figure 4**). For example, when BG_{linReg} is used, the averaged relative difference between the two kriging methods for a single flight is equal to 25% (median = 14%, range = 3% – 62%) for CO_2, 25% (median = 21%, range = 6% – 58%) for CH_4 and 24% (median = 17%, range = 3% – 66%) for CO (Table S2). However, we believe it is most logical to interpolate the individual fluxes, since this interpolation is conducted only for one data set (i.e. calculated point fluxes).

3.2 Improvement of CO_2 fossil fuel and CO emission rate precision through averaging

Table 2 summarizes emission rate results of the nine MBEs obtained from Eq. 1 and 2 (for kriging_F_{ijk} and BG_{linReg}). Individual CO_2, CO, and CH_4 emission rates vary from 10,200 to 20,200 mol/s, 62 to 139 mol/s, and 16 to 189 mol/s, respectively (**Figure 4**). CO_2 and CH_4 emission rates are in the range of emission rates previously reported in Mays et al. (2009) and Cambaliza et al. (2014, 2015) for Indianapolis (from 2,500 to 49,000 mol/s for CO_2 and 12 to 230 mol/s for CH_4). When averaged, our measured emission rates are equal to 14,600 mol/s for CO_2, 108 mol/s for CO and 67 mol/s for CH_4. The average CO_2 emission rate for eight non-growing-season measurements in Mays et al. (2009) was 14,000 (±11,000; 1s) mol/s, for spring and winter of 2008. Our estimate via the Hestia emission model for the November 2014 CO_2 daytime (1200–1600) fossil fuel emission rate for Marion County, which corresponds to the way the downwind transects cut across the city, is 15,100 mol/s, which is indistinguishable from our average. Lauvaux et al. (2016), using an inverse modeling approach, have reported a value of 22,300 mol/s for the non-growing season for a larger 9-county region around Marion County, which they compare to the corresponding 9-county Hestia value of 18,300 mol/s. So, if the inverse model result scales with the relative Hestia

values, the corresponding Lauvaux et al. inverse model value for Marion County would be 18,400 mol/s. For CH_4, our average value, from 43 flights from 2008–2016, is 95 (±21) mol/s at the 95% CL. Although there are few city-wide estimates of GHG emissions, O'Shea at al., (2014) use aircraft-based measurements and the mass balance approach to determine the CO_2, CH_4 and CO emission rates from London, UK. They measured emission rates of 36,000 ± 3300 mol/s, 240 ± 16 mol/s and 220 ± 8 mol/s for CO_2, CH_4 and CO, respectively. These emission rates are considerably greater than those measured for Indianapolis, which is likely due, in part, by the larger population in London (appx 8.6 million) and older infrastructure.

Estimation of the uncertainty for a single MBE (ΔER, see Text S1 in supplementary information, Table S3) was calculated by propagating i) the measurement uncertainties of each term involved in Eq. 1, (i.e. uncertainties in pressure and temperature, which were used to convert into units of mol/m³, uncertainties of the calibration and from the linear fit for background determination and in the wind speed) and ii) the uncertainty of the CBL height due to growth during the experiment. Our uncertainty estimate represents a lower limit of the uncertainty associated with a single MBE, since all the factors known to influence emission rate results, such as entrainment from the free troposphere and interpolation of data points using the kriging approach (Cambaliza et al., 2014), were not accounted for. Relative uncertainties (RSD% = ΔER/ER) calculated from the uncertainty estimate of single MBEs vary between 23% and 91% (average = 44%) for CO_2, 24% and 81% (average = 42%) for CO, and 25% and 153% (average = 67%) for CH_4 (**Table 1**). When emission rates are averaged over the nine MBEs, the SEM95 (representing the expected precision for replication of a nine-point

Table 2: CO_2, CH_4 and CO emission rates from single mass balance experiments and average. CO_2, CH_4 and CO emission rates (mol/s) for single mass balance experiments (MBEs), their respective absolute uncertainties (propagation of uncertainties, results at 95% CL, see supplementary information) and relative precision. The averaged emission rates are reported with the standard error of the mean at 95% CL. The kriging approach was directly applied on fluxes (kriging_F_{ijk}). Background concentrations were calculated from linear functions. DOI: https://doi.org/10.1525/elementa.134.t2

Flight dates in 2014 (mm/dd)	CO_2 emission rate (mol/s)			Precision CO_2	CH_4 emission rate (mol/s)			Precision CH_4	CO emission rate (mol/s)			Precision CO
11/13	13600	±	3000	23%	101	±	30	30%	111	±	26	24%
11/14	11700	±	10700	91%	54	±	44	82%	119	±	68	58%
11/17	19500	±	17000	88%	74	±	110	153%	139	±	110	81%
11/19	20200	±	6240	32%	189	±	100	53%	96	±	57	59%
11/20	13800	±	3700	28%	22	±	13	56%	111	±	41	37%
11/21	14000	±	3900	29%	50	±	14	29%	101	±	29	29%
11/25	14600	±	3500	25%	56	±	14	26%	129	±	34	26%
12/01	14000	±	3400	25%	45	±	11	25%	100	±	31	31%
12/03	10200	±	5400	53%	16	±	24	150%	62	±	21	33%
Average Nov–Dec	**14600**	**±**	**2500**	**17%**	**67**	**±**	**40**	**60%**	**108**	**±**	**17**	**16%**

mean) is equal to 17% for CO_2, 16% for CO, and 60% for CH_4. Although the CH_4 variability is large, the precision of the mean for CO_2 and CO, when calculated from several MBEs performed in a short period of time (so that the emissions can be assumed to be constant), represents a significant improvement compared to the measurement uncertainty estimated for a single MBE measurement (and as shown by Cambaliza et al. (2014) to be 40–50%). This averaging approach represents a considerable improvement compared to the commonly reported uncertainties of GHG emission measurements from urban environments (~50% up to 100%; Trainer et al., 1995; Gurney et al., 2009; Mays et al., 2009; NRC, 2010; Turnbull et al., 2011; Cambaliza et al., 2014, 2015). Since the results for CH_4 are very different, for the same set of flight data, it seems clear that the emission rate variability is substantially different for CH_4.

The proposed averaging method has several potential limitations. All of the MBEs in this study were performed during the late fall when the total CO_2 enhancement from Indianapolis can be attributed to urban fossil fuel-related sources and biogenic CO_2 can be considered negligible (Turnbull et al., 2015). Also during the late fall, the mobile sector is the dominant source of CO emissions (Turnbull et al., 2015). The interpretation of the SEM95 is also based on the assumption that fossil fuel-related CO_2, CO, and CH_4 emissions from Indianapolis are constant during the three-week sampling period. In reality, some daily variability in urban emissions occurs (e.g. due to variability in ambient temperature and subsequent heating and electric power requirements), which may contribute to a larger (poorer) apparent estimated precision. Thus the SEM95 results obtained here are upper limits to the method precision for the period of year considered here, and it is likely that the true method precision is better than reflected in the SEM95 for CO_2 and CO.

3.3 Variability of CH_4 emissions

As discussed above, CH_4 emissions for Indianapolis are considerably more variable day-to-day than are those of CO_2 and CO. The averaged CH_4 emission rate (67 ± 40 mol/s) found for late fall 2014 is close to the emission rate calculated from a bottom-up inventory (57 mol/s, Lamb et al., 2016) built with measurements of selected sources in the city, including natural gas distribution facilities, landfills and waste water treatment facilities. The range of the CH_4 emission rate for the late fall overlaps with but is significantly lower than the range reported for spring-summer 2011 at Indianapolis (135 ± 58 mol/s, averaged from five airborne MBEs; Cambaliza et al., 2015), implying considerable variability for methane.

As shown by Cambaliza et al. (2015), the large enhancement of CH_4 signal from the south side of Indianapolis is attributed to the Southside landfill (SSLF, **Figure 1**), which represents a significant portion of the city-wide CH_4 emissions. We used wind direction data recorded by the BAT probe and the Hybrid Single Particle Lagrangian Integrated Trajectory Model (HYSPLIT, Draxler and Rolph, 2012) to confirm that the large CH_4 enhancements observed in fall 2014 (Figure S5b) can also be

attributed to the SSLF emissions. Cambaliza et al. (2015) determined that the SSLF represented, on average, 33% of the total CH_4 emissions from Indianapolis. Lamb et al. (2016) reported an average CH_4 emission rate of 31 mol/s for the SSLF, using ground-based mobile sampling for inverse plume modeling, and 30 mol/s is obtained from the EPA's GHG Reporting Program. If we assume that the CH_4 average we obtained (67 mol/s) for the city-wide total is representative, the Lamb et al. (2016) value for the SSLF represents 46% of the city-wide total. Variability of CH_4 emissions from landfills is partially dependent on the local climate, which can influence the seasonal oxidation of landfill covers (Xu et al., 2014; Spokas et al., 2015). To better understand the variability of the SSLF emissions, we considered several factors known to influence landfill emissions, such as change in barometric pressure over time, air temperature and precipitation (Xu et al., 2014; Spokas et al., 2015). However, no particular correlations were found between aircraft-determined CH_4 emission rates (the part attributed to the landfill) and average atmospheric temperature recorded at Indianapolis International airport during the experiments (data downloaded from http://www.ncdc.noaa.gov/qclcd/QCLCD, accessed on 10/05/2015), as well as with relative humidity and barometric pressure. Lamb et al. (2016) report that the uncertainty in their determination of the SSLF emission rate is ±30%, or 9 mol/s. In contrast, the SEM95 for the city-wide total we determined is 40 mol/s (60%), i.e. much larger than the SSLF uncertainty, and larger than the average SSLF emission rate reported by Lamb et al. (2016). The total CH_4 variability must, then, derive from CH_4 sources other than the landfill and must be large enough to explain the high variability of the city-wide CH_4 emission rate (SEM95 of 60%). Using the ratio of propane-to-methane ((C_3H_8)/CH_4) concentrations from aircraft flask samples and the slope of the C_3H_8 vs. CH_4 regression, Cambaliza et al (2015) demonstrated that the non-landfill-related CH_4 emissions from Indianapolis represent ~67% of the total CH_4 emissions and could be attributed to the natural gas distribution system. Lamb et al. (2016) estimated a contribution of natural gas sources of 43% from ethane/methane observation coupled with inverse modeling. The authors also found that emissions from natural gas systems can vary by three orders of magnitude, depending on the type of structures in the natural gas system (i.e., pipeline, transmission and storage station, etc.). Emissions from the natural gas system at Indianapolis are potentially due to random small sources (Lamb et al., 2016) and can be highly spatially and temporally variable, which likely explains the variability of CH_4 emission rates we observed in Nov.–Dec. 2014. However, to explain our observed variability, we would need to turn to large GHG contributors, among which are the natural gas sources.

4. Conclusions

From nine mass balance experiments performed in Indianapolis in Nov.–Dec. 2014, we quantified averaged CO_2, CO, and CH_4 emission rates (14,600, 108 and 67 mol/s, respectively). The SEM95 results were equal to 17% for CO_2 and 16% for CO (at the 95% CL), i.e. much lower than for the single-flight precision (~40%) found for Indianapolis (Cambaliza et al., 2014), and for the estimated single-flight uncertainties for the results reported here. We applied averaging to improve estimation of CO_2 and CO urban emission rates during the late fall, when CO_2 and CO emissions are primarily of anthropogenic origin. During the summer, poorer apparent precision would be expected due to the greater CO_2 and CO emissions variability from biogenic emission contributions. A field campaign similar to the one presented in this study (multiple MBEs flown in a short period of time, to allow for an assumption of constant emission rates over the time period) should be performed during the spring and summer seasons to evaluate the averaging method when biogenic CO_2 and CO emissions are not negligible. While the precision obtained in this mass balance experiment is approaching GHG reduction target values, we expect averaged emission rate precisions to be further improved with increasing numbers of airborne MBEs performed over a short period of time. Precisions might also be further improved by either the use of multiple aircraft flying simultaneously, or one aircraft flying at the top of the boundary layer and employing a downward-looking Differential Absorption LIDAR (Dobler et al., 2013; Abshire et al., 2014). Thus the target precision is technically attainable, the main limitation being the cost of the experiment.

Previous aircraft and surface-based measurements have suggested that most of the remaining CH_4 (~2/3 of the city-wide emissions) likely derives from the city's natural gas distribution system (Cambaliza et al., 2015; Lamb et al., 2016). Since the Southside Landfill is a minor component of the total, and the remainder is believed to be derived from the natural gas network, it appears that the variability in our observations of the city-wide total may be attributed to variability in the nature and magnitude of individual leaks in that system. Surface-based measurements of methane, ethane, and $\delta^{13}C$-CH_4 may allow us to differentiate between CH_4 sources, complementing the aircraft-based total emission rate measurement. For this city, evaluation of emission mitigation progress for methane may be difficult because of the large day-to-day variability in the source strength.

Supplemental Files
The supplemental files for this article can be found as follows:

· **Text S1.** Uncertainties for a single MBE.

· **Figure S1.** CRDS calibration.

· **Figure S2.** Flight paths of the aircraft experiments.

· **Figure S3.** Aircraft vertical profiles and estimated boundary layer height.

· **Figure S4.** LIDAR observations.

· **Figure S5.** Downwind CO_2, CH_4 and CO concentrations as a function of distance.

· **Figure S6.** Downwind CO_2, CH_4 and CO fluxes, observations.

· **Figure S7.** Downwind CO_2, CH_4 and CO fluxes after data interpolation (kriging).

· **Table S1.** Meteorological conditions during the field campaign.

· **Table S2.** Emission rates depending on background determination and the kriging method.

· **Table S3.** Uncertainties of individual MBEs.

Acknowledgements
We would like to acknowledge the Purdue University Jonathan Amy Facility for Chemistry Instrumentation (JAFCI) for technical support on the project. We also thank pilots R. M. Grundman and J. N. Poppe, who performed the MBEs with us.

Funding information
This work is part of the Indianapolis Flux Experiment (INFLUX) project, funded by the National Institute of Standards and Technology (NIST), for which we are grateful.

Competing interests
The authors have no competing interests to declare.

Author contibutions
· Project conception and design: AMFH, PBS, BS, MOLC
· Data Acquisition: AMFH, PBS
· Analysis and interpretation of data: AMFH, RMH, PBS, CS, JT, OES, A-EMK, TNL
· Drafting and/or revising the article: AMFH, RMH, PBS, JT, MOLC, OES, A-EMK, TNL, KJD, TL, AK, CS, WAB, RMH, KRG, JW
· Final approval of the version to be published: AMFH, RMH, PBS

References
Abshire, JB, Ramanathan, A, Risis, H, Mao, J, Allan, GR, et al. 2014 Airborne Measurements of CO_2 column concentration and range using a pulsed direct-detection IPDA Lidar. *Remote Sens* **6**: 443–469. DOI: https://doi.org/10.3390/rs6010443

Bergamaschi, P, Corazza, M, Karstens, U, Athanassiadou, M, Thompson, RL, et al. 2015 Top-down estimates of European CH_4 and N_2O emissions based on four different inverse models. *Atmos Chem Phys* **15**: 715–736. DOI: https://doi.org/10.5194/acp-15-715-2015

Bousquet, P, Ciais, C, Miller, JB, Dlugokencky, EJ, Hauglustaine, DA, et al. 2006 Contribution of anthropogénique and natural sources to atmospheric methane variability. *Nature* **443**: 439–443. DOI: https://doi.org/10.1038/nature05132

Cambaliza, MOL, Bogner, JE, Green, RB, Shepson, PB, Harvey, TA, et al. 2017 Field measurements and modeling to resolve m² to km² CH_4 emissions for a complex urban source: An Indiana landfill study, *Elem Sci Anth.*, in press for INFLUX special feature.

Cambaliza, MOL, Shepson, PB, Bogner, J, Caulton, DR, Stirm, B, et al. 2015 Quantification and source apportionment of the methane emission flux from the city of Indianapolis. *Elem Sci Anth.* **3**: 000037. DOI: https://doi.org/10.12952/journal.elementa.000037

Cambaliza, MOL, Shepson, PB, Caulton, DR, Stirm, B, Samarov, D, et al. 2014 Assessment of uncertainties of an aircraft-based mass balance approach for quantifying urban greenhouse gas emissions. *Atmos Chem Phys* **14**: 9029–9050. DOI: https://doi.org/10.5194/acp-14-9029-2014

Carras, JN, Cope, M, Lilley, W and **Williams, DJ** 2002 Measurement and modeling of pollutant emissions from Hong Kong. *Environ Modeling & Software* **17**: 87–94. DOI: https://doi.org/10.1016/S1364-8152(01)00055-X

China's INDC Enhanced actions on climate change: China's intended nationally determined contributions. June 2015. Available at http://www4.unfccc.int/submissions/INDC/Published%20Documents/China/1/China's%20INDC%20-%20on%2030%20June%202015.pdf. Accessed December 2015.

Chu, D 2004 The GLOBEC kriging software package – EasyKrig 3.0; The Woods Hole Oceanographic Institution. Available at http://globec.whoi.edu/software/kriging/easy_krig/easy_krig.html. Accessed May 2015.

Crawford, TL and **Dobosy, RJ** 1992 A sensitive fast-response probe to measure turbulence and heat flux from any airplane. *Boundary Layer Meteor* **59**(3): 257–278. DOI: https://doi.org/10.1007/BF00119816

Crosson, ER 2008 A cavity ring-down analyzer for measuring atmospheric levels of methane, carbon dioxide, and water vapor. *Appl Phys B-Lasers O* **92**: 403–408. DOI: https://doi.org/10.1007/s00340-008-3135-y

Dobler, JT, Harisson, FW, Browell, EV, Lin, B, McGregor, D, et al. 2013 Atmospheric CO_2 column measurements with an airborne intensity-modulated continuous wave 1.57 μm fiber laser lidar. *Applied Optics* **52**(12): 2,874–2,892. DOI: https://doi.org/10.1364/AO.52.002874

Draxler, RR and **Rolph, GD** 2012 HYSPLIT (Hybrid Single-Particle Lagrangian Integrated Trajectory) Model access via NOAA ARL READY website NOAA Air Resources Laboratory, Silver Srping, MD. Available at http://ready.arl.noaa.gov/HYSLIPT.php?.

European commission 2014 Progress towards achieving the Kyoto and EU 2020 objectives. Available at http://ec.europa.eu/transparency/regdoc/rep/1/2014/EN/1-2014-689-EN-F1-1.Pdf.

Forster, P, Ramaswamy, V, Artaxo, P, Berntsen, T, Betts, R, et al. 2007 *Changes in Atmospheric Constituents and in Radiative Forcing. In: Climate Change 2007: The Physical Science Basis. Contribution of Working Group I to the Fourth Assessment Report of the Intergovernmental Panel on Climate Change.* Cambridge University Press, Cambridge, United Kingdom and New York, NY, USA. Solomon, S, Qin, D, Manning, M, Chen, M, Marquis, M, Averyt, KB, Tignor, M and Miller, HL, (eds.).

Garman, KE, Hill, KA, Wyss, P, Carlsen, M, Zimmerman, JR, et al. 2006 An airbone and wind tunnel evaluation of a wind turbulence measurement system for aircraft-based flux measurement. *J Atmos Ocean Tech* **23**: 1696–1708. DOI: https://doi.org/10.1175/JTECH1940.1

Garman, KE, Wyss, P, Carlsen, M, Zimmerman, JR, Stirm, BH, et al. 2008 The contribution of variability of lift-induced upwash to the uncertainty in vertical winds determined from an aircraft platform. *Boundary Layer Meteor* **126**: 461–476. DOI: https://doi.org/10.1007/s10661-013-3517-4

Gioli, B, Carfora, MF, Magliulo, V, Metallo, MC, Poli, AA, et al. 2014 Aircraft mass budgeting to measure CO_2 emissions from Rome, Italy. *Environ Monit Assess* **186**: 2053–2066. DOI: https://doi.org/10.1007/s10661-013-3517-4

Gurney, KR, Mendoza, DL, Zhou, Y, Fischer, ML, Miller, CC, et al. 2009 High resolution fossil fuel combustion CO_2 emission fluxes for the United States. *Environ Sci Technol* **43**: 5535–5541. DOI: https://doi.org/10.1021/es900806c

Gurney, KR, Razlivanov, I, Song, Y, Zhou, Y, Benes, B, et al. 2012 Quantification of fossil fuel CO_2 emissions at the building/street level scale for a large US city. *Environ Sci Technol* **46**: 12194–12202. DOI: https://doi.org/10.1021/es3011282

Hutyra, LR, Duren, R, Gurney, KR, Grimm, N, Kort, EA, et al. 2014 Urbanization and the carbon cycle: Current capabilities and research outlook from the natural sciences perspective. *Earth's future* **2**: 473–495 DOI: https://doi.org/10.1002/2014EF000255

IEA (International Energy Agency) 2008 *World energy outlook.* Available at http://www.worldenergy-outlook.org/media/weowebsite/2008-1994/weo2008.pdf.

IPCC (Intergovernmental Panel on Climate Change) 2014 *Climate Change 2014: Mitigation of Climate Change, Contribution of Working Group III to the Fifth Assessment Report of the Intergovernmental Panel on Climate Change,* Chapter 12, p. 935 [Edenhofer, O, Pichs-Madruga, RR, Sokona, Y, Farahani, E, Kadner, S, Seyboth, K, Adler, A, Baum, I, Brunner, S, Eickemeier, P, Kriemann, B, Savolainen, J, Schlömer, S, von Stechow, C, Zwickel, T and Minx, JC, (eds.)]. Available at http://mitigation2014.org/report/publication/.

Kalthoff, N, Corsmeier, U, Schmidt, K, Kottmeier, C, Fiedler, F, et al. 2002 Emissions of the city of Augsburg determined using the mass balance method. *Atmos Environ* **36**(Supplement 1): 19–S31. DOI: https://doi.org/10.1016/S1352-2310(02)00215-7

Karion, A, Sweeney, C, Kort, EA, Shepson, PB, Brewer, A, et al. 2015 Aircraft-based estimate of total methane emissions from the Barnett shale region. *Environ Sci Technol* **49**: 8124–8131. DOI: https://doi.org/10.1021/acs.est.5b00217

Karion, A, Sweeney, C, Pétron, G, Frost, G, Hardesty, MH, et al. 2013b Methane emissions estimate from airborne measurements over the western United Sates natural gas field, *Geophys Res Lett* **40**: 4393–4397. DOI: https://doi.org/10.1002/grl.50811

Karion, A, Sweeney, C, Wolter, S, Newberger, T, Chen, H, et al. 2013a Long-term greenhouse gas measurements from aircraft. *Atmos Meas Tech* **6**: 511–526. DOI: https://doi.org/10.5194/amt-6-511-2013

Kirschke, S, Bousquet, P, Ciais, P, Saunois, M, Canadell, JG, et al. 2013 Three decades of global methane sources and sinks. *Nature Geosci* **6**: 813–823. DOI: https://doi.org/10.1038/ngeo1955

Lamb, BK, Prasad, K, Cambaliza, MO, Ferrara, T, Lauvaux, T, et al. 2016 n.d. Direct and indirect measurements and modeling of methane emissions in Indianapolis, IN., *Environ. Sci. Technol.*, submitted. DOI: https://doi.org/10.1021/acs.est.6b01198

Lauvaux, TN, Miles, A, Deng, S, Richardson, M, Cambaliza, O, Davis, K, Brian Gaudet, K, Gurney, J, Huang, D, o'Keefe, Y, Song, A, Karion, T, Oda, R, Patarasuk, I, Razlivanov, D, Sarmiento, PB, Shepson, C, Sweeney, J and Turnbull Wu, K 2016 High-resolution atmospheric inversion of urban CO_2 emissions during the dormant season of the Indianapolis Flux Experiment (INFLUX), *J. Geophys. Res. Atmos.*, **121**, DOI: https://doi.org/10.1002/2015JD024473

Le Quéré, C, Moriarty, R, Andrew, RM, Peters, GP, Ciais, P, et al. 2014 Global carbon budget 2014. *Earth Syst Sci Data Discuss* **7**: 521–610. DOI: https://doi.org/10.5194/essdd-7-521-2014

Lind, JA and Kok, GL 1999 Emission strengths for primary pollutants as estimated from an aircraft study of Hong Kong air quality. *Atmos Environ* **33**: 825–831. DOI: https://doi.org/10.1016/S1352-2310(98)00141-1

Marland, G 2008 Uncertainties in accounting for CO_2 from fossil fuels. *J Ind Ecol* **12**: 136–139. DOI: https://doi.org/10.1111/j.1530-9290.2008.00014.x

Marland, G 2012 China's uncertain CO_2 emissions. *Nature Clim Change* **2**: 645 – 646. DOI: https://doi.org/10.1038/nclimate1670

Mays, KL, Shepson, PB, Stirm, BH, Karion, A, Sweeney, C, et al. 2009 Aircraft-based measurements of the carbon footprint of Indianapolis. *Environ Sci Technol* **43**: 7816–7823. DOI: https://doi.org/10.1021/es901326b

McKain, K, Down, A, Raciti, SM, Budney, J, Hutyra, LR, et al. 2015 Methane emissions from natural gas infrastructure and use in the urban region of Boston, Massachusetts. *Proc Nati Acad Sci USA* **112**(7): 1,941–1,956. DOI: https://doi.org/10.1073/pnas.1416261112

Miles, NK, Richardson, SJ, Lauvaux, T, Davis, KJ, Deng, AJ, et al. 2015 May 20. Detection and quantification of atmospheric boundary layer greenhouse gas dry mole fraction enhancements from urban emission: results from INFLUX. NOAA GMD annual meeting; Boulder, CO.

Myhre, G, Shindell, D, Bréon, F-M, Collins, W, Fuglestvedt, J, et al. 2013 *Anthropogenic and Natural Radiative Forcing. In Climate Change 2013: The Physical Science Basis. Contribution of Working Group I to the Fifth Assessment Report of the Intergovernmental Panel on Climate Change.* Cambridge University Press, Cambridge, UK and New York, NY, USA. pp: 659–740. Stocker, TF, Qin, D, Plattner, G-K, Tignor, M, Allen, SK, Boschung, J, Nauels, A, Xia, Y, Bex, V and Midgley, PM, (eds.).

Nisbet, E and Weiss, R 2010 Top-down versus Bottom-up. *Science* **328**(5983): 1241–1243. DOI: https://doi.org/10.1126/science.1189936

NRC 2010 (Committee on Methods for Estimating Greenhouse Gas Emisisons), Verifying Greenhouse Gas Emissions: Methods to Support International Climate Agreements, report n° 9780309152112, The National Academies Press, Washington DC.

Olivier, JGJ, Bouwman, AF, van der Maas, CWM, Berdowski, JJM, Veldt, C, et al. 1996 Description of Edgar Version 2.0: A set of global emission inventories of greenhouse gases and ozone-depleting substances for all anthropogenic and most natural sources on a per country basis and on 1° × 1° grid. RIVM report nr. 771060 002/TNO-MEP report nr. R96/119. National Institute of Public Health and the Environment, Bilthoven, The Netherlands.

O'Shea, SJ, Allen, G, Fleming, ZL, et al. 2014b Area fluxes of carbon dioxide, methane, and carbon monoxide derived from airborne measurements around Greater London: A case study during summer 2012. *J Geophys Res-Atmos* **119**: 4940–4952. DOI: https://doi.org/10.1002/2013JD021269

O'Shea, SJ, Allen, G, Gallagher, MW, et al. 2014a Methane and carbon dioxide fluxes and their regional scalability for the European Arctic wetlands during the MAMM project in summer 2012, *Atmos. Chem. Phys* **14**: 13159–13174. DOI: https://doi.org/10.5194/acp-14-13159-2014

Parrish, DD, Trainer, M, Holloway, JS, Yee, J, Warshawsky, S, et al. 1993 Relationships between ozone and carbon monoxide at surface sites in the North Atlantic region, *J Geophys Res* **103**: 13357–13376. DOI: https://doi.org/10.1029/98JD00376

Peylin, P, Houweling, S, Krol, MC, Karstens, U, Rödenbeck, C, et al. 2011 Importance of fossil fuel emission uncertainties over Europe for CO_2 modeling: model intercomparison, *Atmos Chem Phys* **11**: 6607–6622. DOI: https://doi.org/10.5194/acp-11-6607-2011

President's Climate Action Plan 2013 Available at http://www.whitehouse.gov/sites/default/files/image/president27sclimateactionplan.pdf.

President Obama's Climate Action Plan 2014a Progress Report, Cutting carbon pollution, protecting American communities, and leading internationally. Available at http://www.whitehouse.gov/sites/default/files/docs/cap_progress_report_update_062514_final.pdf.

President's Climate Action Plan 2014b Strategy to reduce methane emissions. Available at http://www.whitehouse.gov/sites/default/files/strategy_to_reduce_methane_emissions_2014-03-28_final.pdf.

Rosenzweig, C, Solecki, W, Hammer, SA and **Mehrotra, S** 2010 Cities lead the way in climate-change action. *Nature* **467**: 909–911. DOI: https://doi.org/10.1038/467909a

Schakenbach, J, Vollaro, R and **Forte, R** 2006 Fundamentals of Successful Monitoring, Reporting, and Verification under a Cap-and-Trade Program. *J of the Air & Waste Management Association* **56**: 1576–1583. DOI: https://doi.org/10.1080/10473289.2006.10464565

Seto, KC, Güneralp, B and **Hutyra, LR** 2012 Global forecast of urban expansion to 2030 and direct impacts on biodiversity and carbon pools. *Proc Natl Acad Sci USA* **109**: 16083–16088. DOI: https://doi.org/10.1073/pnas.1117622109

Spokas, K, Bogner, J, Corcoran, M and **Walker, S** 2015 From California dreaming to California data: challenging historic models for landfill CH_4 emissions. *Elementa* **3**: 000051. DOI: https://doi.org/10.12952/journal.elemena.000051

Sweeney, C, Karion, A, Wolter, S, Newberger, T, Guenther, D, et al. 2015 Seasonal climatology of CO_2 across North America from aircraft measurements in the NOAA/ESRL Global Greenhouse Gas Reference Network. *J Geophys Res-Atmos* **120**: 5155–5190. DOI: https://doi.org/10.1002/2014JD022591

Trainer, M, Ridley, BA, Burh, MP, Kok, G, Walega, J, et al. 1995 Regional ozone and urban plumes in the southeastern United States: Birmingham, a case study. *J Geophys Res* **100**: 18,823–18,834. DOI: https://doi.org/10.1029/95JD01641

Turnbull, JC, Guenther, D, Karion, A, Sweeney, C, Anderson, E, et al. 2012 An integrated flask sample collection system for greenhouse gas measurements. *Atmos Meas Tech* **5**: 2321–2327. DOI: https://doi.org/10.5194/amt-5-2321-2012

Turnbull, JC, Karion, A, Fisher, ML, Faloona, I, Guilderson, T, et al. 2011 Assessment of fossil fuel carbon dioxide and other anthropogenic trace gas emissions from airborne measurements over Sacramento, California in spring 2009. *Atmos Chem Phys* **11**: 705–721. DOI: https://doi.org/10.5194/acp-11-705-2011

Turnbull, JC, Sweeney, C, Karion, A, Newberger, T, Lehman, SJ, et al. 2015 Toward quantification and source sector identification of fossil fuel CO_2 emissions from an urban area: results from the INFLUX experiment. *J Geophys Res Atmos* **120**: 292–312. DOI: https://doi.org/10.1002/2014JD022555

United Nations Framework Convention on Climate Change (UNFCCC) 2010 Report of the Conference of the Partie on its fifteenth session, held in Copenhagen form 7 to 10 December 2009. FCCC/CP/2009/Add.1. Available at http://unfccc.int/resource/docs/2009/cop15/eng/11a01.pdf. Appendix I available at http://unfccc.int/meetings/copenhagen_dec_2009/items/5264.php.

United Nations Framework Convention on Climate Change (UNFCCC) 2015 Adoption of the Paris Agreement. FCCC/CP/2015/L.9/Rev.1. Available at https://unfccc.int/resource/docs/2015/cop21/eng/l09r01.pdf.

Vine, E and **Sathaye, J** 1999 The monitoring, evaluation, reporting and verification of climate change projects. *Mitig Adapt Strateg Glob Change* **4**(1): 43–60. DOI: https://doi.org/10.1023/A:1009651316596

Wennberg, PO, Mui, W, Wunch, D, Kort, EA, Blake, DR, et al. 2012 On the sources of methane to the Los Angeles atmosphere. *Environ Sci Technol* **46**(17): 9,282–9,289. DOI: https://doi.org/10.1021/es301138y

White, WH, Anderson, JA, Blumenthal, DL, Husar, RB, Gillani, NV, et al. 1976 Formation and Transport of Secondary Air Pollutants: Ozone and Aerosols in the St. Louis Urban Plume. *Science* **194**(4261): 187–189. DOI: https://doi.org/10.1126/science.959846

Wunch, D, Wennberg, PO, Toon, GC, Keppel-Aleks, G and **Yavin, YG** 2009 Emissions of greenhouse gases from a North American megacity. *Geophys Res Lett* **36**(L15): 810. DOI: https://doi.org/10.1029/2009GL039825

Xu, L, Lin, X, Amen, J, Welding, K and **McDermitt, D** 2014 Impact of changes in barometric pressure on landfill methane emission. *Global Biogeochem Cy* **28**: 679–695. DOI: https://doi.org/10.1002/2013GB004571

5

Reconciling the differences between a bottom-up and inverse-estimated FFCO$_2$ emissions estimate in a large US urban area

Kevin R. Gurney*, Jianming Liang*, Risa Patarasuk*, Darragh O'Keeffe*, Jianhua Huang*, Maya Hutchins*, Thomas Lauvaux[†], Jocelyn C. Turnbull[‡,§] and Paul B. Shepson[ǁ]

The INFLUX experiment has taken multiple approaches to estimate the carbon dioxide (CO$_2$) flux in a domain centered on the city of Indianapolis, Indiana. One approach, Hestia, uses a bottom-up technique relying on a mixture of activity data, fuel statistics, direct flux measurement and modeling algorithms. A second uses a Bayesian atmospheric inverse approach constrained by atmospheric CO$_2$ measurements and the Hestia emissions estimate as a prior CO$_2$ flux. The difference in the central estimate of the two approaches comes to 0.94 MtC (an 18.7% difference) over the eight-month period between September 1, 2012 and April 30, 2013, a statistically significant difference at the 2-sigma level. Here we explore possible explanations for this apparent discrepancy in an attempt to reconcile the flux estimates. We focus on two broad categories: 1) biases in the largest of bottom-up flux contributions and 2) missing CO$_2$ sources. Though there is some evidence for small biases in the Hestia fossil fuel carbon dioxide (FFCO$_2$) flux estimate as an explanation for the calculated difference, we find more support for missing CO$_2$ fluxes, with biological respiration the largest of these. Incorporation of these differences bring the Hestia bottom-up and the INFLUX inversion flux estimates into statistical agreement and are additionally consistent with wintertime measurements of atmospheric ^{14}CO$_2$. We conclude that comparison of bottom-up and top-down approaches must consider all flux contributions and highlight the important contribution to urban carbon budgets of animal and biotic respiration. Incorporation of missing CO$_2$ fluxes reconciles the bottom-up and inverse-based approach in the INFLUX domain.

Keywords: carbon footprint; carbon flux; fossil fuel CO$_2$

1. Introduction

Anthropogenic carbon dioxide (CO$_2$) emission, primarily from the combustion of fossil fuels, is the largest net annual flux of CO$_2$ to the atmosphere and represents the dominant source of greenhouse gas forcing (*Hansen et al.*, 1998; *LeQuere et al.*, 2013). Anthropogenic CO$_2$ emissions are often used as a near-certain boundary condition when solving total carbon budgets; an endeavor essential to quantifying other components of the carbon cycle and to improving our understanding of the feedbacks between the carbon cycle and climate change (*Gurney et al.*, 2007; *Heimann et al.*, 2008). Similarly, to construct

meaningful projections of greenhouse gas emissions, a mechanistically-based quantification of current emissions is necessary. Finally, greenhouse gas mitigation efforts require improved quantification of fluxes to establish emission baselines, substantiate emission trajectories, and for the identification of efficient, economically-viable greenhouse gas mitigation options (e.g. *Kennedy et al.*, 2010).

All of the motivations for understanding and quantifying fluxes of CO$_2$ are equally applicable to the urban domain, where recent years have seen increasing interest and importance. This interest is driven, in no small part, by the recognition that urban areas currently account for over 70% of energy-related CO$_2$ emissions and are projected to triple in extent between 2000 and 2030 (*Seto* 2012; *IEA*, 2008).

Just as with the larger scales, improved understanding of the carbon flows in cities offers several practical outcomes for urban stakeholders. Quantification of the impacts of mitigation efforts or programs and their effective management remains an important need as more cities agree to voluntary or legislated reduction targets. Similarly

* Arizona State University, Tempe, Arizona, US

[†] Department of Meteorology, Pennsylvania State University, University Park, Pennsylvania, US

[‡] GNS Science, Rafter Radiocarbon Laboratory, Lower Hutt, NZ

[§] National Oceanic and Atmospheric Administration/University of Colorado, Boulder, Colorado, US

[ǁ] Purdue University, West Lafayette, Indiana, US

Corresponding author: Kevin R. Gurney (kevin.gurney@asu.edu)

important are information needs to plan and optimize mitigation strategies. To meet such mitigation targets, action will be taken at local levels where industry functions, consumers live and power is produced. It is at these scales that quantitative information on emissions baselines and mitigation options are most readily needed and it is at the urban landscape scale that knowledge about local mitigation options, costs, and opportunities are the greatest (*Rosenzweig et al.*, 2010; *Fleming and Webber*, 2004; *Salon et al.*, 2010; *Betsill and Bulkeley*, 2006; *Dhakal and Shrestha*, 2010).

The Indianapolis Flux Experiment (INFLUX) experiment emerged from research aimed at quantifying space- and time-explicit fossil fuel carbon dioxide emissions (Hestia) in the city of Indianapolis (*Gurney et al.*, 2012; *Davis et al.*, 2017). The INFLUX effort now includes the original bottom-up quantification system, aircraft-based in situ measurement of CO_2, CH_4, and CO fluxes, and dense, tower-based continuous measurement of mole fraction for CO_2, CH_4, and CO (*Cambaliza et al.*, 2014, 2015; *Heimburger et al.*, 2017; *Miles et al.*, 2016) and flask measurements of CO_2, CH_4, CO, $^{14}CO_2$ and a host of other species (*Turnbull et al.*, 2012; *Turnbull et al.*, 2015). INFLUX has also seen the application of an inverse modeling system that integrates both the bottom-up information, atmospheric observations and atmospheric transport simulation to arrive at an optimal estimate of the total CO_2 flux in an area centered on the City of Indianapolis (*Lauvaux et al.*, 2016). This last research step – the integration of the bottom-up flux estimation with the atmospheric mole fraction measurements and simulated transport – is important in that it paves the way for an information system that integrates multiple approaches to quantifying urban carbon fluxes. Furthermore, these different approaches have complementary strengths – bottom-up estimation is rich with mechanistic and space/time detail but suffers from potential biases in the data and model assumptions used. Atmospheric approaches, by contrast, reliably capture the entire flux but face difficulties in capturing flux detail and remain sensitive to assumptions about atmospheric transport and boundary conditions.

Notable among the recent analysis integrating these two approaches to urban flux estimation, was the difference between the Hestia bottom-up $FFCO_2$ flux estimation of Gurney et al. (2012) and the atmospheric CO_2 inversion result of Lauvaux et al. (2016) in the INFLUX effort. Though the lowest value of the complete atmospheric inversion ensemble range overlapped the upper 2-sigma boundary of the Hestia $FFCO_2$ flux probability distribution, the reference atmospheric inversion and its posterior uncertainty, however, did not. Importantly, the central estimate of the reference inversion was greater than the bottom-up flux estimate by approximately 20% (0.94 MtC) over the eight-month period from September 2012 to April 2013.

Here we consider a simple question regarding the differing estimates of $FFCO_2$ emissions in the INFLUX domain: Can the bottom-up estimation method account for the 0.94 MtC difference between the Hestia $FFCO_2$ flux estimate and that inferred through the atmospheric CO_2 inversion. We consider numerous potential sources

of bias in the Hestia estimation approach to identify the most likely candidates for the difference. This includes examination of an updated version (version 3.0) of the Hestia $FFCO_2$ emissions which made significant changes to the onroad and nonroad emitting sectors. We also consider the possibility of "missing" flux sources – emissions that may be reflected in the mixing ratio measurements but not explicitly included in the prior flux.

We describe our methods in section 2.0 which is mostly a description of the updates to the Hestia Indianapolis $FFCO_2$ emissions data product. In section 3.0, we present the results of our exploration of possible explanations for the difference between the Lauvaux et al. (2016) atmospheric inversion flux estimate and the bottom-up Hestia $FFCO_2$ emissions estimate. In section 4.0 we discuss the most likely candidates that may account for the differences, note the complementary results from the $^{14}CO_2$ monitoring, and discuss methods by which future work can more fully account for biases and missing fluxes, offering some near-term research objectives for further work in the INFLUX effort.

2. Methods
2.1 Hestia-Indianapolis Version 3.0
A new estimate of $FFCO_2$ emissions for Marion County, IN (the location of Indianapolis City) has been generated from the Hestia Project (Hestia-Indianapolis Version 3.0). The previous version (version 2.0) generated a $FFCO_2$ emissions estimate for Marion County and the eight counties surrounding Marion County, but using simpler techniques. The Hestia version 2.0 $FFCO_2$ flux estimate was anchored to the year 2002 and made scaled estimates in all economic sectors (e.g. residential, commercial, industrial, etc.), other than electricity production, for the years 2010–2013 (*Gurney et al.*, 2012). The scaled estimates used statewide fuel sales/consumption statistics from the Department of Energy's Energy Information Agency (DOE EIA). For the larger power plants in the electricity production sector, direct stack monitoring of $FFCO_2$ fluxes were available for all years. For a complete description of the methods employed in the Marion County Hestia version 2.0 $FFCO_2$ flux estimate, see Gurney et al., 2012.

The new version (version 3.0) includes a series of improvements over the version 2.0 estimate. The most important update is the use of the Environmental Protection Agency National Emissions Inventory (NEI) results for the year 2011. This data is relevant to the $FFCO_2$ estimates made for all sectors other than electricity production. Version 3.0 extends the time series to the year 2014 using the same DOE EIA scaling described previously but using the year 2011 as the base year. This offers the opportunity to compare the scaling of version 2.0 to the reported data in version 3.0 for the common year of 2011. Additional improvements were made to the spatial distribution of emitting sources. For example, the onroad sector used an improved road basemap and improved Annual Average Daily Traffic (AADT) data, both of which are used to distribute the county-level estimates of onroad $FFCO_2$ emissions to individual road segments. The improved road basemap had a larger number of individual road classes and a better match to the NEI onroad county-level estimates of $FFCO_2$. The onroad NEI results were driven by the MOVES model as opposed to the NMIM modeling system used in the version 2.0 estimate. MOVES is considered a superior model system for characterizing onroad emissions (*Vallamsundar and Lin*, 2011; *Fujita et al.*, 2012). In version 2.0 the nonroad emissions sector contained no spatial distribution but was evenly spread across Marion County. Version 3.0 employs a series of spatial surrogates derived from EPA data (US EPA; ftp.epa.gov/EmisInventory/surrogates/surrogates_2010). Finally, the point source distribution was improved with more accurate geolocation of point sources, a critical element in linkage to atmospheric modeling.

3. Results

3.1 Hestia version 2.0 and Lauvaux et al. flux inversion

Lauvaux et al. (2016) performed an atmospheric CO_2 inversion for a domain that was centered on the city of Indianapolis (within Marion County) but included the eight counties that surround Indianapolis: Johnson, Morgan, Madison, Hendricks, Shelby, Boone, and Hancock. The inversion generated posterior flux estimates using a five-day moving window between September 2012 and April 2013. The reference case inversion arrived at a posterior flux estimate of 5.5 MtC (one-sigma = ±0.20 MtC). Because the regional atmospheric CO_2 inversion includes assumptions regarding key components of the inversion problem not reflected in either the prior flux or atmospheric measurement uncertainties, the study included a number of sensitivity cases. This resulted in a wider range of posterior flux outcomes which were represented as a numerical span, rather than a probability distribution. The sensitivity cases included variation in the assumed prior error correlation lengths and variation in the time window of observed CO_2 mixing ratios used. The complete ensemble posterior flux for the entire domain ranged from 4.53 MtC to 6.51 MtC.

The Hestia version 2.0 $FFCO_2$ emissions were used as the prior flux in both the Lauvaux et al. (2016) reference case inversion and all the sensitivity cases but one (the case testing the influence of a different prior flux). The Hestia $FFCO_2$ prior flux for the September 2012 to April 2013 period, came to 4.56 MtC/yr, or an 18.7% (0.94 MtC) difference from the inversion reference case posterior flux. Integration of the Hestia version 2.0 $FFCO_2$ flux in its native format over the September to April period, arrives at a total flux of 4.6 MtC/year, slightly higher than the prior used in the inversion experiment (0.04 MtC: 0.9%). The difference is likely due to small inaccuracies commonly encountered in the regridding routines applied to the Hestia values.

Quantification of uncertainty for the Hestia $FFCO_2$ flux data product is particularly difficult since the flux estimation relies to a great extent on self-reported or regulatory-based data sources which are rarely accompanied by uncertainty. Hence, it remains an ongoing effort within the Hestia research to quantify uncertainty and this will be reported with future releases of the Hestia data product. However, in order to supply the atmospheric CO_2

inversion with a required prior flux error, the flux variance assumed at the pixel scale (model grid box) was 60% of the total prior flux in a given pixel based on expert judgement. To arrive at a total domain prior flux uncertainty, a error correlation length of 4km was combined with an urban mask (correlation only occurs between emitting pixels). This resulted in a one-sigma posterior uncertainty for the whole domain of 0.23 MtC. This means that the lower bound of the 2-sigma reference case inversion result (5.10 MtC) does not overlap with the upper bound of the 2-sigma Hestia $FFCO_2$ prior flux value (5.01 MtC).

The Marion County portion of the prior and posterior flux was 2.86 and 3.74 MtC/yr, respectively (a difference of 0.88 MtC/yr), representing a 26% increase between the Hestia $FFCO_2$ prior flux and the posterior flux. The majority of the flux correction (94%) made by the inversion analysis were located within Marion County as opposed to the eight surrounding counties (**Figure 1**). This is not surprising given that the atmospheric monitoring locations were more sensitive to fluxes in Marion County and the magnitude of the fluxes in Marion County were considerably larger than any of the surrounding counties. As the center of commerce in this region and hosting 54% of the population, this result is consistent with expectation. The spatial distribution of the flux correction (*Lauvaux et al.*, 2016, Figures 13 and 15) further confirms this pattern with the dominant flux correction corresponding to the road network and the greater density of residential and commercial buildings in the urban core.

3.2 Difference Hypotheses

As mentioned, the Hestia $FFCO_2$ emissions data product was used as the prior flux within the INFLUX inversion, a necessary constraint given the limitations of the atmospheric CO_2 observational constraint and the uncertainties intrinsic to the atmospheric transport simulation. The role of the prior flux is to offer a reasonable and physically consistent initial flux distribution that the combination of atmospheric observations and atmospheric transport adjust, as needed in the optimization process. One can also consider the inverse result as an important and independent constraint to the generation of a flux data product using the bottom-up technique. In this way, one might examine what potential adjustment to the bottom-up flux construction would be consistent with the inversion result and the internal constraints in the bottom-up data and algorithms. Similarly, there are elements of the measured atmospheric CO_2 that are not intentionally captured in the Hestia $FFCO_2$ emissions data product and these may be explored as an alternative explanation for the 0.94 MtC/yr discrepancy between the two approaches to the flux estimation.

3.2.1 Hestia $FFCO_2$ sectoral error sources

The Hestia $FFCO_2$ emissions for Marion County and the eight counties surrounding Marion County for the eight-month period from September 2012 to April 2013, are dominated by the onroad emission sector (2.24 MtC/yr; 48.2%) followed by the electricity production sector (0.95 MtC/yr; 20.5%), the residential and commer-

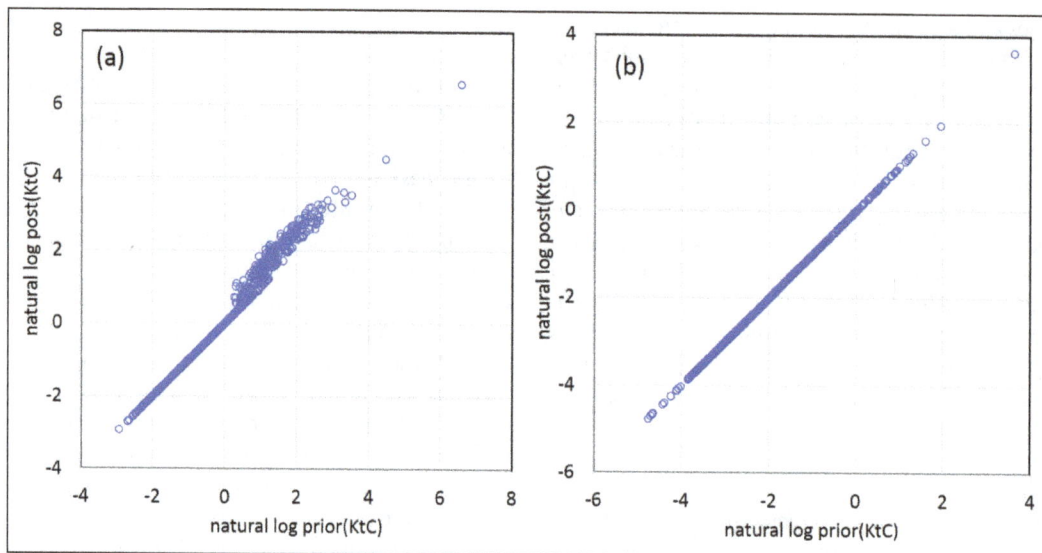

Figure 1: Prior versus posterior gridded CO$_2$ emissions. Scatterplot of the Lauvaux et al. (2016) inversion reference case prior versus posterior CO$_2$ emissions at the scale of 1 kilometer × 1 kilometer gridcells in units of natural log kiloton carbon for **a)** Marion County, Indiana; **b)** eight surrounding counties. DOI: https://doi.org/10.1525/elementa.137.f1

cial buildings (0.62 MtC/yr; 13.4%), the industrial sector (0.45 MtC/yr; 9.7%), and nonroad/rail/airport sectors (0.39 MtC/yr; 9.3%). Here, we consider the three largest of these sectoral divisions, in turn, as candidates for flux bias in the Hestia FFCO$_2$ emissions.

Onroad: As the largest single emitting sector in the domain, onroad FFCO$_2$ emissions bear closer examination as these could be considered a likely candidate for underestimation in the domain-wide Hestia FFCO$_2$ emissions estimate. The approach taken in the Hestia project is to use output of the MOVES 2010b model. It takes county total estimates of vehicle miles traveled (VMT) within each of the nine counties and combines this information with estimates of the onroad fleet of vehicles, their age distribution and fuel economy (*USEPA* 2015) to estimate FFCO$_2$ emissions for each county. County totals for 14 road and six vehicle classes are distributed to the roads based on the road segment-level VMT which is the product of measured average annual daily traffic (AADT) counts and road segment lengths.

Additional estimates of onroad FFCO$_2$ emissions in the INFLUX domain from September 2012 to April 2013 – 2.24 MtC – can be approached using methods independent the Hestia approach. We use the DOE EIA survey-based estimates of 2012 and 2013 "sales/deliveries to onroad consumers" of both diesel and gasoline in the state of Indiana (see **Table 1**) (*USDOE* 2016). We convert these sales/deliveries to carbon using standard heat and carbon content values for No 2 diesel and gasoline (10.07 tCO$_2$/e3gals, 9.12 tCO$_2$/e3gals). We average the two years and extract 8/12 of the total, to capture an equivalent eight months of flux straddling the two years, 2012 and 2013. Finally, we take the statewide proportion of this value according to the US Census 2012/2013 total population of the nine counties in the INFLUX domain, arriving at 1.99 MtC (*US Census*, 2016b).

Because commercial onroad emissions (delivery trucks, interstate commerce, etc.) may be underestimated by a

distribution based on the share of statewide population in the nine counties, we also examine the county share of statewide total retail sales in the nine counties (*US Census*, 2016a). Hence, the state total onroad diesel and gasoline are apportioned to the nine counties based on their share of retail sales. The use of retail sales results in an onroad FFCO$_2$ emissions estimate of 2.25 MtC, a value nearly identical to the Hestia onroad FFCO$_2$ emissions estimate of 2.24 MtC.

Finally, we use a recently published high-resolution onroad FFCO$_2$ emissions data product, DARTE, generated for the United States (*Gately et al.*, 2015). This estimate, though also a bottom-up technique, used a somewhat different approach to onroad emissions, opting to calculate emissions directly from annual average daily traffic (AADT) estimates and statewide proportions of different vehicle classes and their associated travel efficiency. The last year of the DARTE data product is 2012 and has no sub-annual temporal structure. Hence, we used 8/12 of the 2012 estimate within the INFLUX domain and arrive at 2.3 MtC, again, nearly identical to the Hestia onroad FFCO$_2$ estimate of 2.24 MtC. It is worth noting that the two estimates (Hestia and DARTE) have different spatial distribution but similar total domain emissions.

Electricity production: The INFLUX domain includes 12 electricity production facilities that were operational in 2012/2013. Emissions from six of these facilities were retrieved from the Environmental Protection Agency Clean Air Markets Division (CAMD) data reporting, three were retrieved from the Energy Information Administration (EIA) reporting and three reported through the National Emissions Inventory. The relative magnitude of these three sets of emissions reporting were 1.36, 0.001, and 0.003 MtC/yr in the year 2012. As described in Gurney et al. (2016), there are multiple datasets in the United States with independently derived estimates of CO$_2$ emissions from US power plants. Many of these facilities were found to have large differences in monthly estimated

Table 1: Alternative estimates of the onroad FFCO$_2$ emissions for the September 1, 2012 to April 30, 2013 period in the INFLUX domain. DOI: https://doi.org/10.1525/elementa.137.t1

Onroad FFCO$_2$ approach	Estimation technique	Key parameters	Onroad FFCO$_2$ (MtC)	Reference
Hestia	VMT, fleet stats, emission factors	MOVES output, AADT, basemap segment length	2.24	Gurney et al., 2012; USEPA 2015
Statewide fuel	Pop proportion	Onroad gasoline & diesel fuel sales/consumption, 2012/2013 population	1.79	USDOE 2016, US Census 2016
Statewide fuel	Retail sales proportion	Onroad gasoline & diesel fuel sales/consumption, 2012 retail sales	2.25	USDOE 2016, US Census 2016
DARTE	Alternative bottom up		2.31	Gately et al., 2015

FFCO$_2$ emissions when two of the largest datasets were compared (*Gurney et al.,* 2016).

The six facilities reported here as using the CAMD data also report through the EIA allowing for a comparison of reporting. These six facilities account for 99.7% of the electricity production FFCO$_2$ emissions in the INFLUX domain. However, the difference between the CAMD data and the EIA reporting were small, amounting to 0.014 MtC/yr and show the CAMD reporting as the larger of the two, the data source used in the Hestia FFCO$_2$ emissions data product. Hence, the potential for differences is small and in a direction contrary to the hypothesized underestimate.

Hestia Version 3.0: Since the release of the Hestia FFCO$_2$ emissions estimate for the INFLUX inversion effort (version 2.0), updates to a few of the key data sources have become available enabling a version 3.0 of the data product.

Figure 2 presents a pie chart representation of the version 3.0 2011 FFCO$_2$ emissions. As with version 2.0, emissions in Marion County are dominated by the onroad and electricity production sector emissions. The latter is driven by the Harding Street Station, accounting for almost 90% of the electricity production FFCO$_2$ emissions in the year 2011.

Table 2 shows comparison of the Hestia version 2.0 versus version 3.0 FFCO$_2$ annual emissions for Marion County only (the largest county emitter in the INFLUX domain) for the economic sectors and across the 2011–2014 time period. Total FFCO$_2$ emissions for 2011 are nearly unchanged, though there were canceling sectoral changes. In particular, the residential and commercial sector emissions are larger in version 3.0 but the industrial sector emissions were less and compensatorily so. Emissions in 2012 show an increase in total emissions for version 3.0 (from 4.07 to 4.15 MtC/yr) driven primarily by the increase in residential and commercial emissions that were less compensated for by the smaller industrial emissions in version 3.0. The estimate for 2013 decreased by a small amount (from 4.32 to 4.26 MtC/yr) and was mostly driven by the lesser industrial sector emissions. A representative difference for the eight-month period from September 2012 to April 2013 is 0.007 MtC.

It is worth noting that the spatial distribution of the fluxes differs between version 2.0 and 3.0, owing to significant differences in the road basemap used and the new

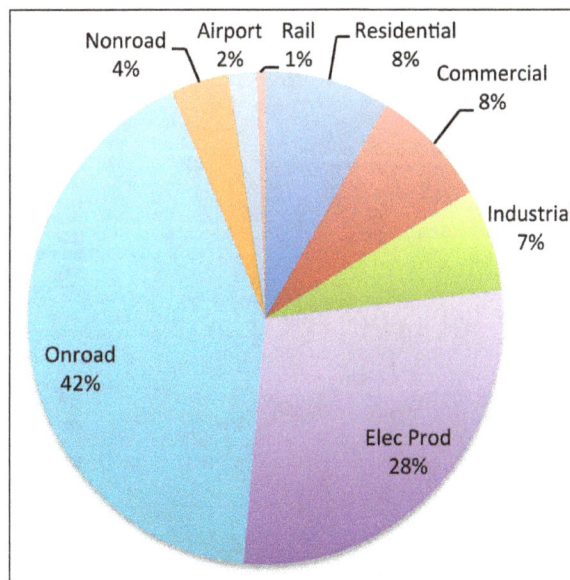

Figure 2: Hestia FFCO$_2$ emissions for Marion County, Indiana. Proportion of the total 2011 Hestia version 3.0 FFCO$_2$ emissions for Marion County, Indiana from each of the eight sectors. DOI: https://doi.org/10.1525/elementa.137.f2

spatial footprint of the nonroad FFCO$_2$ emissions. There is also some spatial difference due to a different proportion of residential, commercial and industrial building emissions. **Figure 3** shows the spatial distribution of FFCO$_2$ emissions in Marion County for the year 2011. The residential, commercial and onroad sectors in addition to the county total are shown. **Figure 4** presents the percentage difference between the version 2.0 and version 3.0 products for the same sectors as **Figure 3**.

3.2.2 Missing CO$_2$ emissions

There are some categories of emissions within the INFLUX domain that are not associated with the combustion of fossil fuel and these could contribute to the 0.94 MtC difference. Because the atmospheric CO$_2$ inversion infers the total CO$_2$ flux within the domain, there is the potential for vegetation and soil carbon exchange to be reflected in the inversion posterior flux estimate but not included in

Table 2: FFCO$_2$ emissions, categorized by sector, comparing Hestia version 2.0 to version 3.0 for Marion County, Indiana, 2011, 2012, and 2013. Units: MtC/yr. DOI: https://doi.org/10.1525/elementa.137.t2

Year/version	Resid	Comm	Ind	Elec Prod	Onroad	Nonroad	Airport	Rail	Total
2011 Version 2.0	0.30	0.29	0.44	1.14	1.70	0.16	0.049	0.010	**4.09**
2011 Version 3.0	0.34	0.33	0.27	1.14	1.70	0.16	0.078	0.026	**4.05**
2012 Version 2.0	0.26	0.27	0.44	1.18	1.70	0.16	0.048	0.011	**4.07**
2012 Version 3.0	0.30	0.38	0.32	1.18	1.70	0.15	0.077	0.029	**4.15**
2013 Version 2.0	0.32	0.32	0.45	1.30	1.70	0.16	0.047	0.011	**4.32**
2013 Version 3.0	0.31	0.32	0.29	1.30	1.72	0.15	0.075	0.099	**4.26**

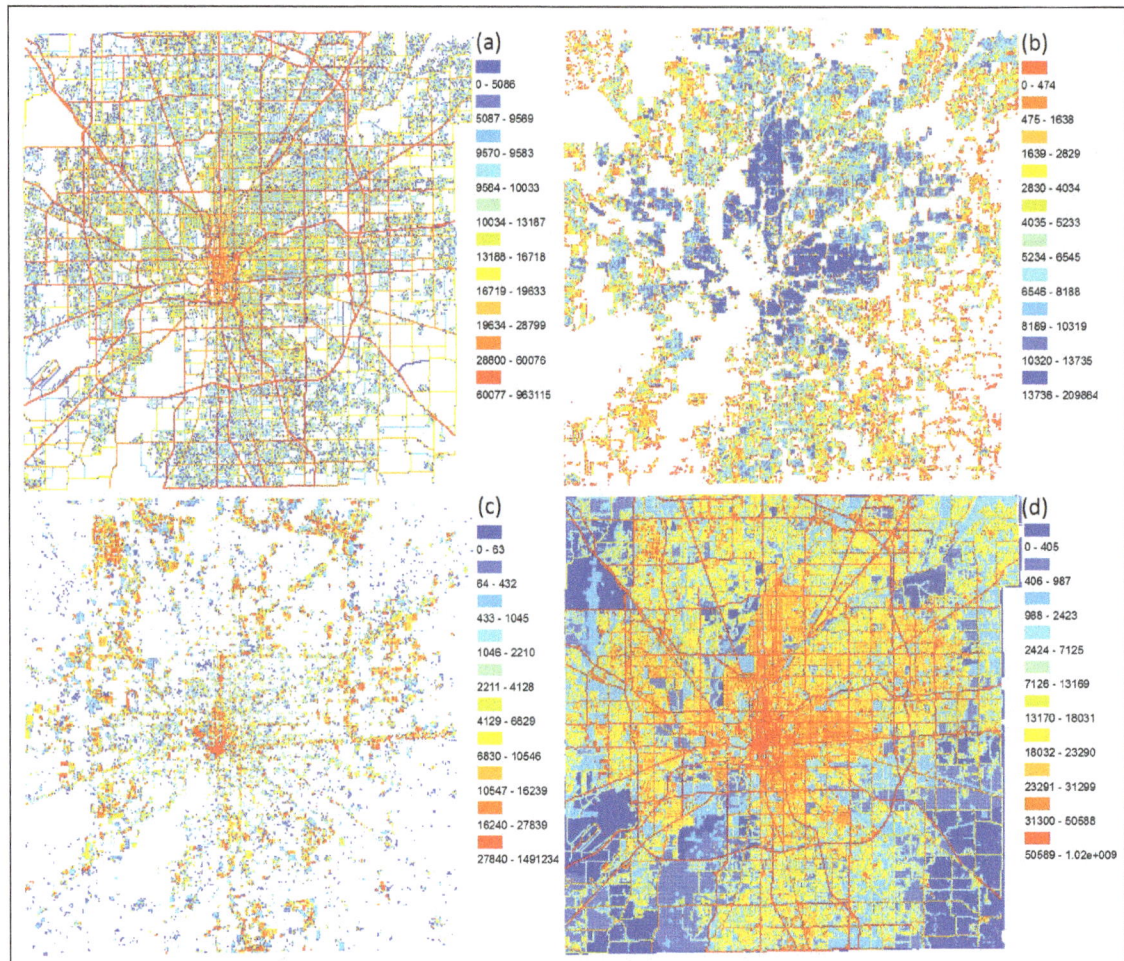

Figure 3: Map of gridded FFCO$_2$ emissions. Marion County, Indiana 2011 FFCO$_2$ emissions from the Hestia version 3.0 gridded at 100 meter × 100 meter resolution for the **a)** onroad sector; **b)** residential sector; **c)** commercial sector; **d)** total emissions. The color bar scales are different in each panel. Units: kgC/yr/grid cell. DOI: https://doi.org/10.1525/elementa.137.f3

the prior flux. Similarly, human/animal respiration occurs within the domain and is most likely driven by imported carbon embedded in food. Hence, these heterotrophic respiration fluxes would be reflected in the atmospheric inversion flux but not included in the Hestia prior FFCO$_2$ flux. We consider both categories of "missing" flux and estimate their magnitudes.

Animal respiration: Respiration from humans and other animals within the INFLUX domain could be a contributor to the difference between the inversion and Hestia

FFCO$_2$ emissions flux estimates. For the purposes of simplicity, we consider only respiration emanating from the human population and domestic pets for which there is statistical information. Wild fauna are not considered. We also assume that all food consumed by humans and pets within the domain is imported from locations outside the domain and the CO$_2$ uptake associated with the vegetation consumed directly or indirectly through consumption of animal products is not accounted for in the atmospheric inversion. Since the eight counties surrounding Marion

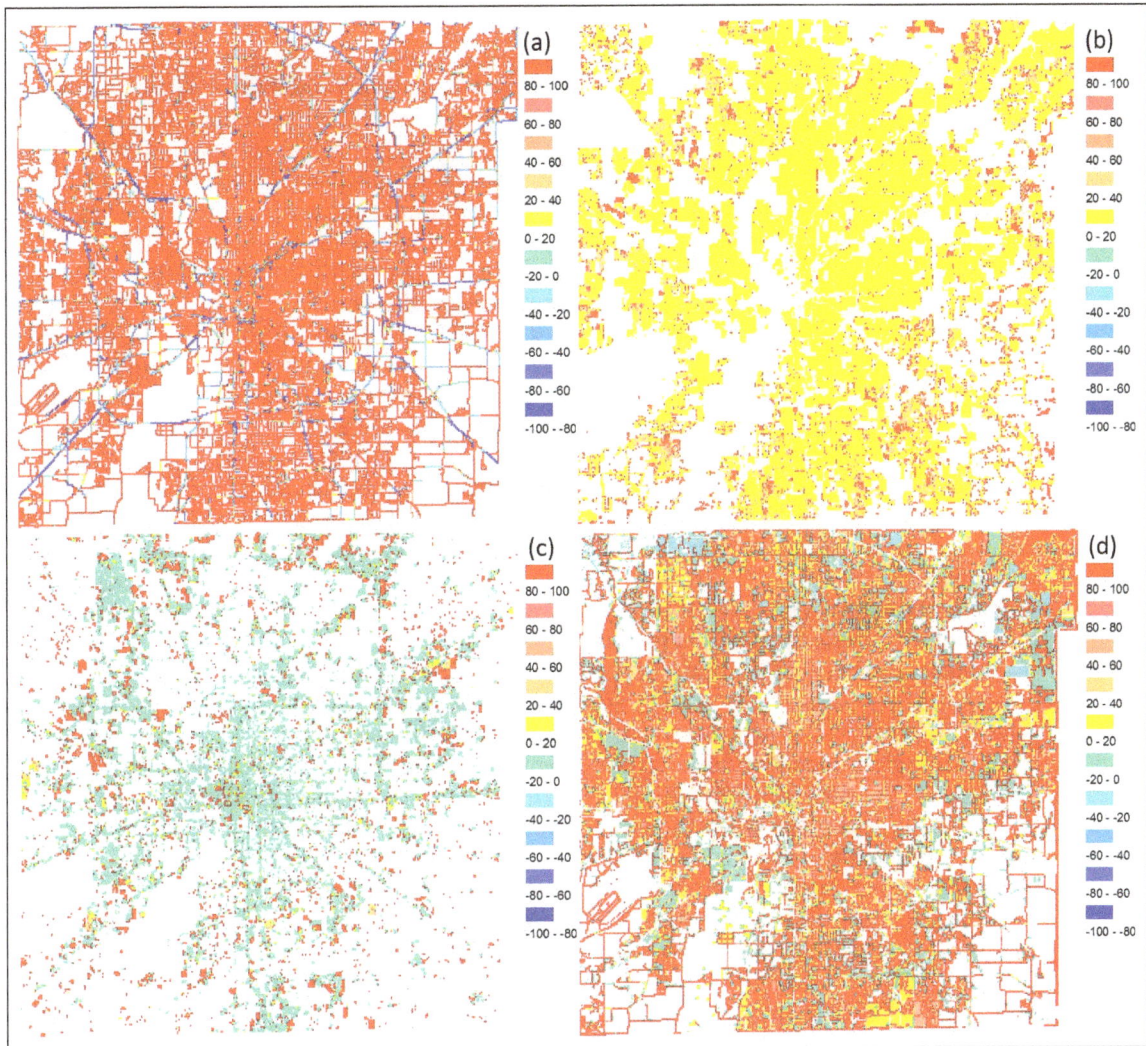

Figure 4: Map of gridded FFCO$_2$ emissions percent difference. Marion County, Indiana 2011 FFCO$_2$ emissions difference between the Hestia version 2.0 and Hestia version 3.0 at 100 meter × 100 meter resolution for the **a)** onroad sector; **b)** residential sector; **c)** commercial sector; **d)** total emissions. Units: percent. DOI: https://doi.org/10.1525/elementa.137.f4

County do engage in agricultural activity, this assumption is flawed to some degree. However, given research indicating that the average travel distance of fruits and vegetables traded in Chicago is greater than 1500 miles, this assumption is reasonable (*Pirog et al.*, 2001).

Assuming an average CO$_2$ exhalation rate of 254 gC/person/day and a 2012/2013 population in the portion of the nine counties in the INFLUX domain of 1,878,546, the average generation of CO$_2$ due to human respiration, scaled to the eight-month interval (8/12), is 0.12 MtC. (*US Census*, 2016; *Prairie and Duarte*, 2007). This is similar in magnitude to a similar estimate made in Turnbull et al. (2015). Assuming dog and cat ownership in the INFLUX domain follows the US national average, 0.22 dog/person and 0.24 cat/person, and the CO$_2$ exhalation rate is roughly 25% that of humans based on mean dog/cat body mass and allometric relationships, an additional 0.014 MtC must be added to human respiration for a total contribution of 0.13 MtC during the September 2012–April 2013 period (*AVMA*, 2012; *Prairie and Duarte*, 2007).

Biotic combustion: Because the Hestia data product only captures combustion of fossil fuel, the combustion of biotic fuel could constitute a source of emissions missing from the prior flux but captured in the measured CO$_2$ mixing ratios. The Covanta Indianapolis Energy facility located in Marion County burns municipal solid waste of biogenic origin to generate electricity. The Hestia FFCO$_2$ emissions data product does not include any emissions derived from biological material. Hence, the difference between the fluxes inferred from monitored CO$_2$, which does include biologically-derived CO$_2$, and the Hestia emissions could be due to this difference. The Covanta facility reported 0.095 MtC during the September 2012 to April 2013 period. Hence, this could be a contributing factor to the 0.94 MtC difference.

A similar category of emissions that is part of the combustion associated with human activities is the use of biotic material for home heating such as woodstoves or wood-burning fireplaces. We approximate this category of emission by using the EPA estimated U.S. total

CO_2 emissions from wood combustion in the residential sector. With a mean U.S. per capita figure, the application of this emission rate to the population within the INFLUX domain comes to approximately 0.05 MtC during the eight-month period in question. This is likely an upper limit on this emission amount since some of the combustion material could be sourced to growth within the INFLUX domain and hence, technically only a portion of the complete gross flux. As with yard/leaf waste that is often burned, there is an offset in time – growth and uptake is temporally separated from the time of combustion. However, the magnitude of this emission amount is too small to be of concern for present purposes.

Biofuel is used within the onroad sector as a component of gasoline. Because the Hestia system estimates onroad $FFCO_2$ using an activity-based approach (i.e. using vehicle miles traveled), this emission of CO_2 is reflected in our $FFCO_2$ emission in the onroad sector, but not tracked separately.

Biosphere respiration: Were the net biosphere exchange to be a positive flux (from the land to the atmosphere), this would result in a positive adjustment to the Hestia $FFCO_2$ emissions estimate. A positive biosphere carbon exchange might be expected from a biological system in which the absolute magnitude of the gross respiration flux exceeded the absolute magnitude of the gross photosynthetic flux. Though not expected during the growing season, where photosynthesis typically dominates the net exchange in mid-latitude vegetation, a positive net exchange may occur during the initial and ending months of the September – April period over which the INFLUX flux inversion operated.

There is little research aimed at quantifying respiration fluxes in urban areas, particularly during months outside the growing season. We draw from three studies that measured urban respiration fluxes in U.S. cities at times of the year coincident with the September-April period of the INFLUX inversion (*Decina et al.*, 2016; *Kaye et al.*, 2005; *Chen et al.*, 2014). The three studies performed measurements in Boston MA, Fort Collins CO, and Baltimore MD. The studies also sampled different urban land cover types but all reflected cover types with either bare soil, grass or forest cover. The fluxes vary by year sampled, month sampled, and land cover type and ranged from near-zero (In December/January) to 4 moles $CO_2/m2/s$. For the rough estimation purposes here we take a conservative estimate from this range of 0.5 – 1 mole $CO_2/m2/s$. We also apply this flux to an estimate of pervious surface area within each of the nine INFLUX counties using an estimate of remotely sensed built-up area (*Pesaresi et al.*, 2015). We assume this area is predominantly grass land cover. Finally, we only apply this respiration flux rate to those days within the September 2012 – April 2013 period for which there were three consecutive days with a 24 hour mean temperature above 32°F and use soil/grass temperature measurements from the West Lafayette IN airport meteorological station (*Indiana State Climate Office*, 2016). Out of these 209 days, 52 days (January 19 – March 11) qualified as having a soil/grass daily mean temperature consistently below 32°F. With a respiration flux of 1 mole $CO_2/m2/s$ we

arrive at a total flux in the domain over this time period of 0.58 – 1.17 MtC.

Indeed, Figure 8 of Lauvaux et al., shows the difference between the Hestia $FFCO_2$ prior flux and the inverted flux at a minimum during the early part of 2013, precisely when the respiration fluxes are at a minimum, coinciding with the lowest ground temperatures of the year.

Figure 5 summarizes both the potential biases in the Hestia $FFCO_2$ emissions estimate and the missing CO_2 fluxes reviewed here. The lower end of the final range of the INFLUX emissions accounted for here in attempting to reconcile the bottom-up and top-down comes to 5.25 MtC while the upper end of the range is 6.12 MtC. The median of the inverse-estimated range, 5.50 MtC, is within the constructed flux range. Though this can only be considered a rough and approximate estimate of the potential differences, the potential difference appears reasonably explained by the hypotheses presented with respiration from the urban biosphere the largest and most significant component in the difference estimate.

4. Discussion and conclusions

This study identifies and quantifies key uncertainties and CO_2 fluxes to account for the disparity between the central estimate of the Lauvaux et al. (2016) atmospheric CO_2 inversion study and the Hestia bottom-up $FFCO_2$ emissions estimate for the INFLUX domain. We examined errors in the Hestia $FFCO_2$ emissions estimate itself and an assessment of missing CO_2 flux sources.

Within the first category, the electricity production sector and the onroad vehicle sector are the most likely candidates given their relative magnitude in the INFLUX domain. Two different reporting streams associated with U.S. power plants show consistency in the INFLUX domain although these two datasets show large differences in some other U.S. locations. Different approaches to estimating onroad vehicle $FFCO_2$ emissions indicate the potential for error. However, of the few alternative approaches to estimating onroad emissions explored here, the Hestia onroad vehicle $FFCO_2$ emissions estimate remains one of the highest, making it less likely to contribute to the 0.94 MtC deficit between the larger inverse estimated flux and Hestia $FFCO_2$ emissions. However, given the challenges of estimating onroad $FFCO_2$ emissions from the bottom-up, the onroad sector must remain a potential source of bias in the comparison.

Three flux categories were explored that were potentially reflected in the inverse-estimated flux but not present in the Hestia $FFCO_2$ flux estimate, by design. The use of biotic fuels in electricity generation is tallied by US agencies but not included in the Hestia system. In the INFLUX domain, the amount of biotic fuel used to generate electricity is small. Similarly, biotic material burned in residential woodstoves and fireplaces is likely too small to be of much consequence. Respiration by both animals and soils/vegetation was considered and of the two, the potential for soil/vegetation respiration is larger by roughly an order of magnitude. Hence, though each of the potential corrections noted here may play a role, the magnitude of soil/vegetation respiration is

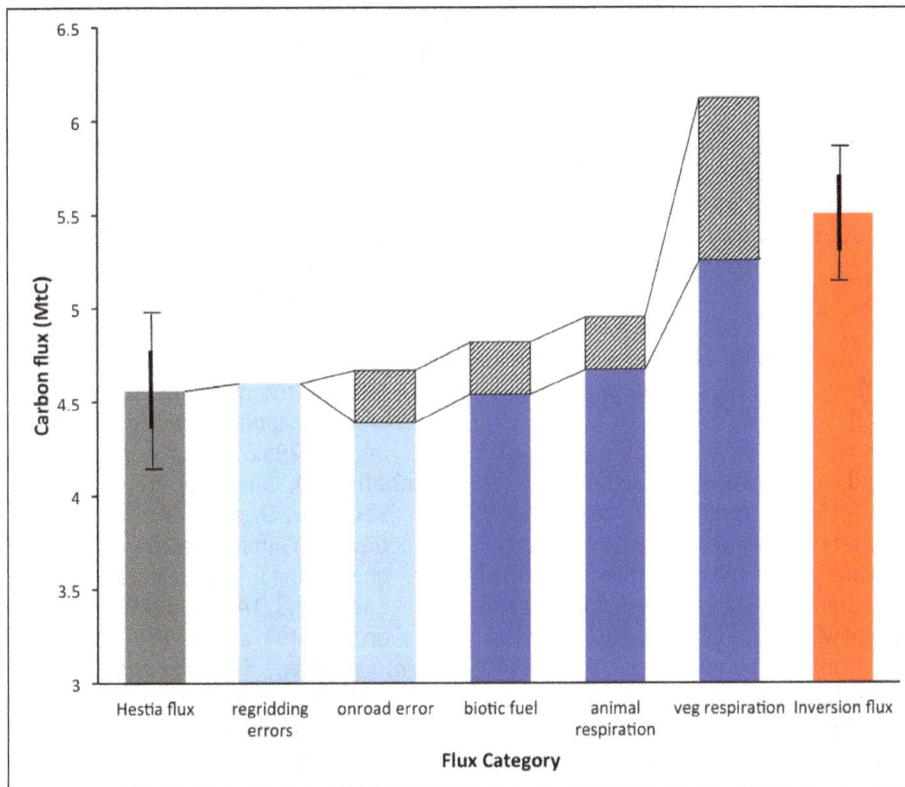

Figure 5: Reconciliation of the INFLUX CO$_2$ emissions. The Hestia FFCO$_2$ flux estimate and uncertainties (thick line: one-sigma; thin line: 2-sigma), the individual CO$_2$ flux reconciliation adjustments, and the reference inversion CO$_2$ flux estimate and uncertainties (thick line: one-sigma; thin line: 2-sigma) for the September 1, 2012 to April 30, 2013 period in the INFLUX domain. The adjustments to the Hestia FFCO$_2$ flux estimate are cumulative from left to right, and the hatched region denotes the range of values associated with the cumulative flux adjustment. Light blue columns represent FFCO$_2$ errors; dark blue columns represent missing CO$_2$ fluxes. Units: MtC. DOI: https://doi.org/10.1525/elementa.137.f5

potentially large enough to singularly explain the disparity. The research community is aware of the potential for soil/vegetation respiration outside the growing season to contribute to observed CO$_2$ mixing ratios, but empirical evidence in the urban domain has been limited. Only recently have there been measurements within mid-latitude urban areas to support the notion that the flux may be large enough to warrant explicit inclusion (*Decina et al.*, 2016; *Kaye et al.*, 2005; *Chen et al.*, 2014). Indeed, the soil/vegetation covered landscape within the urban domain tends to be heavily managed with both water and soil nutrients available throughout the year (*Kaye et al.*, 2006). The potential for soil/vegetation respiration to influence measured CO$_2$ mixing ratios emphasizes the need to consider the fact that the urban biosphere will continue to be active outside times of the year when photosynthesis and/or net uptake is dominant. We conclude that this missing flux can account for nearly all of the discrepancy between the two approaches to within statistical uncertainty, reconciling the bottom-up Hestia FFCO$_2$ estimate with the inversion-based estimate in the INFLUX domain.

Though missing respiration fluxes is the most obvious explanation for the discrepancy, the simplicity of the exploration here does not eliminate a biased bottom-up

emissions data product as a potential factor in the disparity with the inverse-estimated posterior flux. The alternative approaches to estimating the larger contributors to the fossil fuel budget remain limited and hence, cannot be considered conclusive proof that the Hestia data product is not biased. Ideally, improved empirical data and associated uncertainties are required to improve bottom-up estimation of fluxes. Some superior data do exist which would allow for better estimates of FFCO$_2$ emissions from the bottom-up. For example, household utility billing data, though not without measurement uncertainty, offers a direct measured estimate of on-site building emissions. Legal barriers and privacy concerns prevent this information from being shared outside of energy supply utilities and their ratepayers. Arrangements whereby utility billing data is anonymized or aggregated to scales that eliminate privacy concerns should be pursued and standardized (*Pincetl et al.*, 2015). Even representative samples in specific urban domains would improve the estimation algorithms in the building sector.

Onroad vehicle FFCO$_2$ emissions, often the largest single emitting sector in US cities, presents unique data challenges because of its mobile nature. Nevertheless, more comprehensive traffic monitoring and vehicle fleet information from inspection/maintenance recordkeeping could improve the onroad $FFCO_2$ emissions estimate. This information is available in some cities, but standardization and comprehensive coverage is sorely needed.

There are also techniques that assist in making the top-down and bottom-up approaches more consistent in terms of the categorical fluxes estimated. The use of $^{14}CO_2$ measurements as a near-ideal tracer for the fossil fuel component of CO_2 fluxes in urban areas is a powerful way to assist in parsing the budget between the biological and fossil carbon pools (*Miller et al.*, 2012, *Turnbull et al.*, 2006). For linear tracers such as CO_2, inversion systems can incorporate components of CO_2 fluxes as separate tracers, supplying each with unique prior and posterior fluxes (*Enting*, 2002).

Observations of $^{14}CO_2$ have been collected in the INFLUX domain and conclusions elaborated in the work of Turnbull et al. (2015). The results here are consistent with the observed ratios of wintertime urban-enhanced total CO_2 to the fossil fuel-derived component associated with the $^{14}CO_2$ measurements. They found an approximate 20% enhancement of wintertime CO_2 above the fossil fuel-derived CO_2 when using all the measurement towers in the INFLUX domain (*Turnbull et al.*, 2015: **Table 2**, row 5). Though not conclusive proof given the inherent uncertainties and the difficulty of fully eliminating the background inflow of air, the 20% enhancements are nearly identical to the discrepancy between the Hestia $FFCO_2$ flux and the reference inversion posterior flux result and consistent with the reconciliation found here.

Given the importance and estimated magnitude of the biosphere respiration flux as estimated here, an important future task in closing the budget over the INFLUX domain is assessment of the soil/vegetation flux through a combination of direct measurement, land cover and ecosystem modeling.

Acknowledgements
We thank the Environmental Protection Agency for the raw data associated with the 2011 National Emissions Inventory.

Funding information
This work is supported by the National Institute of Standards and Technology grant 70NANB14H321.

Competing interests
The authors have no competing interests to declare.

Author contributions
· Contributed to conception and design: KRG
· Contributed to acquisition of data: JL, DO, RP, MH, PS, JT
· Contributed to modeling results: KRG, JL, JH, RP, TL
· Contributed to analysis and interpretation: KRG
· Drafted and/or revised article: KRG, JT, TL
· Approved the submitted version for publication: KRG

References

American Veterinary Medical Foundation 2012 *U.S. Pet Ownership & Demographics Sourcebook, 2012, American Veterinary Medical Foundation.* U.S. Pet Ownership Statistics. Available at Website: https://www.avma.org/KB/Resources/Statistics/Pages/Market-research-statistics-US-pet-ownership.aspx.

Betsill, MM and **Bulkeley, H** 2006 Cities and the multilevel governance of global climate change. *Global Governance* **12**(2): 141–159.

Cambaliza, OM, Bogner, J, Caulton, DR, Stirm, B, et al. 2015 Quantification and source apportionment of the methane emission flux from the city of Indianapolis. *Elem Sci Anth.* DOI: https://doi.org/10.12952/journal.elementa.000037

Cambaliza, O, Shepson, PB, Caulton, D, Stirm, B, Samarov, D, et al. 2014 Assessment of uncertainties of an aircraft-based mass balance approach for quantifying urban greenhouse gas emissions. *Atmos Chem Phys* **14**(17): 9029–9050. DOI: https://doi.org/10.5194/acp-14-9029-2014

Chen, Y, Day, SD, Shrestha, RK, Strahm, BD and **Wiseman, PE** 2014 Influence of urban land development and soil rehabilitation on soil-atmosphere greenhouse gas fluxes. *Geoderma* **226–227**: 348–353. DOI: https://doi.org/10.1016/j.geoderma.2014.03.017

Davis, K, Lauvaux, T, Gurney, KR, Hardesty, T, Shepson, PB, et al. 2017 The Indianapolis Flux Experiment (INFLUX): A test-bed for anthropogenic greenhouse gas emission measurement and monitoring. *Elem Sci Anth* **5**: 21. DOI: https://doi.org/10.1525/elementa.188

Decina, SM, Hutyra, LR, Gately, CK, Getson, JM, Reinmann, AB, et al. 2016 Soil respiration contributes substantially to urban carbon fluxes in the greater Boston area. *Env Poll* **212**: 433–439. DOI: https://doi.org/10.1016/j.envpol.2016.01.012

Dhakal, S and **Shrestha, RM** 2010 Bridging the research gaps for carbon emissions and their management in cities. *Energy Policy* **38**(9): 4753–4755. DOI: https://doi.org/10.1016/j.enpol.2009.12.001

Fleming, PD and **Webber, PH** 2004 Local and regional greenhouse gas management. *Energy Policy* **32**(6): 761–771. DOI: https://doi.org/10.1016/S0301-4215(02)00339-7

Fujita, EM, Campbell, DE, Zielinska, B, Chow, JC, Lindhjem, et al. 2012 Comparison of the MOVES2010a, MOBILE6.2, and EMFAC2007 mobile source emission models with on-road traffic tunnel and remote sensing measurements. *J Air Waste Manage Assoc* **62**: 1134–1149. DOI: https://doi.org/10.1080/10962247.2012.699016

Gately, CK, Hutyra, LR and **Wing, IS** (2015) Cities, traffic, and CO_2: A multidecadel assessment of trends, drivers, and scaling relationships, PNAS. DOI: https://doi.org/10.1073/pnas.1421723112

Gurney, KR, Ansley, W, Mendoza, D, Seib, B and **Petron, G** 2007 Research needs for process-driven, finely resolved fossil fuel carbon dioxide emissions.

EOS Trans Amer Geophys Union **88**(49): 542–543. DOI: https://doi.org/10.1029/2007EO490008

Gurney, KR, Huang, J and Coltin, K 2016 Bias present in US federal agency power plant CO_2 emissions data and implications for the US clean power plan. *Env Res Lett* **11**: 064005. DOI: https://doi.org/10.1088/1748-9326/11/6/064005

Gurney, KR, Song, Y, Zhou, Y, Benes, B and Abdul-Massih, M 2012 Quantification of fossil fuel CO_2 on the building/street scale for a large US city. *Environ Sci & Tech* **46**: 12194–12202. DOI: https://doi.org/10.1021/es3011282

Hansen, JE, Sato, M, Lacis, A, Ruedy, R, Tegen, I and Matthews, E 1998 Climate forcings in the industrial era. *Proc. Nat. Acad. Sci. USA* **95**(22): 12753–12758. DOI: https://doi.org/10.1073/pnas.95.22.12753

Heimann, M and Reichstein, M 2008 Terrestrial ecosystem carbon dynamics and climate feedbacks. *Nature* **451**: 289–292. DOI: https://doi.org/10.1038/nature06591

Heimburger, AMF, Shepson, PB, Stirm, BH, Susdorf, C, Turnbull, J, et al. 2017 Precision Assessment for the Aircraft Mass Balance Method for Measurement of Urban Greenhouse Gas Emission Rates. *Elem Sci Anth* **5**: 26. DOI: https://doi.org/10.1525/elementa.134

Indiana State Climate Office 2016 Hourly – Purdue Automated. *PAAWS, 1999–present (dataset)*. Available at: http://iclimate.org/data_archive_v3.asp?rdatatype=ph Accessed October 16, 2016.

International Energy Agency 2008 World energy outlook. Paris: Head of communication and information, Office International Energy Agency (EIA).

Kaye, J, Groffman, PM, Grimm, NB, Baker, LA and Pouyat, RV 2006 A Distinct urban biogeochemistry? *Trends in Ecology and Evolution* **21**(4): 192–199. DOI: https://doi.org/10.1016/j.tree.2005.12.006

Kennedy, C, Steinberger, J, Gasson, B, Hansen, Y, Hillman, T, et al. 2010 Methodology for inventorying greenhouse gas emissions from global cities. *Energy Policy* **38**(9): 4828–4837. DOI: https://doi.org/10.1016/j.enpol.2009.08.050

Lauvaux, T, Miles, NL, Deng, A, Richardson, SJ, Cambaliza, MO, et al. 2016 High-resolution atmospheric inversion of urban CO_2 emissions during the dormant season of the Indianapolis flux experiment (INFLUX). *J Geophys Res: Atmos* **121**. DOI: https://doi.org/10.1002/2015JD024473

Le Quéré, C, Andres, RJ, Boden, T, Conway, T, Houghton, RA, et al. 2013 The global carbon budget 1959–2011. *Earth Syst Sci Data* **5**(1): 165–185. DOI: https://doi.org/10.5194/essd-5-165-2013

Miles, NL, Richardson, SJ, Lauvaux, T, Davis, KJ, Deng, A, et al. 2016 Quantification of urban atmospheric boundary layer greenhouse gas dry mole fraction enhancements: Results from the Indianapolis Flux Experiment (INFLUX). *Elem Sci Anth*. In press for INFLUX Special Feature.

Miller, JB, Lehman, SJ, Montzka, SA, Sweeney, C, Miller, BR, et al. 2012. Linking emissions of fossil fuel CO_2 and other anthropogenic trace gases using atmospheric $^{14}CO_2$. *J Geophys Res* **117**(D08): 302. DOI: https://doi.org/10.1029/2011JD017048

Pesaresi, M, Ehrlich, D, Florczyk, AJ, Freire, S, Julea, A, Kemper, T, Soille, P and Syrris, V 2015 GHS built-up grid, derived from Landsat, multitemporal (1975, 1990, 2000, 2014). Available at: http://data.europa.eu/89h/jrc-ghsl-ghs_built_ldsmt_globe_r2015b Accessed October 15, 2016.

Pirog, R, Van Pelt, T, Enshayan, K and Cook, E 2001 Food, fuel, and freeways: An Iowa perspective on how far food travels, fuel usage, and greenhouse gas emissions. *Leopold Center for Sustainable Agriculture: Iowa State University*. Available at: http://www.leopold.iastate.edu.

Rosenzweig, C, Solecki, W, Hammer, SA and Mehrotra, S 2010 Cities lead the way in climate change action. *Nature* **467**: 909–911. DOI: https://doi.org/10.1038/467909a

Salon, D, Sperling, D, Meir, A, Murphy, S, Gorham, R and Barrett, J 2010 City carbon budgets: A proposal to align incentives for climate-friendly communities. *Energy Policy* **38**(4): 2032–2041. DOI: https://doi.org/10.1016/j.enpol.2009.12.005

Seto, KC, Güneralp, B and Hutyra, LR 2012 Global forecasts of urban expansion to 2030 and direct impacts on biodiversity and carbon pools. *Proc. Natl. Acad. Sci. U. S. A.* **109**: 16083–16088. DOI: https://doi.org/10.1073/pnas.1211658109

Turnbull, JC, Miller, JB, Lehman, SJ, Tans, PP, Sparks, RJ and Southon, J 2006 Comparison of $^{14}CO_2$, CO, and SF_6 as tracers for recently added fossil fuel CO_2 in the atmosphere and implications for biological CO_2 exchange: $^{14}CO_2$, CO, and SF_6 as fossil fuel tracers. *Geophys Res Lett* **33**(1): L01817. DOI: https://doi.org/10.1029/2005GL024213

Turnbull, JC, Sweeney, C, Karion, A, Newberger, T, Lehman, SJ, et al. 2015 Toward quantification and source sector identification of fossil fuel CO_2 emissions from an urban area: Results from the influx experiment. *J Geophys Res Atmos* **120**: 292–312. DOI: https://doi.org/10.1002/2014JD022555

Turnbull, J, Guenther, D, Karion, A, Sweeney, C, Anderson, E, et al. 2012 An integrated flask sample collection system for greenhouse gas measurements. *Atmos Meas Tech Disc* **5**: 4077–4097. DOI: https://doi.org/10.5194/amtd-5-4077-2012

United States Census Bureau 2016a American Community Survey (ACS), Data Tables & Tools. Available at: https://www.census.gov/acs/www/data/data-tables-and-tools/index.php.

United States Census Bureau 2016b Annual estimates of the Resident Population, April 1, 2010 to July 1, 2015 (dataset), Population Division, Washington DC. Available at: http://factfinder.census.gov/faces/tableservices/jsf/pages/productview.xhtml?src=bkmk.

United States Department of Energy and Energy Information Administration 2016 (dataset) Available at: www.eia.gov/dnav/pet/pet_cons_prim_dcu_SIN_a.htm.

United States Environmental Protection Agency 2015 2011 National Emissions Inventory, version 2 Technical Support Document, Office of Air Quality Planning and Standards, Air Quality Assessment Division, Emissions Iventory Analysis Group, Research Triangle Park, North Carolina.

Vallamsundar, S and **Lin, J** 2011 MOVES Versus MOBILE: Comparison of Greenhouse Gas and Criterion Pollutant Emissions, *Transportation Research Record: Journal of the Transportation Research Board*, No. 2233, Transportation Research Board of the National Academies, Washington, D.C. 27–35.

6

On the impact of granularity of space-based urban CO_2 emissions in urban atmospheric inversions: A case study for Indianapolis, IN

Tomohiro Oda[*,†], Thomas Lauvaux[‡], Dengsheng Lu[§], Preeti Rao[||], Natasha L. Miles[‡], Scott J. Richardson[‡] and Kevin R. Gurney[¶]

Quantifying greenhouse gas (GHG) emissions from cities is a key challenge towards effective emissions management. An inversion analysis from the INdianapolis FLUX experiment (INFLUX) project, as the first of its kind, has achieved a top-down emission estimate for a single city using CO_2 data collected by the dense tower network deployed across the city. However, city-level emission data, used as *a priori* emissions, are also a key component in the atmospheric inversion framework. Currently, fine-grained emission inventories (EIs) able to resolve GHG city emissions at high spatial resolution, are only available for few major cities across the globe. Following the INFLUX inversion case with a global 1 × 1 km ODIAC fossil fuel CO_2 emission dataset, we further improved the ODIAC emission field and examined its utility as a prior for the city scale inversion. We disaggregated the 1 × 1 km ODIAC non-point source emissions using geospatial datasets such as the global road network data and satellite-data driven surface imperviousness data to a 30 × 30 m resolution. We assessed the impact of the improved emission field on the inversion result, relative to priors in previous studies (Hestia and ODIAC). The posterior total emission estimate (5.1 MtC/yr) remains statistically similar to the previous estimate with ODIAC (5.3 MtC/yr). However, the distribution of the flux corrections was very close to those of Hestia inversion and the model-observation mismatches were significantly reduced both in forward and inverse runs, even without hourly temporal changes in emissions. EIs reported by cities often do not have estimates of spatial extents. Thus, emission disaggregation is a required step when verifying those reported emissions using atmospheric models. Our approach offers gridded emission estimates for global cities that could serves as a prior for inversion, even without locally reported EIs in a systematic way to support city-level Measuring, Reporting and Verification (MRV) practice implementation.

Keywords: Carbon dioxide; emission inventory; urban emissions; atmospheric inversion; INFLUX; geospatial data

1. Introduction

Cities account for more than 70% of global total greenhouse gas (GHG) emissions. Quantifying GHG emissions from cities, which are often the smallest administrative unit, is thus a key challenge towards effective emissions management. Emission inventories (EI) are

* Global Modeling and Assimilation Office, NASA Goddard Space Flight Center, Greenbelt, Maryland, US

† Goddard Earth Sciences Technologies and Research, Universities Space Research Association, Columbia, Maryland, US

‡ Department of Meteorology and Atmospheric Science, The Pennsylvania State University, University Park, Pennsylvania, US

§ Michigan State University, East Lansing, Michigan, US

|| NASA Jet Propulsion Laboratory, Pasadena, California, US

¶ School of Life Sciences, Arizona State University, Tempe, Arizona, US

Corresponding author: Tomohiro Oda (tomohiro.oda@nasa.gov)

a fundamental tool to keep track of emission changes (e.g., national emission inventory (NEI)). However, most cities do not even compile EIs although they have been recognized as practical emission reduction target, even when motivated by international consortiums (e.g., C40 cities climate leadership group). Moreover, EIs are prone to systematic biases from both the emission calculation methodology and the inadequate quality of the underlying activity data (e.g., Guan *et al.*, 2012; Liu *et al.*, 2015). In the absence of a transparent protocol to provide reliable activity data and a robust calibration method, EIs remain uncertain, therefore limited in their ability to measure GHG emission reduction efforts in metropolitan areas (Hutyra *et al.*, 2014). At the country scale (e.g., Kyoto Protocol), EIs aim to determine the level of contribution of various sectors to national carbon budgets thereby supporting the implementation of carbon mitigation for which accurate quantification of emissions is of major importance. The authors believe it is important for the science community to contribute to establishing a framework prefacing the implementation of a complete Monitoring/Reporting/Verification (MRV) practice for cities, guiding stakeholders and emission management policies.

Cities' roles for emission management and emission reduction potential have been identified. However, only few megacities are compiling their EI with the required granularity. Especially, quantification of emissions from cities is preferably done by developing fine-grained bottom-up EI where emission accounting and geolocating are available at the same spatial scale, as done by Gurney et al. (2012) as opposed to most gridded datasets based on disaggregation of national/sectoral emissions (e.g., Andres et al., 1996; Olivier et al., 2005; Janssens-Maenhout et al., 2012; Rayner et al., 2010; Oda and Maksyutov, 2011; Kurokawa et al., 2013; Asefi-Najafabady et al., 2014). The links between human activities and emissions described in a bottom-up framework provide more information on energy use than top-down estimates, which are limited by the ambiguity of mixed source signals in atmospheric observations. However, the development of fine scale EI is often labor intensive and difficult to be completed in a timely manner (i.e., annual basis). In fact, such fine-grained city emission datasets are only available for few locations. Over the continental US, only few EI's have been compiled at the building-level resolution: Indianapolis (Gurney et al., 2012), Los Angeles (Feng et al., 2016), Baltimore (Gurney et al., in preparation) and Salt Lake City (Patarasuk et al., 2016). Furthermore, error quantification and characterization associated with EI's is another emerging issue (e.g., Andres et al., 2016). Especially for fine-grained EI's, uncertainty assessment is non-trivial and involves complex parametric and structural uncertainties (Gurney et al., same issue). Those information should be included in city scale inversion to obtain robust city emission estimates (Lauvaux et al., 2016).

Beyond their original use for city emission accounting, EI is also a key component in top-down methods as they provide a first guess in the optimization problem to help identify the source distribution (Enting, 1995). The use of atmospheric data to verify EI's has been encouraged by several studies (e.g., Nisbet and Weiss, 2010; Pacala et al., 2010) and supported by the analysis of various types of instrumentation (e.g., Kort et al., 2012, Janardanan et al., 2016 for satellite CO_2 data; Basu et al., 2016 for C^{14} radiocarbon data). Recently, an inversion analysis from the Indianapolis Flux experiment (INFLUX) project, as the first of its kind, has achieved a top-down emission estimate for a single city and demonstrated the use of atmospheric CO_2 tower data to constrain urban emissions (Lauvaux et al., 2016). The inversion system used the "Hestia" fine-grained emission dataset (Gurney et al., 2012, data available from http://hestia.project.asu.edu/) as a priori emission and derived emission corrections using atmospheric CO_2 data from the dense tower network within the city domain. The inverse methodology produced 1-km resolution adjustments to the first guess (Hestia) modifying the total emissions by about 20%, a statistically significant change reflecting

possible discrepancies between the two methods including the presence of additional sources beyond anthropogenic emitters (e.g., soil respiration – Gurney et al., 2016). The study also illustrated the impact of assimilating coarser resolution prior emissions taken from the Open-source Data Inventory for Anthropogenic CO_2 (ODIAC) global 1 × 1 km fossil fuel emissions dataset (Oda and Maksyutov, 2011; Oda et al., 2016, data available from http://db.cger.nies.go.jp/dataset/ODIAC/) and its impact on the spatial structures of the emission corrections.

Potentially being applicable to any cities, top down approaches are currently being tested across few metropolitan areas (e.g., Feng et al., 2016), mostly due to the lack of atmospheric GHG networks to constrain city emissions. The deployment of ground-based instruments require an existing infrastructure (i.e. accessible tall towers or high buildings) and expert knowledge to calibrate the instruments (Richardson et al., 2017). Other observing strategies such as future satellite missions (e.g., Orbiting Carbon Observatory-3 – Eldering, 2015; CarbonSAT – Buchwitz et al., 2013; GeoCARB – Polonsky et al., 2014) are currently under development and could provide the required constraint on urban emissions in the near future. In this study, we present the space-based emission field at fine resolution to inform a top-down urban-scale framework. We evaluate the product against an existing fine-grained EI, Hestia, and assess the impact of the fine-scale structures on the posterior emissions estimate. The original ODIAC emissions is a global data set based on disaggregation of national emissions using point source profiles (power plant emission estimates and geolocation) and satellite-observed nighttime lights (e.g., Oda and Maksyutov, 2011). The total emission for the Indianapolis domain taken from ODIAC for a priori was remarkably close to Hestia as shown by Lauvaux et al. (2016), meaning the national emission disaggregation in ODIAC was sufficient for an annual estimate of the whole-city emissions. We present here an improved product at a higher level of granularity with the ambition of achieving the required accuracy in emissions estimates, i.e. sufficient to inform city-scale mitigation policies (i.e. less than 10% annually). However, the emission disaggregation technique using proxy geospatial data, while applicable to the large scale, is limited by the spatial heterogeneity of sources at finer scales. Therefore, proxy data-based emission disaggregation approaches would not work at higher resolutions, especially at the city level when light intensity and population are decorrelated from large emitters. We thus focus on creating better emission spatial structures by determining locations of specific aggregated emission sectors and attempt to make the method applicable to other metropolitan areas.

2. Methods
2.1 Urban emission field
We created a fine-grained emission field from the ODIAC emissions used in Lauvaux et al. (2016). Following the emission disaggregation commonly done in global and region gridded EI studies (e.g. Streets et al., 2003; Janssens-Maenhout et al., 2012; Kuenen et al., 2014), city

emission fields can be approximated by three principal emission type components: point, line and diffused (area) emission sources. **Table 1** shows the sector emission breakdown for Hestia. Values are updated from Gurney *et al.* (2012). It is often fairly straightforward to categorize emissions into few major sectors. For Indianapolis, and likely for many other cities over North America, emissions from transportation can account for a major fraction of the city total (about half – or 49% – for Indianapolis). In the original ODIAC emissions, power plant emissions, which are often the major emitting sector at the national scale, are already distributed using geolocation of power plants taken from CARMA (www.carma.org) (Oda and Maksyutov, 2011; Oda *et al.*, 2017). The transportation sector emissions are distributed as a diffused source. Thus, we preserved the power plant emission information from the ODIAC dataset and disaggregated the non-point source emissions (total minus point source emissions) using geospatial datasets. We used both the global road network data and satellite-data driven surface imperviousness data at 30 × 30 m resolution to generate a final product at a spatial scale similar to Hestia. We distributed the residual (non-point emissions) using the Global Roads Open Access Data Set (gROADS) v1 developed by the SocioEconomic Data and Applications Center (SEDAC) (CIESIN/ITOS/ University of Georgia, 2013, http://sedac.ciesin.colum-bia.edu/data/set/groads-global-roads-open-access-v1) for transportation sector emissions (i.e. line source emissions) and used the satellite-data driven 30 m surface imperviousness data (National Land Cover Dataset (NLCD) 2011 http://www.mrlc.gov/nlcd2011.php) for diffused source emissions (i.e. area source emissions). ODIAC does not distinguish emissions from the different sectors as emission estimates are based on country scale fuel consumption statistics (Oda *et al.*, 2017). In this study, we calculated the fraction of transportation emissions using Hestia

(see **Table 1**). The sectoral emission approach is applicable to any city assuming that sectoral total estimates are available. If not, an average of sectoral contributions from other cities across the country should provide a fairly similar distribution. The impervious surface used here indicate four levels of development (high, medium, low and open space, see **Figure 1**), but the four categories are aggregated to one as the surface imperviousness does not directly inform CO_2 emission sectors (e.g. industrial, residential and commercial), but potential locations for area sources. We thus used population data taken from Census (www.census.gov for the year 2011) to create spatial gradient on sector emission areas indicated by gROAD data and impervious data. The use of population is a classic proxy for human emissions (e.g., Andres *et al.*, 1996) even applied for transportation emission (e.g., Olivier *et al.*, 2002) as population and traffic density are highly correlated. The use of population data is therefore a reasonable approach as a first order approximation. We found a difference of 0.3% in total emissions when projecting our 1 × 1 km ODIAC into the impervious surface data fields (30 m resolution). We corrected the iODIAC emissions of the difference by adjusting the entire field.

2.2 INFLUX urban inversion system

The flux inversion analyses in this study were done using the urban high-resolution atmospheric CO_2 inversion system developed by Lauvaux *et al.* (2016). The urban inversion system is built around the Weather Research Forecasting model coupled with Chemistry (WRF-Chem) modified for passive tracers described as Lauvaux *et al.* (2012). The version of WRF model used in Lauvaux *et al.* (2016) has Four Dimensional Data Assimilation (FDDA) capability and the World Meteorological Organization (WMO) observations were assimilated in order to simulate atmospheric CO_2 concentration with the best accurate meteorological

Table 1: A summary of annual total sectoral emissions indicated by Hestia. Values are updated from Gurney *et al.* (2012). DOI: https://doi.org/10.1525/elementa.146.t1

Hestial emission sector	Type	Emissions (tC/yr)	Share (%)
OnRoad	Line	3,360,000	49.2%
Electricity Production	Point	1,362,000	19.9%
Industrial NonPoint	Area	492,000	7.2%
NonRoad	Area	477,000	7.0%
Residential NonPoint	Area	458,000	6.7%
Commercial NonPoint	point	369,000	5.4%
Industrial Point	Point	188,000	2.8%
Airport	Point	82,000	1.2%
Commercial Point	point	25,000	0.4%
Railroad	Line	21,000	0.3%
Total	–	6,835,000	100.0%

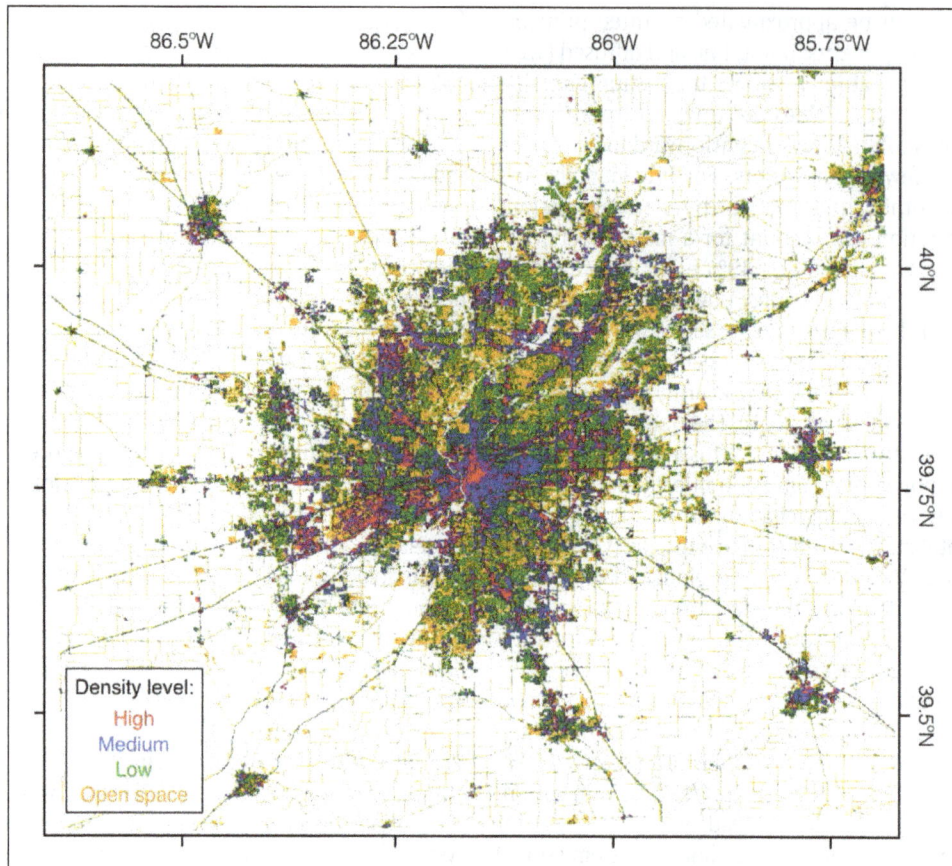

Figure 1: Impervious data over Indianapolis, IN. Data were taken from the National Land Cover Dataset (NLCD, http://www.mrlc.gov/nlcd2011.php). The impervious data indicate four different levels of developed surface: high intensity (red), medium intensity (blue), low intensity (green) and open-space. According to the NCDC categorization (http://www.mrlc.gov/nlcd06_leg.php), high density indicates 80–100% imperviousness, medium indicates 50–79%, low indicates 20–49% and open space indicates less than 20%. DOI: https://doi.org/10.1525/elementa.146.f1

conditions (Deng *et al.*, 2017). Lauvaux *et al.* (2016) used three WRF model grid configurations in nested mode (9km, 3km and 1km, see **Figure 1** of Lauvaux *et al.*, 2016). This study focuses on the Indianapolis metropolitan area that is defined by 87 × 87 grid points at 1 km resolution. The urban inversion system employs the Lagrangian Particle Dispersion Model (LPDM) described by Uliasz (1994) as an adjoint model for the WRF-Chem model. Lagrangian particles are released from CO_2 observation locations and transported backward in time to yield the contributions from surface fluxes and boundary contributions. As in Lauvaux *et al.* (2016), we used CO_2 data from nine towers of the INFLUX network, all of them operational over the period September 2012 to April 2013 (Miles *et al.*, 2017). The system assimilates CO_2 data and solves for 5-day corrections to surface anthropogenic emissions over the dormant season during which the biospheric contribution is small (about 5% of the total CO_2 emissions, reported by Turnbull *et al.*, 2015). Additional modeling details are available in Lauvaux *et al.* (2016).

We will evaluate the different prior emissions by computing the final mismatch in CO_2 mixing ratios referred here as goodness-of-fit after inversion, both over the whole city and for each individual tower site. Because prior error covariances are also constructed according to

the prior emissions, the goodness-of-fit depends on the distribution of sources across the inversion domain and their associated errors. The error variances will be a function of the emissions for each pixel whereas the error covariances will correspond to an exponentially decaying function assuming a correlation length scale of 4 km between urban pixels (similar to Lauvaux *et al.*, 2016). We note here that inverse emissions depend on the a priori but the relative performances will reflect the consistency between atmospheric data and the different prior emission products. Therefore, higher correlations between the posterior mixing ratios and the observations are evidences of a better agreement between the prior emissions and the true fluxes.

3. Results and discussions

3.1 Impervious data as a proxy for diffused sources

Figure 1 shows the impervious surface data over Indianapolis. We extracted three categories that indicate the level of development (high, medium and low) and a category for open space. According to NCDC categorization (http://www.mrlc.gov/nlcd06_leg.php), high density indicates 80–100% imperviousness, medium indicates 50–79%, low indicates 20–49% and open space indicates less than 20%. Although a single category is unlikely to correspond to one particular emissions sector, the city structures are clearly depicted with developed areas and open spaces, the major road transport network (e.g., beltway and interstate highways) and blocky patterns in residential areas. Compared to the spatial structures of ODIAC (see **Figure 2a**), the use of impervious data significantly reduces the mapping error by distributing the emissions over well-identified urban areas rather than smoothed zones overlapping with non-emitting areas. The impervious data might be able to identify particular emission sectors, but no clear relationship between the imperviousness categories and emission sectors can be established. In this study, we aggregated the four imperviousness categories and used them with population density maps as a proxy for diffused emissions.

3.2 30 × 30 m improved ODIAC emission field (iODIAC)
The 30 × 30 m improved emission field (iODIAC) and the other fields are shown in **Figure 2**. The emission gradients over the areas depicted by the impervious surface data were driven by population. Thanks to the use of 1 × 1 km gridded population data, the blocky features are visible across the area (see **Figure 2b**). As expected, the emission mapping error is significantly reduced in iODIAC field compared to ODIAC, with iODIAC field being more closely related to Hestia, although emission gradients are modeled rather than being determined by sectoral information. We present a quantitative assessment of the iODIAC emissions in the following section by performing inversions over the city and by computing statistical metrics to evaluate the improved representation of urban CO_2 emissions.

3.3 Inversion results
Table 2 shows the summary of the 8-month inverse estimates over Indianapolis. Assuming Hestia is the best estimate of Indianapolis CO_2 emissions, the nightlight-based disaggregation emissions from ODIAC (only for non-point sources) are performing reasonably

well for a middle-size city like Indianapolis. When the inversion was performed using the 30 × 30 m improved emission field (iODIAC) as a priori, the inverse estimate differed by only 0.4 Mt/yr over 8 months (about 8% of the total emissions) compared to the Hestia-based inversion. The inversion result with ODIAC was slightly closer to the Hestia inversion result by 0.2 MtC/yr, within the uncertainty range of 0.6 MtC/yr. The spatial structure of the prior emissions has an indirect impact on the inverse emissions. Because the error variances are scaled with prior emissions, specific areas or points may be more or less susceptible to adjustments. Therefore, the differences in the total emissions will depend on the presence of sources near the observation locations which defines the degree of freedom of the prior emissions (i.e. error variances of the prior emissions). Overall, the sharp spatial emission gradients in iODIAC affected the whole-city inverse emissions producing a lower estimate over the entire period (lower by 0.17 MtC/yr). Assuming that iODIAC emissions represent the urban area more accurately than ODIAC, this result shows the sensitivity of the top-down estimate to the fine-scale structure as described by the prior emissions.

Figure 3 shows the prior and posterior emission fields for the three inversion cases, i.e. inv_{Hestia}, inv_{ODIAC} and inv_{iODIAC} emissions. Although the total inv_{iODIAC} emission estimate differs from inv_{Hestia}, the two inverse emission distributions shared major spatial patterns especially with high emissions. The correlation with Hestia was increased from 46% to 52% and the Mean absolute Error (MAE) was reduced by 14% compared to inv_{ODIAC}. Both statistics are significant considering that the increased resolution of iODIAC artificially decreases the correlation (i.e. increases the MAE) due to misplacements of larger gradients in iODIAC. Smoother structures in ODIAC tend to have better correlations, attributable to smaller spatial gradients. We also note here that the power plant emissions were removed to avoid artificially high correlation values (the three maps share identical power plant information).

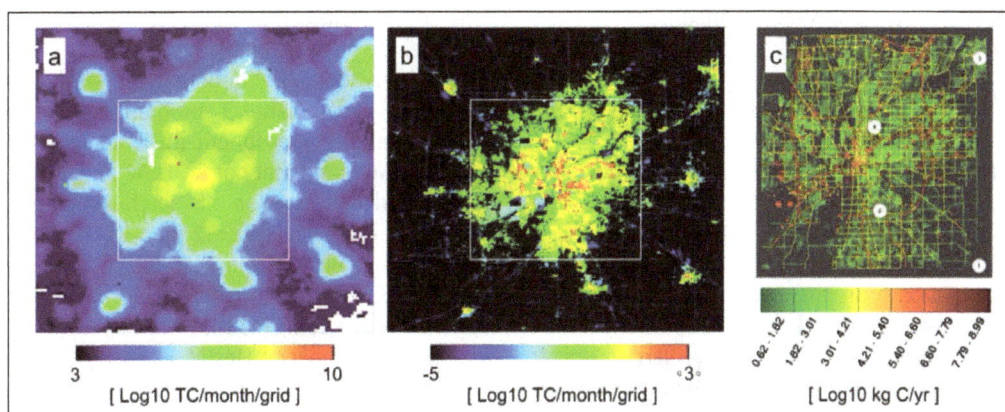

Figure 2: A comparison of three emission fields: 1 × 1 km ODIAC **(a)**, 30 × 30 m improved ODIAC emission field (iODIAC) **(b)**, and Hestia **(c)**. The Figure 2a and 2b indicate the same area as Figure 1. The Hestia emission map (Marion county only) was adopted from Gurney *et al.* (2012). The numbers on the Hestia map indicate focused emission zones in Gurney *et al.* (2012). The original high-resolution image is available at the Hestia project web page (http://hestia.project.asu.edu/index.shtml). The white box on the ODIAC (a) and iODIAC (b) roughly indicates the Hestia domain in Figure 2c. DOI: https://doi.org/10.1525/elementa.146.f2

Table 2: A summary of three inversion results with different prior emission fields. Values are the total emissions from the study domain, given in the unit of MtC/yr. DOI: https://doi.org/10.1525/elementa.146.t2

Prior emission	Hestia* (MtC/yr)	ODIAC* (MtC/yr)	iODIAC – this study (MtC/yr)
A priori	4.56	4.12	4.15
A posteriori	5.5	5.3	5.13

*Values are taken from Lauvaux *et al.* (2016).

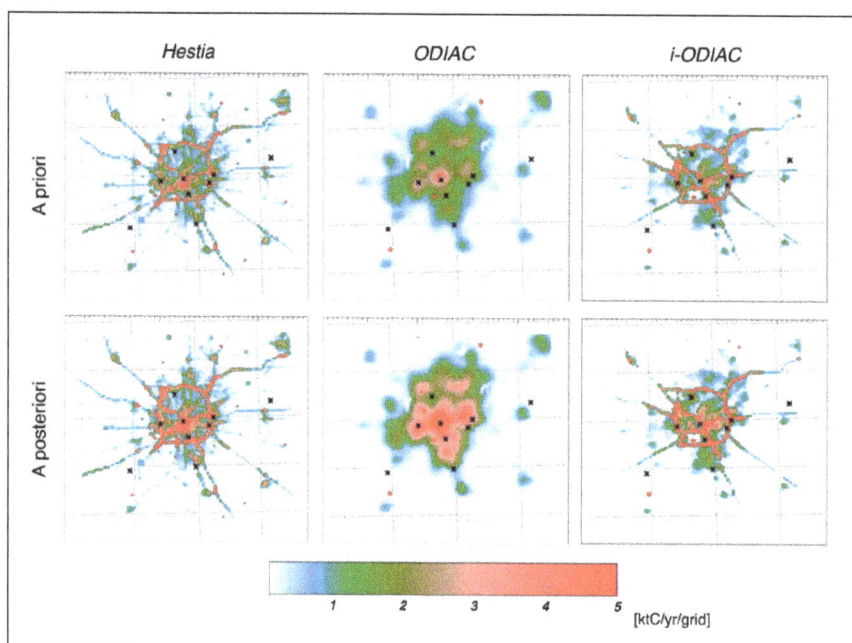

Figure 3: Spatial distributions of a priori (upper) and a posteriori (lower) emissions over Indianapolis, IN. Emission corrections were obtained at a 1 × 1 km resolution. Values are given in the unit of ktC/yr/grid. DOI: https://doi.org/10.1525/elementa.146.f3

Figure 3 illustrates the high resemblance between Hestia and iODIAC (upper row, left and right panels) compared to the smoothed pattern of emissions in ODIAC (upper row, middle panel). The inversion shows more diffuse emission corrections when using ODIAC (lower row, middle panel), while emission adjustments are guided by the spatial patterns in the iODIAC prior field and the error variances constructed accordingly (lower row, right panel).

In **Figure 4**, the temporal variations in the posterior emissions are shown. As shown in previous inversion cases by Lauvaux *et al.* (2016), atmospheric data constrain the temporal variability while prior emissions have no significant impact on the inverse 5-day variations. The inverse results confirm that while spatial information remains a limiting factor despite the large number of towers over the city, temporal variations in the emissions being primarily constrained by observations rather than a priori information. Therefore, the lack of diurnal and sub-monthly variability in iODIAC is overcome by the observational constraint. This result is discussed further in Section 4.3 with potential implications for the development of future high resolution EI's.

We calculated the model-observation mismatch for the three inversion cases as a measure of the goodness-of-fit

before and after inversion. Because the prior errors are fairly similar over the whole city, this result illustrates the capability of the inversion to fit the observed mole fractions and therefore the quality of the prior. If the prior structures are inconsistent with the gradients in the atmospheric observations, the goodness-of-fit will not improve after the inversion. **Table 3** summarizes the values calculated from all the atmospheric measurements used in the inversion. We found that both iODIAC and inv$_{iODIAC}$ showed smaller model-observation mismatch compared to ODIAC and inv$_{ODIAC}$ emissions (–0.382 ppm vs. –0.487 ppm after inversion, and –0.819 ppm vs. –1.05 ppm before inversion), with iODIAC being further away than the Hestia case. This result confirms that iODIAC emission distribution is closer to that of Hestia, allowing the inversion to improve the fit to the atmospheric observations, which indirectly confirms a better distribution of the posterior emissions. The authors would like to highlight that, unlike the Hestia case, weekly to diurnal temporal patterns were not applied to neither ODIAC nor iODIAC.

We further looked at model-observation mismatch for each tower assimilated in the inversion. **Figure 5** shows the model-observation mismatch on a per-tower basis. In this analysis, only the posterior fit was used. The fit

Table 3: A prior and posterior model-observation mismatch. Values are calculated from all the measurements used in the inversion. Values are given in the units of ppm. DOI: https://doi.org/10.1525/elementa.146.t3

Prior emissions	Hestia (ppm)	ODIAC (ppm)	iODIAC – this study (ppm)
A priori	−0.769	−1.05	−0.819
A posteriori	−0.279	−0.487	−0.382

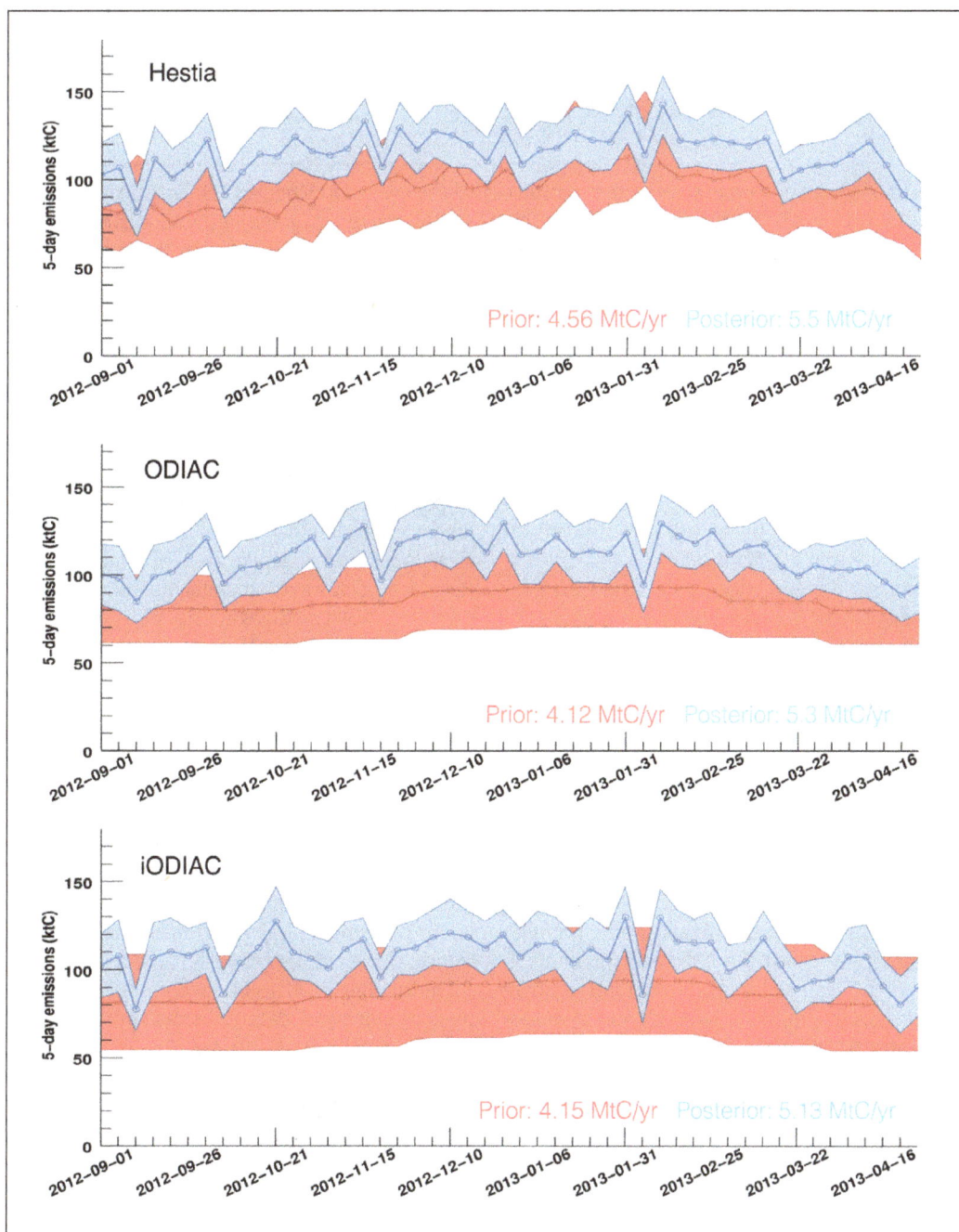

Figure 4: Time series of a priori (in pink) and a posteriori (in blue) emissions of CO_2 aggregated over the Indianapolis inversion domain (indicated in Figure 1) using Hestia (top), ODIAC (middle), and iODIAC (bottom) as prior emissions. DOI: https://doi.org/10.1525/elementa.146.f4

of Hestia emissions are available in Miles *et al.* (same issue). Here we only consider the fit to the posterior emissions. This analysis revealed that the posterior model-observation goodness-of-fit are similar or even better with inv_{iODIAC} compared to inv_{Hestia} emissions for most of the sites, except for sites #04 and #12 which are located on the south side of the city. For the site within the belt-way (site #02, 03, 05, 07, and 10) where the emissions

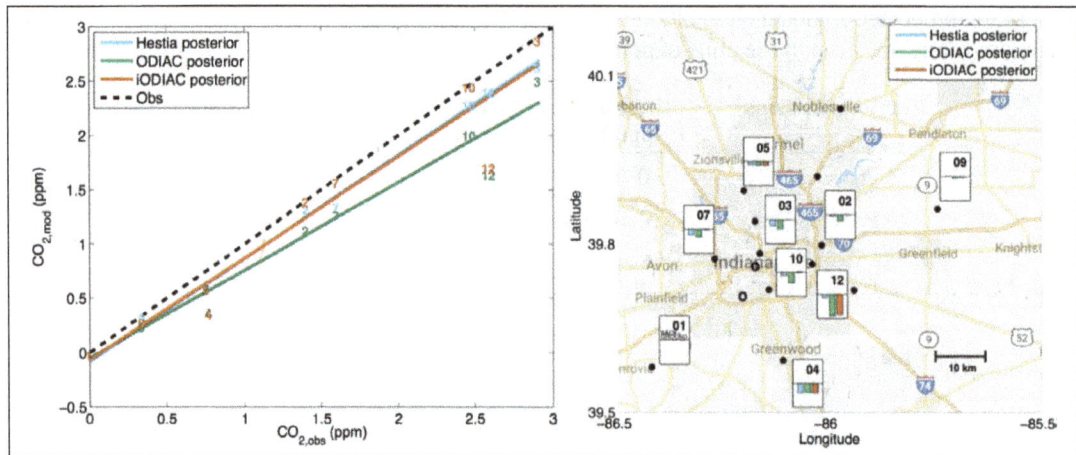

Figure 5: Posterior model-observation mismatch at nine INFLUX towers. Left: Modeled CO_2 enhancement as a function of observed CO_2 enhancement for the Hestia prior (blue), ODIAC posterior (green), iODIAC posterior (orange), and the 1-to-1 line (black dashed). Numbers indicate data point for individual sites. Right: Posterior model-observation mismatch at nice INFLUX towers for the Hestia posterior (blue), ODIAC posterior (green), iODIAC posterior (orange). The site number is shown in the upper right corner of each plot and the black circles indicate the locations of the sites. Open black circles indicate the location of the power plants within the city. The y-axis for each plot extends from -1 to $+1$ ppm CO_2. The observation are not available for this time period at Site 06, 08, 11 and 13. DOI: https://doi.org/10.1525/elementa.146.f5

are most intense, the iODIAC case outperformed the other two cases. For site #04, the model-observation fit is similar for the three cases, indicating a missing adjustment in all cases. For site #12, the inv_{ODIAC} and inv_{iODIAC} model-observation differences are much larger than for the Hestia case.

4. Current limitations and future perspectives

Given the use of generic geospatial data that are available globally, our downscaling approach is applicable to any city in a systematic and timely manner, although the accuracy of the disaggregation method could vary due to errors/biases from larger scale EIs and/or solely due to the potential regional errors/biases in emission disaggregation. The use of very high-resolution satellite-driven data such as impervious surface data for emission mapping can be computationally expensive. For similar studies over multiple cities, the collection of impervious data for urban emissions only represents a small fraction of the surface of the globe which decreases significantly the amount of data and processing of such application. As pointed out by Lauvaux *et al.* (2016), error quantification and characterization for city scale inversion is often extremely difficult to implement due to the lack of information and the computational expense when considering large volume of data in EI's. Our approach could also provide a limited but meaningful opportunity to perform error quantification and characterization by providing alternative emission field to be compared. Thus, the authors believe that emission downscaling approach will help informing city emissions in a global framework for city top-down MRV, especially with future space-based carbon-observing missions. Here we discuss current limitations and future perspectives of this study in a context of city MRV implementation.

4.1 Emission information

As pointed out earlier, the lack of EI reported by cities is a fundamental, limiting factor in city MRV. Although the authors believe that development of a fine-grained EI such as Hestia is an ideal way to accurately quantify city emissions and inform top down methods in a city MRV framework, emission accounting for cities via compilation of EIs needs to be more commonly available and following existing guidelines, such as the Global Protocol for Community-Scale Greenhouse Gas Emission Inventories (GPC, http://www.ghgprotocol.org/city-accounting). With sector-specific information, more accurate emission modeling can be implemented instead of making crude assumptions about sectoral contributions (e.g., applying national-level sectoral distributions or averaged city sectoral fractions to every city). Spatially defined EIs or geolocation information will also greatly support the introduction of the complexity and the diversity of anthropogenic sources in the resulting emission field at fine scale.

The quality of EI is often correlated with the goodness of statistical data collected from various institutions or directly from private organizations (e.g., Olivier and Peters, 2002; Marland, 2008; Andres *et al.*, 2012). Most of the countries that are thought to be producing lower quality EIs are unlikely to be able to compile high-accuracy EIs at the city scale. Collecting accurate data at large scales for aggregated EIs (e.g., national and province levels) remains more practical than city-scale emissions. Therefore, the construction of fine-grained top down estimates to support city-scale EIs is an attractive solution to produce more accurate estimates in any country, and possibly offer a monitoring of the reported emissions, consistent with estimates from larger scales. As an example, Guan *et al.* (2012) reported a 1 Gt CO_2 difference between estimates based on national and province level statistics in China.

4.2 Disaggregation (Mapping) error

Initially, the agreement between ODIAC and Hestia total emissions suggests that the downscaling approach can give us a reasonable estimate for whole-city emissions (within 10%). However, disaggregation (mapping) error can be more significant when moving to higher spatial resolutions. Especially at very high spatial resolution, source locations have to be determined rather than estimated or approximated using proxy data. As seen in the emission pattern, ODIAC provides maps of CO_2 emissions over areas that are unlikely to be emitting (see **Figure 2**). Other than the resolution mismatch (1 km vs. 30 m), the underlying nightlight data used in ODIAC, provided by the Defense Meteorological Satellite Program (DMSP) Operational Linescan System (OLS) nightlight data (https://ngdc.noaa.gov/eog/dmsp. html), have known limitations (e.g., Elvidge et al., 2013). The authors are working on applying new nightlight environmental product developed from data collected by Visible Infrared Imaging Radiometer Suites (VIIRS) on Suomi National Polar-orbiting Partnership (NPP) satellite (Román and Stokes, 2015) to the ODIAC emission model (Oda et al., 2017). There are a number of improvements in VIIRS over the previous instrument which will mitigate the mapping error originating from the use of current nightlight data.

Although the satellite-driven data used in this study for downscaling (e.g., nightlights and impervious surface data) turned out to be useful for determining source regions within a city, nightlights intensity, or development density in impervious surface data, does not fully explain any emission spatial gradients within the emitting area. In this study, we used population data to model the spatial emission gradient. In future study, we will examine the impact of emission gradients on the posterior emission estimates constrained by other proxies, which could be a source of bias in the current inversion setup.

Given the absence of other EI estimates, the evaluation of biases in the emission field remains unachievable. However, geolocation information used to map the emissions can be addressed from various data sources. Although emissions estimates could be significantly biased for sources such as power plants and transportation, we could determine the precision of the geolocation at a minimum (e.g., locations of power stacks and road networks). This first step is critical for city-scale inversions because atmospheric data are unlikely to determine the locations of large sources within the city limits. The verification of intense sources is also limited to few proxies such as public information from Google Map/Earth. However, the limited numbers of large point sources remain manageable within each city compared to the national scale EIs (e.g., Oda and Maksyutov, 2011). This type of error/uncertainty has been discussed in other studies (e.g., Oda and Maksyutov, 2011; Woodard et al., 2014)

4.3 Time profiles

In this study, we focused on the impact of spatial emissions distributions on the inverse emissions without including any temporal variations in the a priori beyond monthly time scale (except Hestia). The seasonality in ODIAC is taken from estimates made by the Carbon Dioxide Information Analysis Center (CDIAC) at Oak Ridge National Laboratory (Oda et al.,

2016). The CDIAC seasonality is based on national monthly fuel statistics, rather than subnational (e.g., state) monthly statistics. Thus, the actual subnational seasonality might be different. According to GPC inventory guidelines, future products may include an annual (i.e. 12 month) inventory. The development of monthly emissions would greatly improve the current level of information in EIs. Climatology may also be used for modeling purposes such as Nassar et al. (2010). The response to environmental conditions and human events (e.g., regular weekday/weekends vs. holidays) should be detectable and therefore quantifiable, if applied. Overall, the authors would like to highlight that the inversion with iODIAC was able to show a very good match with the atmospheric observations comparable to Hestia inversion case over an 8-month period. Future work will aim to assess the impact of temporal profiles in the emissions relative to the impact of finer spatial distributions.

4.4 Error specification

The lack of the error quantification/characterization in the fine-grained emission dataset was discussed by Lauvaux et al. (2016). As mentioned earlier, many sources of uncertainties can affect the emissions and need to be carefully considered depending on the flux resolution (e.g., time and space) of interest. Most of the emission datasets are based on disaggregation of emissions (e.g., CDIAC, EDGAR) where proxy data are used at many different levels. The proxy data are used to approximate the spatial emissions and thus are usually not appropriate at urban scales where individual processes are identifiable. Emission intercomparison may not be highly meaningful but given the lack of physical measurements or EIs constructed at comparable spatial resolutions, model intercomparison remains valid. In the current inversions, the absence of definition for emissions errors is critical, impairing the ability of top down methods (Lauvaux et al., 2016). Given the relatively good performance of iODIAC and the presence of detailed spatial structures, the assessment of emissions errors is a critical objective for urban inversions to improve both the distribution and the total emissions of the city.

5. Conclusions

We present the first space-based emission field at fine resolution to inform a top-down urban-scale framework. Following the INFLUX inversion case with a global 1 × 1 km ODIAC fossil fuel CO_2 emission dataset as a prior, we further improved the 1 × 1 km emission field from the global ODIAC dataset to describe higher levels of emission granularity at the city-scale such as roads and point sources, often missing in coarser resolution products. We approached city emission fields with three types of geometrical objects to represent the principal emission sector components: point, line and diffused (area) emission sources. While preserving the point source information in the ODIAC dataset, we disaggregated the non-point source emissions using geospatial dataset such as global road network data and satellite-data driven surface imperviousness data to generate a 30 × 30 m resolution emission field, comparable to the spatial scale of Hestia. Our disaggregation theoretically can be applied to any global cities and provide an emission estimate with spatial distributions even EI are not compiled locally. The posterior emission estimate summed over the whole city was about 5.1 MtC/yr and remains statistically similar to the previous inversion using ODIAC (5.3 MtC/yr, as reported by Lauvaux et al., 2016). However, the inversion with the 30 × 30 m emission field yielded flux corrections with major spatial patterns matched with those of the inverse using a state-of-the-art building-level emission product, and the optimized model-observation mismatches were similar across the city despite the absence of hourly variability in the prior emissions.

Although emission disaggregation is not often the best approach to inform emissions at a high spatial resolution, our result showed that the use of the geospatial data allowed us to improve the prior emission spatial structure within the city and the potential for providing city emissions where fine-grained emissions data are not available. Beyond the simple mapping of GHG emissions, we quantify here the indirect gain of information by using better-informed a priori emissions, further increasing the potential of the top down approach. This combined approach is particularly useful as fine-grained emission products like Hestia are rarely available for a vast majority of the large metropolitan areas across the globe. Currently, city scale emissions are reported for some cities within local climate action such as Compact of Mayors (https://www.compactofmayors.org/). If we were to start with such activities using atmospheric information, the reported EI (often without spatial distributions) needs to be disaggregated, in order to be incorporated into models. Our method offers a potential approach to a global verification system of city emissions (MRV) using a disaggregation method and an atmospheric inversion system at the urban scale. Given the availability of generic geospatial data, our approach could provide fine-scale city emissions in various locations as future CO_2 observations from ground-based or space missions become more systematically available.

Acknowledgements

The authors would like to thank Junmei Tang for providing her expertise to data processing of geospatial data.

Funding information

TO is supported by NASA Carbon Cycle Science program (Grant #NNX14AM76G). TL is supported by the National Institute for Standards and Technology (INFLUX project #70NANB10H245) and the National Oceanic and Atmospheric Administration (grant #NA13OAR4310076).

Competing interests

The authors have no competing interests to declare.

Author contributions

· TO and TL conceived and designed the study, implemented the analysis and drafted the manuscript.
· NLM and SJR acquired atmospheric CO_2 data.
· KRG developed and provided Hestia emission dataset.
· PR and DL contributed to geospatial data processing. All the authors contributed to improve the manuscript.

References

Andres, RJ, Boden, TA, Bréon, F-M, Ciais, P and **Davis, S,** et al. 2012 A synthesis of carbon dioxide emissions from fossil-fuel combustion, *Biogeosciences* **9**: 1845–1871. DOI: https://doi.org/10.5194/bg-9-1845-2012

Andres, RJ, Boden, TA and **Higdon, DM** 2016 Gridded uncertainty in fossil fuel carbon dioxide emission maps, a CDIAC example, *Atmos. Chem. Phys.* **16**: 14979–14995. DOI: https://doi.org/10.5194/acp-16-14979-2016

Asefi-Najafabady, S, Rayner, PJ, Gurney, KR, McRobert, A, Song, Y, et al. 2014 A multiyear, global gridded fossil fuel CO_2 emission data product: Evaluation and analysis of results. *J. Geophys. Res. Atmos.* **119**: 10213–10231. DOI: https://doi.org/10.1002/2013JD021296

Buchwitz, M, Reuter, M, Bovensmann, H, Pillai, D, Heymann, J, et al. 2013 Carbon Monitoring Satellite (CarbonSat): assessment of atmospheric CO_2 and CH_4 retrieval errors by error parameterization. *Atmos. Meas. Tech.* **6**: 3477–3500. DOI: https://doi.org/10.5194/amt-6-3477-2013

Center for International Earth Science Information Network − CIESIN − Columbia University, and Information Technology Outreach Services − ITOS − University of Georgia 2013 Global Roads Open Access Data Set, Version 1 (gROADSv1). Palisades, NY: NASA Socioeconomic Data and Applications Center (SEDAC). DOI: https://doi.org/10.7927/H4VD6WCT

Deng, A, Lauvaux, T, Davis, KJ, Gaudet, BJ, Miles, N, Richardson, SJ, Wu, K, Sarmiento, DP, Hardesty, RM, Bonin, TA, Brewer, WA and **Gurney, KR** 2017 Toward reduced transport errors in a high resolution urban CO_2 inversion system, *Elem Sci Anth.* In press for the INFLUX Special Feature.

Eldering, A 2015 The OCO-3 Mission: Overview of Science Objectives and Status, 2015 American Geophysical Union (AGU) Fall Meeting, San Francisco.

Elvidge, CD, Baugh, K, Zhizhin, M and **Hsu, FC** 2013 Why VIIRS data are superior to DMSP for maping nighttime lights, *Proc. the Asia-Pacific Advanced Network* **35**: 62–69. DOI: https://doi.org/10.7125/APAN.35.7

Enting, IG, Trudinger, CM and **Francey, RJ** 1995 A synthesis inversion of the concentration and $d_{13}C$ atmospheric CO_2, *Tellus B* **47**: 35–52. DOI: https://doi.org/10.3402/tellusb.v47i1-2.15998

Feng, S, Lauvaux, T, Newman, S, Rao, P, Ahmadov, R, et al. 2016 Los Angeles megacity: a high-resolution land–atmosphere modelling system for urban CO_2 emissions, *Atmos. Chem. Phys.* **16**: 9019–9045. DOI: https://doi.org/10.5194/acp-16-9019-2016

Guan, D, Liu, Z, Geng, Y, Lindner, S and Hubacek, K 2012 The gigatonne gap in China's carbon dioxide inventories. *Nature Climate Change* **2**: 672–675. DOI: https://doi.org/10.1038/nclimate1560

Gurney, K, Razlivanov, I, Song, Y, Zhou, Y, et al. 2012 Quantification of fossil fuel CO_2 emission on the building/street scale for a large US city. *Environ. Sci. & Technol.* **46**: 12194–12202. DOI: https://doi.org/10.1021/es3011282

Gurney, KR, Liang, J, Patarasuk, R, O'Keeffe, D, Huang, J, et al. 2016 Reconciling the differences between a bottom-up and inverse-estimated $FFCO_2$ emissions estimate in a large US urban area, *Elem Sci Anth.* In press for the INFLUX Special Feature.

Hutyra, L, Duren, R, Gurney, KR, Grimm, N, Kort, E, et al. 2014 "Urbanization and the carbon cycle: Current capabilities and research outlook from the natural sciences perspective", *Earth's Future.* DOI: https://doi.org/10.1002/2014EF000255

Janardanan, R, Maksyutov, S, Oda, T, Saito, M, Kaiser, JW, Ganshin, A, Stohl, A, Matsunaga, T, Yoshida, Y and Yokota, T 2016 Comparing GOSAT observations of localized CO_2 enhancements by large emitters with inventory-based estimates, *Geophys. Res. Lett.* **43**. DOI: https://doi.org/10.1002/2016GL067843

Janssens-Maenhout, G, Dentener, F, Van Aardenne, J, Monni, S, Pagliari, V, et al. 2012 EDGAR-HTAP: a Harmonized Gridded Air Pollution Emission Dataset Based on National Inventories. Ispra (Italy): *European Commission Publications Office*; 2012. JRC68434, EUR report No EUR, **25**: 299–2012, ISBN 978-92-79-23122-0, ISSN 1831-9424.

Kort, EA, Frankenberg C, Miller, CE and Oda, T 2012 Space-based observations of megacity carbon dioxide. *Geophys. Res. Lett.* **39**: 17–22. DOI: https://doi.org/10.1029/2012GL052738

Kuenen, J, Visschedijk, A, Jozwicka, M and Gon, H 2014 TNO-MACC_II emission inventory; a multi-year (2003–2009) consistent high-resolution European emission inventory for air quality modelling, *Atmos Chem Phys* **14**(20): 10963–10976. DOI: https://doi.org/10.5194/acp-14-10963-2014

Kurokawa, J, Ohara, T, Morikawa, T, Hanayama, S, Janssens-Maenhout, G, et al. 2013 Emissions of air pollutants and greenhouse gases over Asian regions during 2000–2008: Regional Emission inventory in ASia (REAS) version 2, *Atmos. Chem. Phys.* **13**: 11019–11058. DOI: https://doi.org/10.5194/acp-13-11019-2013

Lauvaux, T, et al. 2016 High-resolution atmospheric inversion of urban CO_2 emissions during the dormant season of the Indianapolis Flux Experiment (INFLUX), *J. Geophys. Res. Atmos.* **121**: DOI: https://doi.org/10.1002/2015JD024473

Liu, Z, et al. 2015 Reduced carbon emission estimates from fossil fuel combustion and cement production

in China, *Nature* **524**: 335–8. DOI: https://doi.org/10.1038/nature14677

Marland, G 2008 Uncertainties in Accounting for CO_2 From Fossil Fuels. *J. Industrial Ecology* **12**: 136–139. DOI: https://doi.org/10.1111/j.1530-9290.2008.00014.x

Miles, NL, et al. 2017 Detectability and quantification of atmospheric boundary layer greenhouse gas dry mole fraction enhancements in an urban landscape: Results from the Indianapolis Flux Experiment (INFLUX) *Elem Sci Anth.* In press for the INFLUX Special Feature.

Nassar, R, Jones, DBA, Suntharalingam, P, Chen, JM, Andres, RJ, et al. 2010 Modeling global atmospheric CO_2 with improved emission inventories and CO_2 production from the oxidation of other carbon species. *Geosci. Model Dev.* **3**: 689–716. DOI: https://doi.org/10.5194/gmd-3-689-2010

Nisbet, E and Weiss, R 2010 Top-down versus bottom-up, *Science* **328**: 1241–1243. DOI: https://doi.org/10.1126/science.1189936

Oda, T and Maksyutov, S 2011 A very high-resolution (1 km × 1 km) global fossil fuel CO_2 emission emission inventory derived using a point source database and satellite observations of nighttime lights. *Atmos. Chem. and Phys.* **11**: 543–556. DOI: https://doi.org/10.5194/acp-11-543-2011

Oda, T, Maksyutov, S and Andres, RJ 2017 The Open-source Data Inventory for Anthropogenic CO_2 (ODIAC) fossil fuel emission model version 3.0 (ODIAC v3.0), in preparation for *Earth Syst. Sci. Data.*

Olivier, JGJ 2002 On the Quality of Global Emission Inventories: Approaches, Methodologies, Input Data and Uncertainties, PhD thesis University Utrecht, ISBN 90-393-3103-0.

Olivier, JGJ, Aardenne, JAV, Dentener, FJ, Pagliari, V, Ganzeveld, LN and Peters, JAHW 2005 Recent trends in global greenhouse gas emissions: Regional trends 1970–2000 and spatial distribution of key sources in 2000, *J. Integr. Env. Sci.* **2**: 81–99. DOI: https://doi.org/10.1080/15693430500400345

Olivier, JGJ and Peters, JAHW 2002 Uncertainties in global, regional, and national emissions inventories. In *Non-CO_2 greenhouse gases: Scientific understanding, control options and policy aspects,* Van Ham J, Baede, APM, Guicherit, R and WIlliams-Jacobse, JFGM (eds.). Springer, New York, USA, 525–540.

Pacala, SW, et al. 2010 Verifying Greenhouse Gas Emissions: Methods to Support International Climate Agreements. *Committee on Methods for Estimating Greenhouse Gas Emissions; National Research Council,* National Academy of Sciences, 124 pp.

Patarasuk, P, Gurney, KR, O'Keeffe, D, Song, Y and Huang, J, et al. 2016 Application of high-resolution fossil fuel CO_2 emissions quantification to urban climate policy in Salt Lake County, Utah USA, *Urban Ecosystems.* DOI: https://doi.org/10.1007/s11252-016-0553-1

Polonsky, IN, O'Brien, DM, Kumer, JB, O'Dell, CW and the geoCARB Team 2014 Performance of a geostationary mission, geoCARB, to measure CO_2, CH_4 and CO column-averaged concentrations, *Atmos. Meas. Tech.* **7**: 959–981. DOI: https://doi.org/10.5194/amt-7-959-2014

Rayner, PJ, Raupach, MR, Paget, M, Peylin, P and **Koffi, E** 2010 A new global gridded data set of CO_2 emissions from fossil fuel combustion: Methodology and evaluation. *J. Geophys. Res.* **115**(D19): 306, DOI: https://doi.org/10.1029/2009JD013439

Richardson, SJ, Miles, NL, Davis, KJ, Lauvaux, L, Martins, DK, et al. 2017 CO_2, CH_4, and CO tower in situ measurement network in support of the Indianapolis FLUX (INFLUX) Experiment. *Elem Sci Anth.* In press for the INFLUX Special Feature.

Román, MO and **Stokes, EC** 2015 Holidays in Lights: Tracking cultural patterns in demand for energy services. *Earth's Future.* DOI: https://doi.org/10.1002/2014EF000285

Streets, D, et al. 2003 An inventory of gaseous and primary aerosol emissions in Asia in the year 2000, *J. Geophys. Res.,* **108**(D21): 8809. DOI: https://doi.org/10.1029/2002JD003093

Turnbull, JC, et al. 2015 Toward quantification and source sector identification of fossil fuel CO_2 emissions from an urban area: Results from the influx experiment. *J. Geophys. Res. Atmos.* **120**: 292–312. DOI: https://doi.org/10.1002/2014JD022555

Uliasz, M 1994 Lagrangian particle modeling in mesoscale applications, in Environmental Modelling II, Zanetti, P (ed.). *Computational Mechanics Publications,* pp. 71–102.

Woodard, D, Branham, M, Buckingham, G, Hogue, S, Hutchins, M, et al. 2014 A spatial uncertainty metric for anthropogenic CO_2 emissions, Greenhouse Gas Measurement and Management (2014), **2–4**: 139–160. DOI: https://doi.org/10.1080/20430779.2014.1000793

7

A large source of dust missing in Particulate Matter emission inventories? Wind erosion of post-fire landscapes

N. S. Wagenbrenner*, S. H. Chung[†] and B. K. Lamb[†]

Wind erosion of soils burned by wildfire contributes substantial particulate matter (PM) in the form of dust to the atmosphere, but the magnitude of this dust source is largely unknown. It is important to accurately quantify dust emissions because they can impact human health, degrade visibility, exacerbate dust-on-snow issues (including snowmelt timing, snow chemistry, and avalanche danger), and affect ecological and biogeochemical cycles, precipitation regimes, and the Earth's radiation budget. We used a novel modeling approach in which local-scale winds were used to drive a high-resolution dust emission model parameterized for burned soils to provide a first estimate of post-fire PM emissions. The dust emission model was parameterized with dust flux measurements from a 2010 fire scar. Here we present a case study to demonstrate the ability of the modeling framework to capture the onset and dynamics of a post-fire dust event and then use the modeling framework to estimate PM emissions from burn scars left by wildfires in U.S. western sagebrush landscapes during 2012. Modeled emissions from 1.2 million ha of burned soil totaled 32.1 Tg (11.7–352 Tg) of dust as PM_{10} and 12.8 Tg (4.68–141 Tg) as $PM_{2.5}$. Despite the relatively large uncertainties in these estimates and a number of underlying assumptions, these first estimates of annual post-fire dust emissions suggest that post-fire PM emissions could substantially increase current annual PM estimates in the U.S. National Emissions Inventory during high fire activity years. Given the potential for post-fire scars to be a large source of PM, further on-site PM flux measurements are needed to improve emission parameterizations and constrain these first estimates.

Keywords: wildfire; dust; PM10

1. Introduction

Wildfire smoke is the largest source of primary $PM_{2.5}$ (particulate matter with aerodynamic diameter less than 2.5 mm) and the third largest source of PM_{10} (PM with aerodynamic diameter less than 10 mm) in the United States (EPA, 2011). Air quality degradation from wildfire smoke is a widespread issue in the western U.S. (Jaffe et al., 2008). This work investigates wind erosion of burned soils as an additional fire-related source of atmospheric PM. In contrast to wildfire smoke, post-fire PM (mineral dust and ash, hereafter referred to as dust) is not as conspicuous an issue since wind erosion events are highly intermittent in time and space and the sources tend to be in remote areas. Yet, there is evidence (Miller et al., 2012; Sankey et al., 2009; Wagenbrenner et al., 2013; supplemental material) that

post-fire landscapes of the U.S. Great Basin (delineated by the gray-colored area in **Figure 1**) are major contributors to atmospheric dust.

The intense heat of a wildfire can penetrate surface soils increasing erodibility by destroying naturally occurring soil crusts (Ford and Johnson, 2006), increasing soil water repellency (Ravi et al., 2007), and decreasing aggregate stability (Varela et al., 2010). Fires in the Great Basin typically burn quickly and intensely, often consuming all vegetation due to the arid conditions and flammability of the fuels. This leaves dry, loose, bare soil and ash exposed to high winds, also characteristic of this region (Jewell and Nicoll, 2011). Post-fire dust sources are among the strongest atmospheric dust sources reported (Wagenbrenner et al., 2013) and may contribute substantially to the global dust budget.

PM source identification and quantification is important for accurate characterization of atmospheric PM. Atmospheric PM impacts human health (Dockery and Pope, 1994), visibility (Hyslop, 2009), snowpack dynamics including snowmelt (Skiles et al., 2012), snow chemistry (Rhodes et al., 2010), and avalanche danger (Summit County, 2014), ecological and biogeochemical cycles (Field

* US Forest Service, Rocky Mountain Research Station, Missoula Fire Sciences Laboratory, Missoula, Minnesota, United States

† Laboratory for Atmospheric Research, Department of Civil and Environmental Engineering, Washington State University, Pullman, Washington, United States

Corresponding author: N. S. Wagenbrenner
(nwagenbrenner@fs.fed.us)

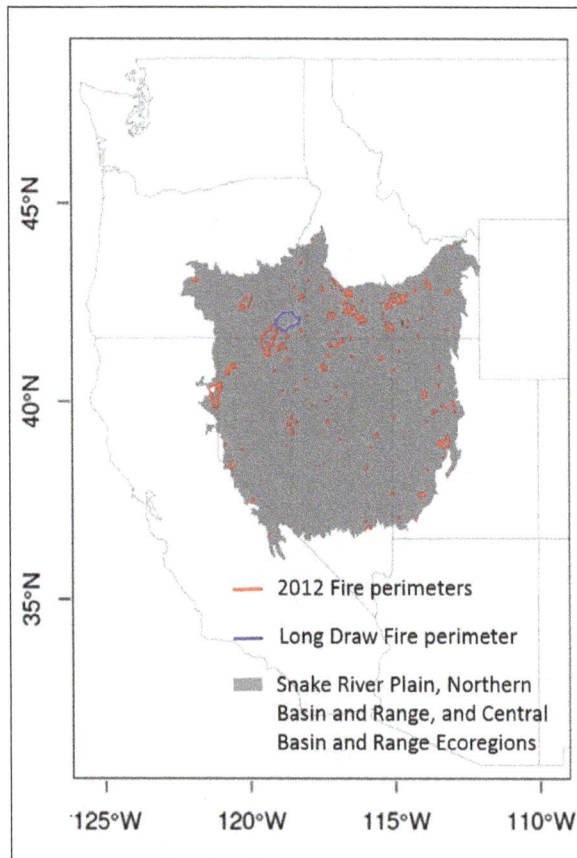

Figure 1: Areas burned by wildfires in the Snake River Plain, Northern Basin and Range, and Central Basin and Range Level III Ecoregions during 2012. The area burned by the Long Draw Fire is outlined in blue. DOI: https://doi.org/10.1525/elementa.185.f1

et al., 2010), precipitation regimes (Sassen et al., 2003), and the Earth's radiation budget (Tegen et al., 1996). But quantification of PM emissions is difficult, particularly for dust, due to the spatial, temporal, and compositional variability of the sources (e.g., Ginoux et al., 2012).

The lack of information regarding dust sources, particularly small-scale or quasi-permanent sources, has hampered understanding and modeling of the global dust cycle (e.g., Ginoux et al., 2012). The best characterized dust sources are topographic depressions containing deep alluvial sediment deposits in arid regions (Prospero et al., 2002). Small-scale and transient sources have been recognized as significant contributors to the global dust budget, but difficulties in identifying emission 'hot spots' have limited their investigation and inclusion in emission inventory estimates (Zender et al., 2003b; Okin et al., 2011). Post-fire sources pose an especially difficult case since they are highly transient, with emissions occurring intermittently for maybe just a year post-fire (Wagenbrenner et al., 2013), and different locations are burned each year.

Additionally, there are potentially specific concerns for post-fire dust. Post-fire dust likely exhibits different optical, chemical, and mechanical properties than typical mineral dust due to the composition of the post-fire material (a mix of ash and soil particles). Little is known regarding

how size distributions and compositional makeup may differ between dust from post-fire sources and mineral dust originating from agricultural or desert sources. Post-fire field measurements suggest that post-fire dust likely contains a higher fraction of $PM_{2.5}$ than typical mineral dust (Wagenbrenner et al., 2013), likely due to the presence of ash in the surface soil.

In this work we focus on the U.S. Great Basin, where a large number of wildfires occur each year and the landscapes are susceptible to post-fire erosion (Miller et al., 2012; Sankey et al., 2009; Wagenbrenner et al., 2013). Furthermore, dust from the Great Basin has substantial influences on water resources in the U.S. Intermountain West due to dust-snow-albedo effects on snowmelt (Deems et al., 2013; Skiles et al., 2012). The spread of invasive species has increased fuel loadings across the U.S. Great Basin, led to more frequent and severe fires (Balch et al., 2013), and presumably increased post-fire erodibility and PM emissions. Post-fire emissions may provide a positive feedback on the invasive species-altered fire regime loop in this region as the dust emitted from burned areas also contains stores of native plant seedbanks (e.g., Chambers and MacMahon, 1994), which further reduces competition for invasive species.

In this work, we present a new PM emission model based on measurements from Wagenbrenner et al. (2013). To our knowledge, Wagenbrenner et al. (2013) is the only study to-date to report PM vertical flux measurements from a fire scar. The fire scar in Wagenbrenner et al. (2013) was burned by the 2010 Jefferson Fire northwest of Idaho Falls, ID (northeast portion of the area delineated in gray in **Figure 1**). The fire, driven by high southwesterly winds, burned over 100,000 acres of grass and sagebrush in just days. The fire consumed essentially all vegetation and left behind dry, bare soil that was highly erodible. Dust emissions were visible immediately after the fire and elevated emissions were frequently measured until nearly one year post-fire, when vegetation began to re-establish. Additional details can be found in Wagenbrenner et al. (2013).

The objectives of this work are to (1) demonstrate the ability of a new modeling framework to simulate emission and transport of PM during a large post-fire dust event and (2) use the new post-fire PM emission model to provide a first estimate of the contribution of western U.S. fire scars to atmospheric PM. We first provide a case study to demonstrate the behavior of a post-fire wind erosion event and the ability of the new modeling approach to capture this behavior. Then we present a broader emissions modeling study to provide the first estimate of the annual contribution of PM emissions from post-fire dust sources.

2. Methods
2.1. Models
WindNinja (Forthofer et al., 2014) was used to model the local wind field and PM emissions from the fire scars. WindNinja is a diagnostic microscale wind model that accounts for local mechanical and thermal effects of the underlying terrain on the flow field. Wagenbrenner et al. (2016) found that WindNinja downscaling improved surface wind forecasts in complex terrain.

WindNinja uses a semi-empirical PM emission model (Draxler et al., 2001) parameterized for burned soil. Vertical flux of PM_{10} is calculated in WindNinja as:

$$F_v = \frac{K\rho}{g}u_* \; u_*^2 \left(-u_{*t}^2\right)$$ (1)

where K is the PM_{10} release factor, ρ is air density, u_* is friction velocity, and u_{*t} is threshold friction velocity.

The PM_{10} release rate and threshold friction velocity were developed from post-fire field measurements following a 2010 wildfire in the Snake River Plain ecoregion (Wagenbrenner et al., 2013). The PM_{10} release rate and threshold friction velocity were 0.0007 m^{-1} and 0.22 ms^{-1} (**Table 1**). We assumed no re-vegetation and dry, uniform soil conditions representative of the burn scar in Wagenbrenner et al. (2013). Emissions outside of the burn perimeters were assumed to be negligible. The emission model does not account for soil moisture or texture. Given the severe impact fire has on the surface soil and vegetation and the lack of other post-fire PM measurements available, we believe this is a reasonable approximation for an initial attempt to quantify post-fire PM emissions.

Friction velocity is calculated in WindNinja based on a log profile normal to the ground.

$$u_* = \frac{U \; \kappa}{\ln(\frac{z-d}{z_0})}$$ (2)

$U(z)$ is the wind speed at height z above the ground, κ is the von Karman constant, d is the zero-plane displacement, and z_0 is the roughness parameter. First, the gridded winds are computed in WindNinja. $U(z)$ is then computed in each cell of a near-surface layer of the computational mesh using a rotated coordinate system where the z-axis is normal to the ground. Finally u_* is calculated in each cell of the near-surface layer. d is set to 0.0 m, κ is set to 0.4, and z_0 is set to 0.01 m, representative of smooth bare soil (**Table 1**).

The transport and dispersion for the case study were simulated using the AIRPACT-3 regional air quality modeling system for the Pacific Northwest (Chen et al., 2008; Chung et al., 2012). AIRPACT-3 uses the Weather Research and Forecasting (WRF) model (Skamarock et al., 2008) for meteorology and the Community Multiscale Air Quality (CMAQ) model (Byun and Schere, 2006) for chemistry and transport. CMAQ simulates the transport and removal from the atmosphere of PM by wet and dry deposition according to meteorology simulated by WRF. The default AIRPACT-3

Table 1: Parameters used in the PM emission model. DOI: https://doi.org/10.1525/elementa.185.t1

Parameter	Symbol	Value	Units
von Karman constant	κ	0.4	–
Zero-plane displacement	d	0.0	m
Roughness parameter	z_0	0.01	m
PM_{10} release rate	K	0.0007	m^{-1}
Threshold friction velocity	u_{*t}	0.22	$m\,s^{-1}$

modeling system includes all major known sources of PM in the region but, similar to other regional air quality models, does not include post-fire PM emissions, which were added for this work. Based on the measurements of Wagenbrenner et al. (2013), 60% of post-fire PM_{10} emissions was assigned to coarse-mode PM (PMC or particulate matter with diameters between 2.5 and 10 μm) and the remainder was assigned to $PM_{2.5}$. Post-fire PM_{10} is calculated as the difference in modeled PM_{10} concentrations from CMAQ simulations with and without post-fire dust emissions.

2.2. A case study
The vast expanse of the Great Basin is sparsely populated, which means fewer people to observe or be immediately affected by post-fire dust events. One of the best documented (Video S1) cases of a population-impacting event occurred on August 5, 2012 when a large dust plume originating from the Long Draw fire scar (**Figure 1**) in southeast Oregon caused an exceedance of the PM_{10} National Ambient Air Quality Standard (NAAQS) 150 km downwind in Boise, Idaho. This case study was used to investigate the feasibility of simulating the onset and evolution of a post-fire dust event with a regional air quality model linked with a high-resolution dust emission model.

HYSPLIT backward trajectories originally suggested the Long Draw fire scar as a potential source of the dust plume (e.g., **Figure 2**). Inspection of NEXRAD base reflectivity in the area at the time of the event further indicated the dust was likely emitted from the burn scar as high winds associated with near-by thunderstorms passed over the area and moved toward Boise, ID (supplemental material).

We simulated hourly PM emissions and transport for 24 hours beginning at midnight LT on August 4. Hourly surface winds from a nearby Remote Automated Weather Station (Grassy Mountain RAWS; www.raws.dri.edu) were used to drive the emission model. The fire perimeter was extracted from the Monitoring Trends in Burn Severity (MTBS84) dataset (Eidenshank et al., 2007). We compared modeled PM concentrations to observed PM data from Boise and Nampa provided by the Idaho Department of Environmental Quality (http://airquality.deq.idaho.gov).

2.3. Annual emissions estimate
Burn perimeters from wildfires that occurred during 2012 in the Snake River Plain, Northern Basin and Range, and Central Basin and Range Level III Ecoregions (Omernik and Griffith, 2014) were investigated in this study (**Figure 1**). Fire perimeters and start dates were extracted from the MTBS84 dataset. Only wildfires larger than 1000 acres were included in the analysis. There were 120 burn perimeters covering 1.2 million ha. The year 2012 was chosen because at the time that this work was initiated it was the most recent complete archive of annual fire perimeters and because it included the Long Draw Fire, which we had already chosen to investigate as a case study. This was a particularly active fire year in the U.S. Great Basin (1.2 million ha burned in 2012 vs. 319,651 ha in 2011 and 255,504 ha in 2010) and thus, estimates from this work are representative of a high fire activity year.

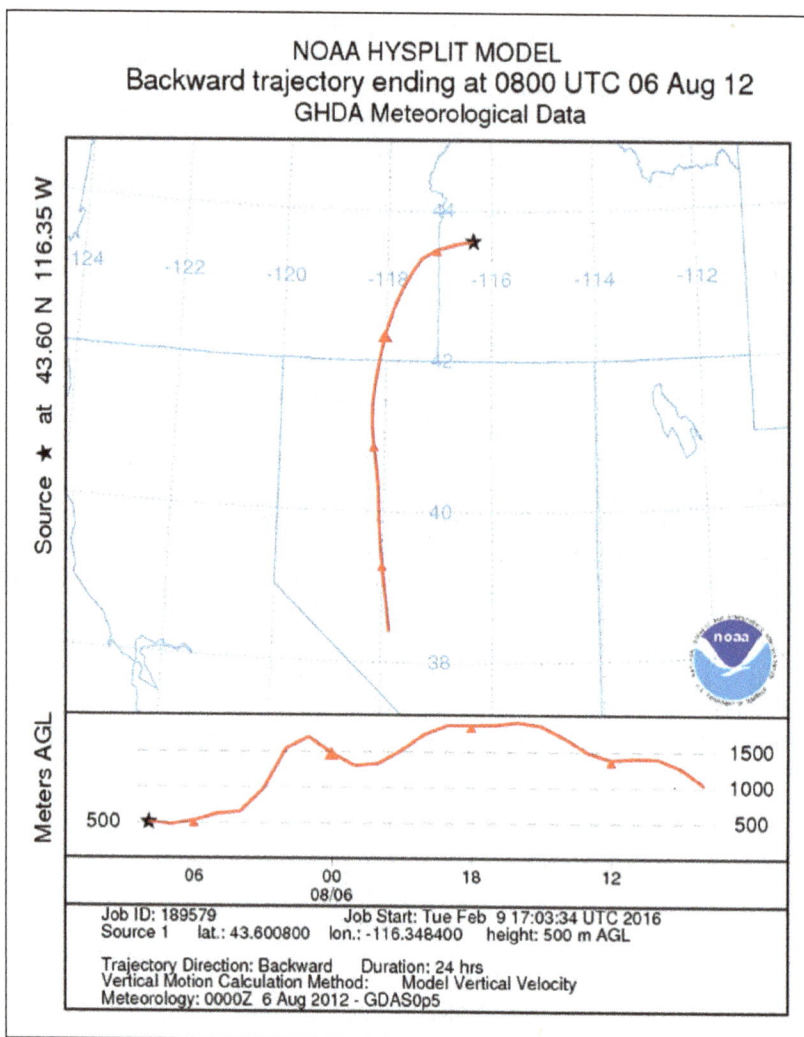

Figure 2: Hysplit backward trajectory for the August 5 dust event observed in Boise, Idaho. The modeled end time is adjusted (offset of 6 hours based on the WRF-CMAQ results) to account for underestimation of wind speed in the meteorological model. DOI: https://doi.org/10.1525/elementa.185.f2

Archived 12-km North American Model (NAM) meteorology was used as input to WindNinja for the emissions estimate work. NAM meteorology was chosen because of its relatively high spatial resolution for regional-scale meteorological modeling and because it was readily obtainable from the National Centers for Environmental Prediction for our temporal and spatial extents. Temporal resolution, determined by NAM forecast output frequency, was one hour. The horizontal grid resolution of the PM emission model was on the order of 100 m but varied according to the domain extent for each fire.

We modeled hourly emissions for each burn scar in **Figure 1** for one year post-fire. The hourly simulations started 24 hours after the fire start date and continued through December 31, 2012, or until the first snowfall, whichever was earlier. Since the grass and sagebrush fuels in the ecoregions investigated in this study typically burn very quickly and accurate temporal fire perimeter information is not available, we assumed the final burn perimeter existed the day following the fire start date. Periods of

precipitation and 24 hours after each precipitation event were omitted from the analysis period.

3. Characterization and quantification of post-fire dust

3.1. A case study: 5–6 August, 2012 dust event in Boise, Idaho

The modeled emissions at the peak of the wind event are shown **Figure 3a**. The modeled emission map clearly shows emission "hot spots" within the burn perimeter due to the mechanical effects of the terrain on the local flow field. The hot spots correspond to areas of higher friction velocities (e.g., due to wind speed-up over a ridge). The modeled emissions varied from 0 to 140 mg m^{-2} s^{-1} over the domain. The highest predicted emissions were on a ridge oriented from northwest to southeast in the southwest portion of the domain. Although we do not have on-site emission measurements to corroborate the modeled emission pattern, the idea of emission hot spots related to terrain modification of the local wind field has been documented in other studies (e.g., Goosens and Offer, 1997).

Figure 4b shows the dust plume approaching Boise, ID around 1800 LT on 5 August. Transport and dispersion modeling captured the general plume trajectory (**Figure 3c**). The modeled transport suggested that PM was lofted high into the atmosphere not far downwind from the burn scar. **Figure 3c** shows high surface concentrations over the burn scar, but lower surface concentrations immediately downwind. The modeled plume eventually covered most of Idaho and extended into parts of Washington, Montana, and Canada (**Figure 3c**). The high PM concentrations along the Idaho-Montana border are due to interception of the dust plume by the high-elevation Bitterroot Mountains (**Figure 3c**).

Observed peak concentrations were 2650 $\mu g\ m^{-3}$ in Boise at 21:00 (LT) and 2397 $\mu g\ m^{-3}$ in Nampa at 20:00 (LT) on 5 August (**Figure 3d**). Elevated concentrations persisted at both locations for about 8 hours. Modeled surface concentrations in Boise and Nampa were lower than, but on the order of the measured concentrations (**Figure 3d**).

Modeled peak concentrations were 962 $\mu g\ m^{-3}$ in Boise at 12:00 on 6 August and 2040 $\mu g\ m^{-3}$ in Nampa at 03:00 (LT) on 6 August. This indicates that the modeled PM emission rate for this event was reasonable.

The timing of the modeled peak concentrations was delayed due to an issue with the modeled meteorology used to drive the transport model. The WRF meteorological model underestimated the high wind speeds associated with the thunderstorm outflows which drove the emission and local transport of PM (discussed in Section 2.2). This underestimation of wind speed increased the residence time of the modeled plume and is at least partially responsible for the high modeled PM concentrations in Boise and Nampa following the concentration peaks. The 24-hr average concentration was still underestimated for Boise (483 $\mu g\ m^{-3}$ observed vs. 279 $\mu g\ m^{-3}$ modeled), but was over-predicted for Nampa (385 $\mu g\ m^{-3}$ observed vs. 561 $\mu g\ m^{-3}$ modeled) during 17:00 5 August to 17:00 6 August.

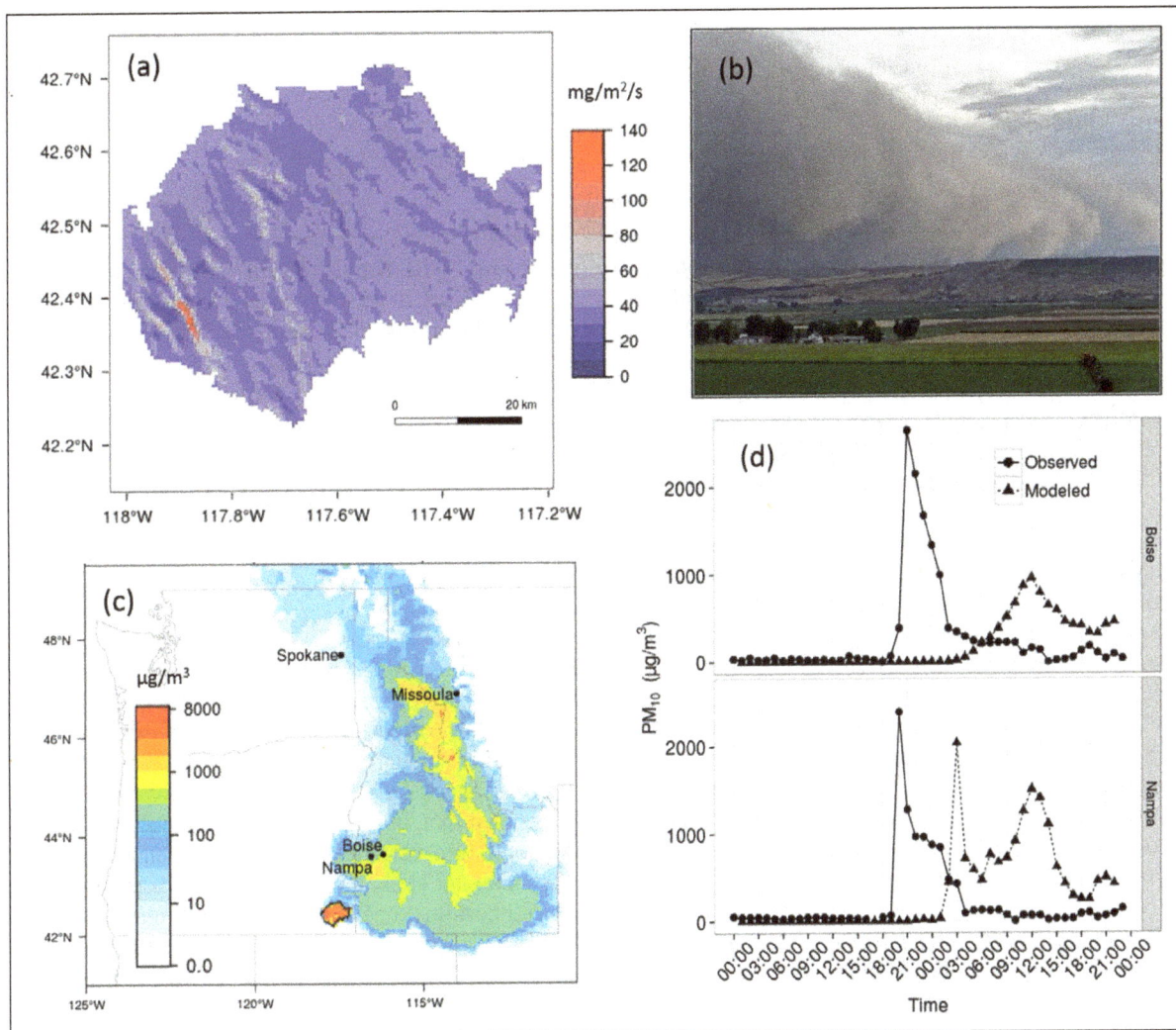

Figure 3: **Modeled emissions from the Long Draw Fire. (a)** A haboob originating from the Long Draw Fire and approaching Boise, Idaho on August 5, 2012. **(b)** Photo credit: ktvb.com. Modeled PM_{10} concentrations during the haboob event on August 5, 2012 **(c)** Modeled and observed surface concentrations in Boise and Nampa (~ 34 km west of Boise), Idaho. **(d)** Observed data are provided by the Idaho Department of Environmental Quality (http://airquality.deq.idaho.gov). DOI: https://doi.org/10.1525/elementa.185.f3

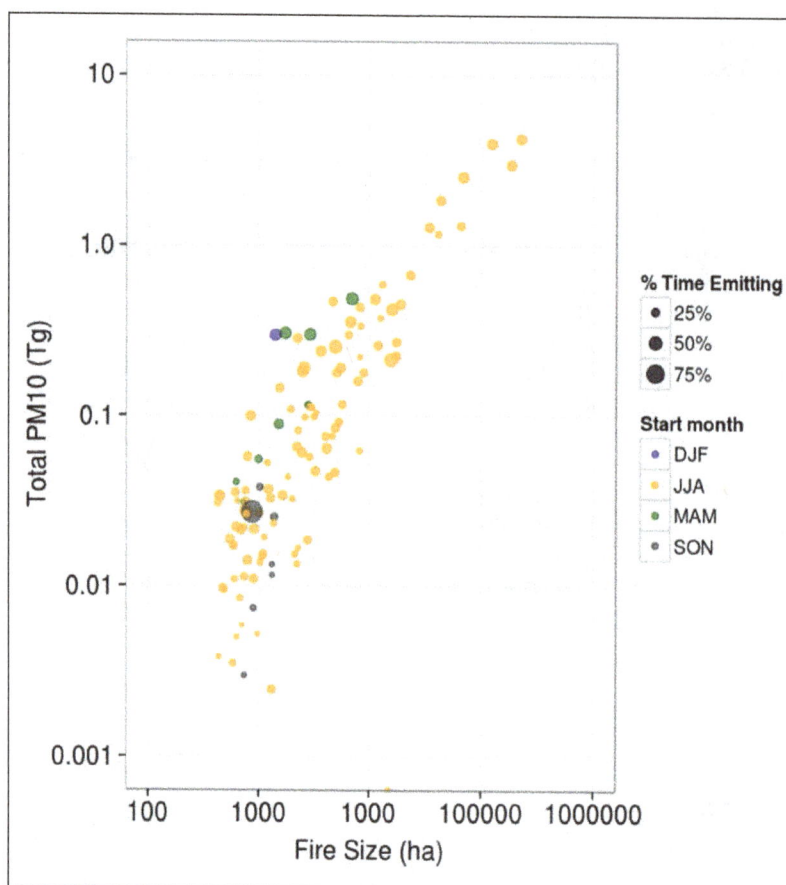

Figure 4: Total PM$_{10}$ emitted from areas burned by wildfire during 2012. The size of the circle indicates the percent of time (from the day after the fire to the end of the year) during which at least one pixel was emitting PM$_{10}$. The color of the circle indicates the start month of the fire. DOI: https://doi.org/10.1525/elementa.185.f4

The general behavior of the post-fire dust event was captured in this modeling effort. WindNinja simulated reasonable PM emission levels when driven by local observed winds, based on comparisons with observed surface concentrations in Boise and Nampa. The dust transport modeling showed relatively good performance for PM surface concentrations although timing was off due to issues with the modeled meteorology. The modeled PM accumulated surface loadings were 32% lower and 53% higher than the observed surface loadings at Boise and Nampa, respectively, and were 64 and 145 times higher than the modeled PM loadings in the absence of PM emissions from the Long Draw fire scar. This case study demonstrates the potential for widespread impacts of post-fire wind erosion as well as the feasibility of accounting for post-fire dust within a regional air quality modeling framework.

3.2. An estimate of annual post-fire PM emissions
In this section we describe the modeled annual emissions from the fire scars in **Figure 1**. The modeling approach is described in Section 2.3. Total emitted PM$_{10}$ was 32.1 Tg (11.7–352 Tg) including 40% or 12.8 Tg (4.68–141 Tg) of which was estimated as PM$_{2.5}$. The uncertainty ranges in parentheses are based on emission model uncertainties described in Section 3.3. These estimates suggest post-fire

landscapes could be the largest source of PM$_{10}$ and PM$_{2.5}$ in the continental U.S. and increase annual total PM$_{10}$ and PM$_{2.5}$ emissions by 171% (62%–1872%) and 231% (84%–2545%), respectively (**Table 2**), during a high fire activity year. PM$_{10}$ emissions among the burned areas ranged from 0.003 to 21.0 kg m^{-2}, with a mean of 3.49 kg m^{-2} and a median of 2.83 kg m^{-2}.

Total emitted PM$_{10}$ scaled with the size of the burned area (**Figure 4**). The largest fires occurred during JJA, which is the driest season in the Great Basin region; as such, the main window for post-fire dust emissions extended from mid-summer through late fall, depending on the timing of winter precipitation. However, previous field studies have shown that, in some cases, post-fire dust sources can also contribute PM after snowmelt the following spring (Wagenbrenner et al., 2013) and in some cases more than one year post-fire, depending on re-establishment of vegetation in the burned area (Miller et al., 2012). Our simulations did not take this post-winter emission period into account.

The percent of time emitting (PTE) was calculated for each fire as the number of hours from the start date of the fire to the end of the year divided by the number of hours during which at least one model grid cell was emitting PM$_{10}$. The PTE ranged from 0.06 to 91%, with a median of 23%. The strongest source areas had PTEs around 30%,

Table 2: Annual PM emission estimates[a]. DOI: https://doi.org/10.1525/elementa.185.t2

Region	PM$_{10}$ Tg	PM$_{2.5}$ Tg
Global[b]		
Mineral dust	1220–1540	308
CONUS[c]		
All sources	**18.8**	**5.54**
Mineral dust (non-agriculture)	9.95	1.15
Agriculture	4.08	0.81
Wildfire smoke	2.29	1.93
Other	2.44	1.65
Basin ecoregions		
Wildfire smoke[d]	0.12	0.10
Post-fire dust[e]	**32.1**	**12.8**
	(11.7–352)	**(4.68–141)**

[a]Basin ecoregions includes the Snake River Plain, Northern Basin and Range, and Central Basin and Range Level III Ecoregions defined in Omernik and Griffith (2014).
[b]Ginoux et al., 2012 and Zender et al., 2003a;
[c]2011 National Emission Inventory (EPA, 2011);
[d]Urbanski et al. 2011;
[e]This study.

which suggests that the largest contributions to total PM emissions were from occasional high-wind events, rather than persistent, moderate-strength winds (**Figure 4**).

3.3. Emission estimate uncertainties

Primary sources of uncertainty in the emission model are the estimated threshold friction velocity and PM$_{10}$ release factor. The values for these parameters were chosen based on the work in Wagenbrenner et al. (2013). Our modeling approach assumes that all fire scars in the U.S. Great Basin emit PM in the same manner as the Jefferson Fire scar emitted PM in Wagenbrenner et al. (2013). Wagenbrenner et al. (2013) was the first (and to our knowledge, still the only) study to report vertical dust fluxes from a fire scar; thus, we believe these are the best estimates for post-fire emission parameters at this time. The values used in the current study are within the range reported in Wagenbrenner et al. (2013).

Wagenbrenner et al. (2013) reported two values for threshold friction velocity: 0.20 m s^{-1} for the fall and 0.55 m s^{-1} for the spring period following snowmelt. A constant value of 0.22 m s^{-1} was used in this work. This value is close to the fall value reported in Wagenbrenner et al. (2013). We think this is appropriate, as the increase in threshold friction velocity at the site in Wagenbrenner et al. (2013) was likely induced by accumulated snow and subsequent snowmelt. The burned areas simulated in this work, however, never experienced snow cover and thus, should be more representative of the fall conditions. Additionally, inspection of the modeled friction velocities in this work indicated that a threshold friction velocity of 0.55 m s^{-1} would have given unrealistic results. For example, the largest friction velocity modeled on the Long Draw fire was 0.54 m s^{-1}. That means that no PM$_{10}$ would have been emitted from that site with a threshold friction

velocity of 0.55 m s^{-1}, which is not correct. Increasing (decreasing) the threshold friction velocity by 10% to 0.24 m s^{-1} (0.20 m s^{-1}) decreases (increases) the estimated total emitted PM$_{10}$ to 27.9 Tg (37.4 Tg).

Wagenbrenner et al. (2013) reported average PM$_{10}$ release rates for three pre-revegetation dust events. The value used in this work (0.0007 m^{-1}) is between the late fall and early spring values reported in that study. Increasing (decreasing) the PM$_{10}$ release factor to the highest (lowest) pre-revegetation event-averaged value of 0.0066 m^{-1} (0.0003 m^{-1}) reported in that study increases (decreases) the estimated total emitted PM$_{10}$ to 303 Tg (13.8 Tg). Assuming the actual PM$_{10}$ release factor was within the pre-revegetation range reported in Wagenbrenner et al. (2013) and the actual threshold friction velocity was within 10% of the fall value reported in the 2013 study, total emitted PM$_{10}$ is estimated as 32.1 Tg with a range of 11.7–352 Tg.

For the annual emissions work, we assume uniform, dry, bare soil with characteristics similar to that of the Jefferson Fire scar within all burn perimeters. Essentially, we assume that the impact of the fire overwhelms differences in soil characteristics, including soil texture and moderate changes in soil moisture. Given the lack of PM measurements from fire scars and the documented severe effect of wildfire on surface soils (Ford and Johnson, 2006; Ravi et al., 2007; Varela et al., 2010), we believe this is a reasonable assumption. We do not account for non-erodible areas, such as rocky outcroppings; however, we assume these areas would also not be susceptible to fire and so would not be found within the burn perimeters investigated in this work (at least not at scales larger than the resolution used to map the fire perimeters). We assume the final fire perimeter exists 24 hours after the start date of the fire based on Wagenbrenner et al. (2013). The Jefferson Fire reported in that study burned 109,000 acres in roughly two days in late July 2010. Dust emissions were visible the day after the fire was contained and persisted for almost a year post-fire. This included emissions after precipitation events and following a 3 month period of snow. Emissions did not tail off in Wagenbrenner et al. (2013) until vegetation began to grow back at the site.

Additional evaluation of the PM emission model is desirable; however, lack of on-site post-fire PM flux measurements prohibits direct evaluation of the emission parameterization. This limits model evaluations to indirect metrics, such as downwind atmospheric PM concentrations, in-situ or remotely-sensed observations of dust plumes, or derived parameters such as AOD. We conducted additional analyses of the largest predicted events using Air Quality System (AQS) PM data, HYSPLIT modeling, and satellite imagery (S4). Results confirm that post-fire dust is a potential source of atmospheric PM, but also highlight the limitations associated with using existing networks and remotely-sensed data for dust source attribution. Ultimately, additional on-site measurements of post-fire PM flux are needed to constrain the emission estimates.

4. Conclusions

The Long Draw case study demonstrated the ability of simulating a large post-fire dust event with a high-resolution dust emission model linked with a regional air quality modeling system. The onset and general dynamics of the observed plume were captured by the modeling framework. The emission model predicted emission hot spots within the fire scar that corresponded to areas of higher friction velocities. It was necessary to drive the emission model with on-site wind observations in order to capture the high winds associated with nearby thunderstorms. These high winds were underestimated by the meteorological model used to drive transport and dispersion of the emitted PM, which resulted in mistimed and underestimated peak model concentrations downwind. Despite these issues with the forecast meteorology used to drive the transport model, modeled 24-hr average concentrations (279 μg m^{-3} at Boise and 561 μg m^{-3} at Nampa) were on the order of the observed 24-hr average concentrations (483 μg m^{-3} at Boise and 385 μg m^{-3} at Nampa).

Our 2012 emissions estimate indicates that Great Basin fires could generate more PM$_{10}$ and PM$_{2.5}$ as post-fire dust than all fires combined in the continental U.S. (CONUS) release in smoke plumes during high fire activity years. While this estimate is large, it seems plausible given the source strength reported for the Jefferson Fire scar in Wagenbrenner et al. (2013) and for the Long Draw Fire scar in this case study. The rangeland landscapes of the Great Basin support less biomass per unit area than other landscapes susceptible to fire (e.g., forests). Additionally, less biomass available for fire consumption results in less PM as smoke. Rangeland fires are also some of the largest wildfires in the U.S. since fire spreads quickly through the dry sagebrush and grass fuels and relatively topographically simple terrain (compared to mountainous areas) of this region. Eight of the wildfires in 2012 were larger than 10,000 ha and the three largest fires burned more than 100,000 ha. Larger burned areas result in more exposed soil and ultimately more PM as dust.

Given the estimated substantial contribution of post-fire landscapes to the PM inventory and the likelihood that these sources will persist and possibly grow in the future, it is important to better quantify and characterize post-fire PM emissions. This study provided a first estimate of annual post-fire emissions for a high fire activity year; however, many assumptions were made due to lack of PM measurements in post-fire environments. This work largely relied on the assumption that fire scars in the Great Basin emit PM in the same manner as the Jefferson Fire scar reported in Wagenbrenner et al. (2013). Future studies should focus on narrowing the uncertainty around emission estimates and better characterizing post-fire dust sources. We project that post-fire dust could be an important PM source in steppe ecoregions around the world, such as the Eurasian Steppe belt and the cold Patagonian steppe.

Supplemental Files

The additional files for this article can be found as follows:
- **Supplemental file 1: Video S1.** Video footage of the August 5, 2012 dust event.

- **Supplemental ile 2: Figure S1.** MODIS Terra Satel-lite imagery of a post-fire dust event.

- **Supplemental ile 3: Figure S2.** Base reflectivity measured by NEXRAD at Boise, ID (KCBX).

- **Supplemental file 4: Text S1.** Analysis with AQS data.

Acknowledgements

Thanks to Charles McHugh of the Missoula Fire Science Laboratory for providing the fire perimeter data and reviewing an early version of the manuscript. Thanks to Shawn Urbanski of the Missoula Fire Sciences Laboratory for providing fire emissions inventory data and reviewing this manuscript. Thanks to the Northwest International Air Quality Environmental Science and Technology (NW-AIRQUEST) Consortium for funding the AIRPACT air quality forecasting system.

Funding Information

This work was supported by the USDA Forest Service Rocky Mountain Research Station.

Competing Interests

The authors have no competing interests to declare.

Contributions

- Performed the dust emission simulations, carried out the analysis, and wrote the paper: NW
- Performed the transport and dispersion simulation: SC
- Conceived of the central idea: BL
- Helped with the analysis: SC, BL
- Edited the paper: SC, BL

References

Balch, J K, Bradley, B A, D'Antonio, C M and **Gomez-Dans, J** 2013 Introduced annual grass increases regional fire activity across the arid western USA (1980–2009). *Glob Change Biol* **19**: 173–183. DOI: https://doi.org/10.1111/gcb.12046

Byun, D and **Schere, K L** 2006 Review of the governing equations, computational algorithms, and other components of the Models-3 Community Multiscale Air Quality (CMAQ) modeling system. *Appl Mech Rev* **59**: 51–77. DOI: https://doi.org/10.1115/1.2128636

Chambers, J C and **MacMahon, J A** 1994 A day in the life of a seed: movements and fates of seeds and their implications for natural and managed systems. *Annu Rev Ecol Sys* **25**: 263–292. DOI: https://doi.org/10.1146/annurev.es.25.110194.001403

Chen, J, Vaughan, J, Avise, J, O'Neill, S and **Lamb, B** 2008 Enhancement and evaluation of the AIRPACT ozone and PM2.5 forecast system for the Pacific

Northwest. *J Geophys Res* **113**. DOI: https://doi.org/10.1029/2007JD009554

Chung, S H, Herron-Thorpe, F L, Lamb, B K, VanReken, T M, Vaughan, J K, et al. 2012 Application of the Wind Erosion Prediction System in the AIR-PACT regional air quality modeling framework. *Trans ASABE* **56**: 625–641. DOI: https://doi.org/10.13031/2013.42674

Deems, J S, Painter, T H, Barsugli, J J, Belnap, J and Udall, B 2013 Combined impacts of current and future dust deposition and regional warming on Colorado River Basin snow dynamics and hydrology. *Hydrol Earth Syst Sci* **17**: 4401–4413. DOI: https://doi.org/10.5194/hess-17-4401-2013

Dockery, D W and Pope III, C A 1994 Acute respiratory effects of particulate air pollution. *Annu Rev Public Health* **15**: 107–132. DOI: https://doi.org/10.1146/annurev.pu.15.050194.000543

Draxler, R R, Gillette, D A, Kirkpatrick, J S and Heller, J 2001 Estimating PM10 air concentrations from dust storms in Iraq, Kuwait and Saudi Arabia. *Atmos Environ* **35**: 4315–4330. DOI: https://doi.org/10.1016/S1352-2310(01)00159-5

Eidenshank, J, Schwind, B, Brewer, K, Zhu, Z, Quayle, B, et al. 2007 A project for monitoring trends in burn severity. *Fire Ecol* **3**: 3–21. DOI: https://doi.org/10.4996/fireecology.0301003

EPA 2011 National Emissions Inventory. http://www.epa.gov/ttn/chief/net/2011inventory.html. Accessed 09.04.15.

Field, J P, Belnap, J, Breshears, D D, Neff, J C, Okin, G S, et al. 2010 The ecology of dust. *Front Ecol Environ* **8**: 423–430. DOI: https://doi.org/10.1890/090050

Ford, P L and Johnson, G V 2006 Effects of dormant- vs. growing-season fire in short grass steppe: biological soil crust and perennial grass responses. *J Arid Environ* **67**: 1–14. DOI: https://doi.org/10.1016/j.jaridenv.2006.01.020

Forthofer J M, Butler B W and Wagenbrenner, N S 2014 A comparison of three approaches for simulating fine-scale surface winds in support of wildland fire management. Part I. Model formulation and comparison against measurements. *Inter J Wild Fire* **23**: 969–981. DOI: https://doi.org/10.1071/WF12089

Ginoux, P, Prospero, J M, Gill, T E, Hsu, N C and Zhao, M 2012 Global-scale attribution of anthropogenic and natural dust sources and their emission rates based on MODIS Deep Blue aerosol products. *Rev Geophys* **50**(3). DOI: https://doi.org/10.1029/2012RG000388

Goosens, D and Offer, Z Y 1997 Aeolian dust erosion on different types of hills in a rocky desert: wind tunnel simulations and field measurements. *J Arid Environ* **37**: 209–229. DOI: https://doi.org/10.1006/jare.1997.0282

Hyslop, N P 2009 Impaired visibility: the air pollution people see. *Atmos Environ* **43**: 182–195. DOI: https://doi.org/10.1016/j.atmosenv.2008.09.067

Jaffe, D, Hafner, W, Chand, D, Westerling, A, Spracklen, D 2008 Interannual variations in PM$_{2.5}$ due to wildfires in the western United States. *Environ Sci Technol* **42**: 2812–2818. DOI: https://doi.org/10.1021/es702755v

Jewell, P W and Nicoll, K 2011 Wind regimes and Aeolian transport in the Great Basin, USA. *Geomorphology* **129**: 1–13. DOI: https://doi.org/10.1016/j.geomorph.2011.01.005

Miller, M E, Bowker, M A, Reynolds, R L and Goldstein, H L 2012 Post-fire land treatments and wind erosion – Lessons from the Milford Flat Fire, UT, USA. *Aeol Res* **7**: 29–44. DOI: https://doi.org/10.1016/j.aeolia.2012.04.001

Okin, G S, Bullard, J E, Reynolds, R L, Ballantine, J-A C, Schepanski, K, et al. 2011 Dust: Small-scale processes with global consequences. *Eos Trans AGU* **92**: 241–242. DOI: https://doi.org/10.1029/2011EO290001

Omernik, J M and Griffith, G E 2014 Ecoregions of the conterminous United States: Evolution of a hierarchical spatial framework. *Environ Manag* **54**: 1249–1266. DOI: https://doi.org/10.1007/s00267-014-0364-1

Prospero, J M, Ginoux, P, Torres, O, Nicholson, S E and Gill, T E 2002 Environmental characterization of global sources of atmospheric soil dust identified with the Nimbus 7 Total Ozone Mapping Spectrometer (TOMS) absorbing aerosol product. *Rev Geophys* **40**. DOI: https://doi.org/10.1029/2000RG000095

Ravi, S, D'Odorico, P, Zobeck, T M, Over, T M and Collins, SL 2007 Feedbacks between fires and wind erosion in heterogeneous arid lands. *J Geophys Res* **G04007**. DOI: https://doi.org/10.1029/2007JG000474

Rhodes, C, Elder, K and Greene, E 2010 The influence of an extensive dust event on snow chemistry in the southern Rocky Mountains. *Arctic Antarctic, Alpine Res* **42**: 98–105. DOI: https://doi.org/10.1657/1938-4246-42.1.98

Sankey, J B, Germino, M J and Glenn, N F 2009 Aeolian sediment transport following wildfire in sagebrush steppe. *J Arid Environ* **73**: 912–919. DOI: https://doi.org/10.1016/j.jaridenv.2009.03.016

Sassen, K, DeMott, P J and Poellet, M R 2003 Saharan dust storms and indirect aerosol effects on clouds: CRYSTAL-FACE results. *Geophys Res Lett* **30**. DOI: https://doi.org/10.1029/2003GL017371

Skamarock, W C, Klemp, J B, Dudhia, J, Gill, D O, Barker, D M, et al. 2008 A description of the Advanced Research WRF Version 3. NCAR technical note NCAR/TN/u2013475+ STR.

Skiles, S M, Painter, T H, Deems, J S, Bryant, A C, et al. 2012 Dust radiative forcing in snow of the Upper Colorado River Basin: 2. Interannual variability in radiative forcing and snowmelt rates. *Water Resour Res* **48**: W07522. DOI: https://doi.org/10.1029/2012WR011986

Summit County 2014 Avalanche danger. http://www.summitdaily.com/news/10886504-113/dust-snow-avalanche-colorado. Accessed 02.04.15.

Tegen, I, Lacis, A A and Fung, I 1996 The influence on climate forcing of mineral aerosols from disturbed soils. *Nature* **380**: 419–422. DOI: https://doi.org/10.1038/380419a0

Urbanski, S P, Hao, W M and Nordgren, B 2011 The wildland fire emission inventory: western United States emission estimates and an evaluation of uncertainty. *Atmos Chem Phys* **11**: 12973–13000. DOI: https://doi.org/10.5194/acp-11-12973-2011

Varela, M E, Benito, E and Keizer, J J 2010 Effects of wildfire and laboratory heating on soil aggregate stability of pine forests in Galicia: the roles of lithology, soil organic matter content and water repellency. *Catena* **83**: 127–134. DOI: https://doi.org/10.1016/j.catena.2010.08.001

Wagenbrenner, N S, Germino, M J, Lamb, B K, Robichaud, P R and Foltz, R B 2013 Wind erosion from a sagebrush steppe burned by wildfire: measurements of PM_{10} and total horizontal sediment flux. *Aeol Res* **10**: 25–36. DOI: https://doi.org/10.1016/j.aeolia.2012.10.003

Wagenbrenner, N S, Forthofer, J M, Lamb, B K, Shannon, K S and Butler, B W 2016 Downscaling surface wind predictions from numerical weather prediction models in complex terrain with WindNinja. *Atmos Chem Phys Discuss.* DOI: https://doi.org/10.5194/acp-2015-761

Zender, C S, Bian, H and Newman, D 2003a Mineral dust entrainment and deposition (DEAD) model: Description and 1990s dust climatology. *J Geophys Res* **107**. DOI: https://doi.org/10.1029/2002JD002775

Zender, C S, Newman, D and Torres, O 2003b Spatial heterogeneity in Aeolian erodibility: Uniform, topographic, geomorphic, and hydrologic hypotheses. *J Geophys Res* **108**. DOI: https://doi.org/10.1029/2002JD003039

8

Ten-year chemical signatures associated with long-range transport observed in the free troposphere over the central North Atlantic

B. Zhang[*,†], R. C. Owen[*,‡], J. A. Perlinger[*], D. Helmig[§], M. Val Martín[||], L. Kramer[*,¶],
L. R. Mazzoleni[*] and C. Mazzoleni[*]

Ten-year observations of trace gases at Pico Mountain Observatory (PMO), a free troposphere site in the central North Atlantic, were classified by transport patterns using the Lagrangian particle dispersion model, FLEXPART. The classification enabled identifying trace gas mixing ratios associated with background air and long- range transport of continental emissions, which were defined as chemical signatures. Comparison between the chemical signatures revealed the impacts of natural and anthropogenic sources, as well as chemical and physical processes during long transport, on air composition in the remote North Atlantic. Transport of North American anthropogenic emissions (NA-Anthro) and summertime wildfire plumes (Fire) significantly enhanced CO and O_3 at PMO. Summertime CO enhancements caused by NA-Anthro were found to have been decreasing by a rate of 0.67 ± 0.60 ppbv/year in the ten-year period, due possibly to reduction of emissions in North America. Downward mixing from the upper troposphere and stratosphere due to the persistent Azores-Bermuda anticyclone causes enhanced O_3 and nitrogen oxides. The $d[O_3]/d[CO]$ value was used to investigate O_3 sources and chemistry in different transport patterns. The transport pattern affected by Fire had the lowest $d[O_3]/d[CO]$, which was likely due to intense CO production and depressed O_3 production in wildfire plumes. Slightly enhanced O_3 and $d[O_3]/d[CO]$ were found in the background air, suggesting that weak downward mixing from the upper troposphere is common at PMO. Enhancements of both butane isomers were found during upslope flow periods, indicating contributions from local sources. The consistent ratio of butane isomers associated with the background air and NA-anthro implies no clear difference in the oxidation rates of the butane isomers during long transport. Based on observed relationships between non-methane hydrocarbons, the averaged photochemical age of the air masses at PMO was estimated to be 11 ± 4 days.

Keywords: Long-range transport patterns to Azores; Long-term observations of ozone and ozone precursors; Non-methane hydrocarbon aging

1. Introduction

The central North Atlantic Ocean is a remote region where transport of pollutants from North America, Europe, and Africa has been observed. The meteorology in this region is usually controlled by a persistent mid-latitude anticyclone, known as the Azores-Bermuda High. Its strength and location change seasonally and interannually, causing

synoptic transport patterns to vary, bringing air masses from different continents. There has been much effort to study the tropospheric chemistry and long-range transport over the North Atlantic. In the 1990s, the North Atlantic Regional Experiment (NARE, (Fehsenfeld et al., 1996)) was conducted to investigate transport of CO and O_3 from North America. It was found that transport of air pollutants has a significant influence on air composition over the North Atlantic Ocean. During the International Consortium for Atmospheric Research on Transport and Transformation (ICARTT, Fehsenfeld et al., 2006) experiment, observations including ground and aircraft measurements were collected over North America, the Azores and Europe to study the impacts of long-range transport crossing the North Atlantic. The more recent BORTAS campaign focused on remarkable long-range transport of wildfire emissions to the North Atlantic (Parrington et al., 2012). The majority of the studies for

* Atmospheric Sciences, Michigan Technological University, Houghton, US

† National Institute of Aerospace, Hampton, US

‡ Environmental Protection Agency, Research Triangle Park, US

§ Institute of Arctic and Alpine Research, University of Colorado, Boulder, US

|| Chemical and Biological Engineering, University of Sheffield, Sheffield, UK

¶ University of Birmingham, Birmingham, UK

Corresponding author: B. Zhang (bzhang3@mtu.edu)

the North Atlantic Ocean focused on transport events or short-term observations. The chemical climatology and the roles of different transport patterns have not been discussed based on long-term observations.

Pico Mountain Observatory (PMO), in the Azores, Portugal, is an island site established in 2001 for the purpose of studying North American outflow (Honrath et al., 2004). The elevation of PMO is 2225 m asl, so the site allows monitoring long-range transport in the free troposphere for most of the time. Tracer gases have been observed at PMO since 2001. Data collected until now provide precious long records that are very useful to investigate different transport patterns and the impacts on air composition over the central North Atlantic.

Studies of chemical transformation during transport can rely on relationships of simultaneously-observed chemicals in downwind regions. $d[O_3]/d[CO]$ (the linear regression slope between O_3 and CO concentrations) has been used as a measure of O_3 enhancement in downwind regions of continental sources (e.g., Parrish et al., 1993). Variations in the slope observed in previous studies reflected differences in air composition and differing O_3 chemistry. A $d[O_3]/d[CO]$ value of 0.3 was found to be a signature value for rural sites in eastern North America, and was concluded to be a result of mixing of fresh pollution emissions and aged air mass (Chin et al., 1994). Cooper et al. (2001) investigated $d[O_3]/d[CO]$ in different airstreams associated with mid-latitude cyclones, and suggested that the slope could be different depending on air composition and photochemistry. Honrath et al. (2004) found a higher summertime $d[O_3]/d[CO]$ (~1.0) in the free troposphere over the central North Atlantic and suggested potential O_3 production during transport from North America. A previous study (Zhang et al., 2014) investigated the evolution of $d[O_3]/d[CO]$ during selected transport events of pollution plumes from North America to PMO in a semi-Lagrangian view, using data from PMO and a chemical transport model. It was found that photochemical CO loss can also contribute to increases in $d[O_3]/d[CO]$ in long-range transport, so the higher $d[O_3]/d[CO]$ at PMO compared to places closer to emission sources (i.e., over the U.S.) were due to not only photochemical ozone production but also CO loss. For the comparisons of $d[O_3]/d[CO]$ at a given location, such as PMO, the differences in CO loss for pollution impacted transport should not be significant. Zhang et al. (2014) estimated 10% and 12% CO loss due to photochemistry for two transport events of polluted plume having significant different OH concentrations and transport pathways. Thus, differences in observed $d[O_3]/d[CO]$ at PMO should mainly reflect varying emission strength and ozone chemistry for the pollution impacted transport.

Non-methane hydrocarbons (NMHC) mainly undergo reactions with OH radicals in the troposphere, and their relative reaction ratios are useful tools to determine the photochemical age of transported air masses. Lighter NMHC usually react more slowly than heavier ones. Consequently, concentrations of lighter NMHC are still measurable after long-range transport. The differences in NMHC decay rates produce specific patterns in the observed mixing ratios, which have been used to reflect photochemical ages of air masses (Mckeen and Liu, 1993). The natural logarithm of [n-butane]/[ethane] versus that of [propane]/[ethane] has been used as a photochemical clock of air masses (Parrish et al., 1992). The butane isomer ratio, [i-butane]/[n-butane], has been used to investigate oxidation pathways by OH, ocean-emitted chlorine atom, and other oxidants (Hopkins et al., 2002).

In this work, a ten-year record of trace gas observations at PMO was used to study relationships between air composition and transport patterns. The objectives of this work are to provide a classification of major transport patterns to PMO that are characterized by different pollution sources and transport pathways, and to investigate the associated chemical signatures. We investigated O_3 and O_3 precursors in spring (April and May), summer (June, July and August) and fall (September and October). We only focused our analyses on months from April to September, because data from PMO were most available during the periods.

2. Methods
2.1. PMO measurements and research
PMO was established in the summer of 2001 on top of Pico Mountain (38.47° N, 28.40° W, 2225 m a.s.l.) in the Azores Islands, Portugal. The elevation of PMO in the central North Atlantic ocean makes it an excellent site for conducting research on gaseous and particle species transported in the free troposphere. Since the establishment of PMO, studies have been conducted to understand the impacts of pollution transport, photochemistry, and meteorology on the air composition at this baseline site. Honrath et al. (2004) and Owen et al. (2006) focused on CO and O_3 observations in the earlier years, and specifically investigated the enhancements due to transport from North America. Both Lapina et al. (2006) and Val Martín et al. (2006) studied impacts of fire emissions on enhancement of air pollutants at PMO. Val Martín et al. (2008b) focused on a three-year period of nitrogen oxides observation and studied the seasonal variations and sources of nitrogen oxides. Helmig et al. (2008) and Honrath et al. (2008) assessed NMHC levels observed at PMO, and used NMHC data to investigate oxidation chemistry and plume aging over the North Atlantic. Aerosol properties and composition have also been measured to investigate aerosol radiative effect and aging during long-range transport (Fialho et al., 2005; Dzepina et al., 2015; China et al., 2015). A recent work by Helmig et al. (2015) discussed observed seasonal cycles and the degree of photochemical aging of NMHC at PMO using data collected during 2004–2014. The study here uses ten years of measurements (2001–2010) of CO, O_3, nitrogen oxides ($NO_x = NO + NO_2$), total reactive nitrogen (NO_y), and NMHC to study the chemical signatures of different transport patterns (data coverage is shown in **Figure 1**). Data used in this work are archived at http://instaar.colorado.edu/pico/pico_archive/default.html. Details about the instrumentation of individual species are summarized in Honrath et al. (2004) (CO and O_3), Val Martín et al. (2006) (NO_x and NO_y), Kleissl et al. (2007) (meteorology), Tanner et al. (2006) (NMHC), and Helmig et al. (2008) (NMHC).

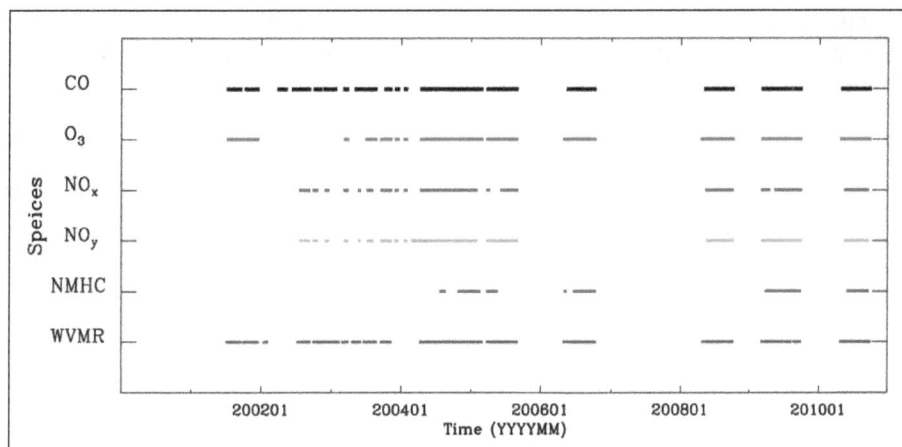

Figure 1: Availability of trace gas observations at PMO during 2001–2010. DOI: https://doi.org/10.1525/elementa.194.f1

2.2. Transport and CO tracer simulations

The Lagrangian particle dispersion model FLEXPART (version 8.2; (Stohl et al., 1998)) was used to build a ten-year archive of transport trajectories for PMO. Backward simulations were launched every three hours from PMO. Meteorology fields from the European Center for Medium-Range Weather Forecasting (ECMWF), featured with 3-hourly temporal resolution, 1° horizontal resolution, and 60 vertical levels, were used to drive simulations for 2001–2004. Meteorology fields of the Global Forecast System (GFS) and its Final Analysis (FNL), featured with the same temporal and horizontal resolutions, but 26 vertical levels, were used to drive FLEXPART for the years 2005–2010. The output was saved in a grid with a horizontal resolution of 1° latitude by 1° longitude, and eleven vertical levels from the surface to 15,000 m a.s.l. A "retroplume" obtained in FLEXPART backward simulations describes upwind distributions of residence time (introduced with details in supplemental information (SI)), which is a useful tool to study transport pathways. Retroplumes are multiplied with CO emission inventories from the Emissions Database for Global Atmospheric Research (EDGAR version 3.2 (Olivier and Berdowski, 2001)) and the Global Fire Emissions Database (GFED v3.1, daily averaged fire emissions (Mu et al., 2011)) to estimate influence from anthropogenic and wildfire sources, respectively. The product is hereafter called "FLEXPART_CO" (examples are given in Figure S1). It has been found that GFED v3.1 misses biomass burning emissions from small fires, but the bias due to such misses was estimated to be only 9% in annual total emissions in the temperate to boreal North America (Randerson et al., 2012). Not all of the missing emissions can be transported to PMO, so the chance of missing "Fire" events should be even lower and is believed to have minor impacts on the analyses in this work.

Uncertainties in transport pathways simulated by FLEXPART can be due to the parameterizations representing temporally and spatially unresolved transport processes (Stohl et al., 2011). In terms of vertical transport processes, boundary layer mixing and updrafts in convection are both considered in FLEXPART. Time-varying PBL height determines the vertical mixing of air parcels. Failure to capture the daily PBL peak can lead to overestimation of surface concentrations. In FLEXPART, PBL height is calculated using the Richardson number concept based on the wind and temperature fields given in the meteorology (Vogelezang and Holtslag, 1996). The meteorology fields used in this work had a temporal resolution of 3 hours, and at least one local afternoon time was included. Subgrid variables, including topography and land use, can also affect vertical mixing. Oversimplified topography may overlook shear stress and sensible heat, creating biases in vertical distributions of tracers. Another highly parameterized sub-grid process is cloud convection. FLEXPART redistributes air parcels vertically in convection-activated grids using the approach of Emanuel and Emanuel and Zivkovic-Rothman (1999), which determines air parcel displacement in up- and down-drafts based on temperature and humidity fields. The convection scheme is not necessarily the same scheme used in driving meteorology. These parameterizations have been fully tested and validated using surface and in situ measurements (Stohl et al., 1998; Brioude et al., 2013); therefore we do not believe that these associated uncertainties are significant enough to change the conjectures and conclusions of this paper.

2.3. Classification of transport patterns

In this section, we list the constraints used to determine different transport patterns. Helmig et al. (2015) also studied long-term transport patterns to Pico in different seasons by using HYSPLIT modeling. They found that fast transport was associated with shorter photochemical ages as indicated by observed ln[propane]/[ethane]. Here, we intend to further explore the transport types. By analyzing FLEXPART products and meteorological data, we identified several major transport patterns in the ten-year record of trace gas observations. The constraints are discussed in the following and summarized in **Table 1**.

2.3.1. Air mass origin

The primary consideration in categorizing the transport is where it originates. We considered transports originating from North America, Europe and Africa, downward mixing from the upper troposphere and lower stratosphere (UTLS), and North Atlantic background air. Transport origin was determined based on the distributions of upwind residence

Table 1: Constraints to define transport patterns, abbreviations of the transport patterns, and the estimated occurrence frequencies in spring (April and May), summer (June, July, and August), and fall (September and October). DOI: https://doi.org/10.1525/elementa.194.t1

Transport Patterns	Abbrev.	Constraint*	Occurrence Frequency ** %		
			Spring	Summer	Fall
North American flow	NA	RT over North America in 0–5 km a.s.l. greater than 50% for at least 1 day	40	30	32
NA affected by anthropogenic emissions	NA-anthro	North America anthropogenic FLEXPART_CO contribution greater than 15 ppbv, fire <15 ppbv	16	15	13
NA-anthro lifted during transport	NA-anthro-lifted	1. Same as NA-anthro	4.0	2.7	2.5
		2. RT in a vertical range 0–2.5 km reduced 30% within 24 hour during 0–15 days upwind			
NA-anthro within or close to MBL	NA-anthro-low	1. Same as NA-anthro	3.1	1.7	<1.0
		2. RT within 0–2.5 km greater than 40% from 0–5 days upwind			
NA-anthro quick transport	NA-anthro-young	1. Same as NA-anthro	6.7	4.3	4.0
		2. FLEXPART_CO age from anthropogenic emissions less than 7 days			
NA-anthro aged transport	NA-anthro-aged	1. Same as NA-anthro	6.2	5.3	4.5
		2. FLEXPART_CO age from anthropogenic emissions greater than 10 days			
Wildfire affected	Fire	Wildfire FLEXPART_CO contribution greater than 15 ppbv, Anthropogenic <15 ppbv	<1.0	7.3	2.6
Fire quick transport	Fire-young	1. Same as Fire	<1.0	2.5	1.0
		2. FLEXPART_CO age from wildfire emissions less than 7 days			
Fire slow transport	Fire-aged	1. Same as Fire	<1.0	3.0	1.3
		2. FLEXPART_CO age from wildfire emissions greater than 10 days			
European flow	EU	RT over Europe greater than 50% for at least 1 day	1.8	<1.0	<1.0
African flow	AF	RT over Afreica greater than 50% for at least 1 day	<1.0	<1.0	2.6
North Atlantic free troposphere background	NATL	RT over North Atlantic in 0–5 km a.s.l. greater than 70% through 0–10 days upwind	17	19	8.3
Downward mixing from UTLS	Upper	RT less than 40% in 0–5 km through 2–12 days upwind	2.1	2.3	4.1
Upslope flow	Upslope	Simulated PBL height greater than altitude of Pico or DSL less than PBL height	24	13	15

* RT = residence time of FLEXPART retroplumes.
** Spring = April–May; Summer = June–August; Fall = September–October Unclassified time percentages are 22%, 39%, and 40 % for spring, summer, and fall, respectively.

time (RT) over the North Atlantic Ocean and the surrounding continents. **Figure 2a** provides the regional fraction of upwind RT averaged for all the retroplumes simulated for the ten years, which shows the averaged RT fractions in different regions at upwind days. During 0–3 days upwind, retroplume RT mainly distribute over the North Atlantic. The North American RT contribution started to appear between 1–2 days upwind, indicating the extreme shortest transport time from the North American continent to PMO. The average North American RT reached its maximum 7–8

days upwind at 22%. We defined the retroplumes that had greater than 50% RT distributed in the low-middle troposphere (0–5 km a.s.l.) over North America for a period longer than one day as "North American transport", with an assigned abbreviation of "NA". The altitude limitation was applied for targeting ground emissions because air originating from the upper troposphere would certainly have a different chemical composition. The upper limit of 5 km was set much higher than the typical boundary layer height to take into account surface air masses lofted by convec-

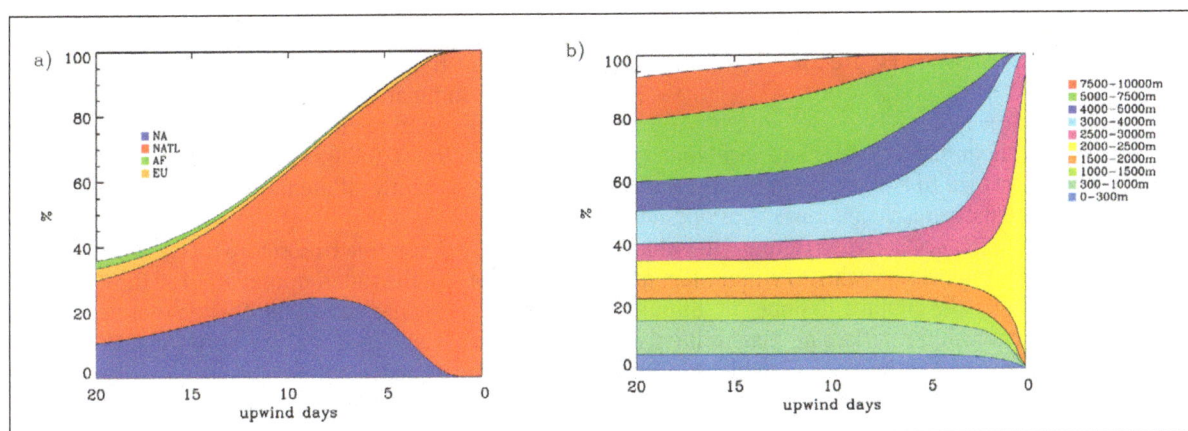

Figure 2: Averaged upwind residence time distributions over different geographical regions **(a)** and vertical levels (%) **(b)** for 3 hourly retroplumes in 2010 simulated by FLEXPART. Note that not all transport sources and heights are included in the legends, and therefore total percentages are less than 100% in the upwind days. NA: North America; NATL: North Atlantic; AF: Africa; EU: Europe. DOI: https://doi.org/10.1525/elementa.194.f2

tion over continents. Using even higher levels (i.e., 7.5 km or 10 km) would have certainly captured deep convection but would also have increased the probability of including free troposphere air that was decoupled from surface emissions. The North Atlantic free troposphere background, "NATL", was constrained by a higher percentage (70%) of RT and a longer time limit (10 days) within 0–5 km a.s.l. over the North Atlantic in order to reduce the probability of overlapping with air masses originating from the continents. It is worth mentioning that we did not exhaustively classify all observations into a specific transport pattern. Instead, we determined the observations with clear transport characteristics and paid less attention to transport that had multiple origins and complicated pathways.

2.3.2. Impact of emission type and air mass age
FLEXPART_CO was used in this study to determine the periods when PMO was affected by two types of pollution emissions, i.e., anthropogenic and wildfire emissions (see details in SI). We used a FLEXPART_CO mixing ratio enhancement of 15 ppbv (compared to 80–100 ppbv as a typical background CO at PMO) as the cutoff in determining the periods affected by fresh emissions. For example, if FLEXPART_CO from biomass burning emission was greater than 15 ppbv for a particular time, then the period was considered as a fire event. The 15 ppbv used here equals the standard deviation of observed half-hourly CO for the ten-year period. We used the same anthropogenic emission inventory in FLEXPART for all the years in order to keep a consistent threshold for identification of affected periods, although the actual emissions of CO and reactive nitrogen oxides over North America have decreased in recent decades. The simulated FLEXPART_CO shows a wide range from 3 ppbv (small emission influence) to more than 150 ppbv (large emission influence). An example of FLEXPART_CO originating from different continents and types of emission is given in Figure SI1a.

FLEXPART_CO was also used to estimate the age of continental emissions after long-range transport by averaging the time elapsed between CO emission pick-up and the arrival at PMO in the model (see SI for details). Honrath et al. (2004) found that transport from North America that

causes significant changes in CO at PMO usually takes 5 to 7 days. Thus, we defined the transport retroplumes that had average FLEXPART_CO ages less than 7 days as short transport and those that were older than 10 days as aged transport. A one-month FLEXPART_CO age spectrum time series is shown in Figure SI1b for demonstration purpose. This constraint can be combined with the determined air mass origin to further distinguish transport patterns. For example, aged transport of North American anthropogenic emissions is referred to as "NA-anthro-aged". In previous work, this method was used to assess the influence of wildfire emission and transport pathway (Val Martín et al., 2008a; Dzepina et al., 2015; China et al., 2015).

2.3.3. Transport height
We also defined a few subcategories determined by transport at different heights or that experienced RT redistribution in height. Kleissl et al. (2007) reported that the MBL height at Pico was typically 800–1500 m simulated by the Global Data Assimilation System (GDAS, 2014). We studied the GDAS data again for the ten-year period and found the MBL occasionally reduced to 500 m or reached up to 2000 m. For such estimates, we defined a "low" transport pattern which had over 40% of its RT below 2.5 km during the transport period. This height was chosen because it was the first vertical level in the configured FLEXPART that reaches above 2 km. In order to capture the transport scenario, in which pollution emission was lifted by deep convection or by warm conveyor belts, we defined the "lifted" scenario when the RT was reduced by at least 30% in the 0–2.5 km within 24 hours during the simulated upwind transport. The fraction of RT change during lifting refers to the findings in a climatology study of warm conveyor belts over 15 years (Eckhardt et al., 2004). The last transport scenario defined by height was downward mixing, which brought air from the UTLS. Retroplumes having less than 40% of their RT within 0–5 km during 2–12 upwind days were defined as downward mixing having the abbreviation "Upper". In this definition, most of the air mass subsides from above 5 km to the level of PMO in the last few days of transport. The last two days were not considered because 100% of the RT should be at the

level of PMO (2–2.5 km) when the simulated air parcels approached PMO as shown in **Figure 2b**.

2.3.4. Potential upslope flow

Local emissions from human activities and vegetation on Pico Island can be carried to PMO by upslope flow, which may lead to different chemical signatures. Upslope flow occurs through mechanically-forced lifting, in which strong synoptic winds blowing against the mountain are deflected upwards by the mountain slope, or through buoyant forcing, in which the surface air mass is lifted as a result of solar heating and release of latent heat. The frequency of upslope flow was found to be less than 20% at PMO from May to September by a micrometeorological study conducted during summertime in 2004 and 2005 (Kleissl et al., 2007). In our study, we examined potential upslope flow events for the ten-year period. For mechanically lifted upslope flow, the height of a dividing streamline (DSL) for the flow climbing upwards to PMO was calculated by using the method described in Sheppard (1956) and the wind speed profiles in the ECMWF and GFS/FNL meteorological datasets. DSL heights indicate the lowest level of air that can reach the top of Pico Mountain after consumption of the mechanical energy and, if any, latent heat. If the calculated DSL was lower than PBL height, upslope flow to PMO was considered as having occurred. We also considered deep convection caused by strong solar heating over the Island. We obtained PBL heights at PMO from GDAS ($1°$ latitude by $1°$ longitude resolution) for the ten-year period and compared those with the height of PMO. When estimated PBL height exceeded the altitude of PMO, upslope flow was also considered as having occurred. In addition, we compared water vapor mixing ratios (WVMR) at PMO to the WVMR interpolated at the PMO elevation in the radiosonde measurements at Terceira (http://esrl.noaa.gov/raobs/). Similar to the approach used by Ambrose et al. (2011), we calculated the seasonal means of WVMR at 2225 m at the radiosonde site, Terceira. We determined, for each season, the driest portion of the PMO WVMR that derived equivalent seasonal means to the means at 2225 m above Terceira (within 0.01 g/kg). These subsets were considered as WVMR of free troposphere. The remaining data, and associated periods, were considered as anabatic upslope flows for each season, in addition to the two upslope flow situations determined above.

3. Results and discussions

3.1. Transport patterns and associated chemical signatures

The estimated occurrence frequencies of determined transport patterns are summarized in **Table 1**. The occurrence frequencies and chemical signatures for each transport pattern were calculated independently by applying the listed constraint(s) to the entire ten-year observations and model results. Thus, some categories are not exclusive. The largest data overlap between two transport patterns was less than 10% for NATL and Upper. The results are discussed together with observed chemical signatures in this section. We focus on the transport patterns with

occurrence frequencies greater than 1%. The chemical signatures of trace gases determined for each transport pattern are summarized in **Table 2**. For the comparisons of chemical signatures in the following discussions, the Welch's t-test ($\alpha = 0.05$) was used to determine if the values were significantly different from each other.

3.1.1. North Atlantic free troposphere background

Figure 3a shows the superimposed RT of all the FLEXPART retroplumes classified as NATL in the ten-year period. The RT of NATL concentrated over the North Atlantic and was mostly isolated from continental emissions due to the constraints applied in Section 2. The occurrence frequencies of NATL were 17%, 19%, and 8.3% in spring, summer, and fall, respectively. The highest frequency in summer was likely due to the strengthened Azores-Bermuda anticyclone over the North Atlantic, which tends to constrain air to the central North Atlantic.

Due to the constraint implemented to classify NATL, we viewed this transport pattern as a flow of aged air that has been isolated from direct transport of continental sources for a long enough time such that the chemical signatures represent the background air composition of PMO. The average observed CO during NATL periods from the ten-year record were 113, 83, and 82 ppbv for spring, summer, and fall, respectively (**Table 2**). CO has a tropospheric lifetime of a few months, so the seasonal variations reflect the northern hemisphere seasonal CO background, which is driven by enhanced emission in cold seasons and accelerated oxidation in summer (Logan et al., 1981). Such maxima of CO in spring have been found at other remote sites (e.g., Macdonald et al. (2011) and for the Northern Hemisphere in general (Worden et al., 2013). The average O_3 of NATL were 39, 31, and 33 ppbv for spring, summer, and fall, respectively. These values are similar to the previously reported O_3 background at Bermuda (30–40 ppbv (Li et al., 2002)), but lower than the eastern U.S. mean O_3 in spring (47 ppbv) and summer (46 ppbv) (Cooper et al., 2012), reflecting an ozone destruction in the remote North Atlantic background air. The relatively lower O_3 in summer and fall, compared with spring, was likely the result of increased photochemical destruction of O_3 due to conversion of O_3 to OH when water vapor was more abundant. A springtime O_3 maximum has been found in multiple locations in the free troposphere. The causes of the maximum seem location dependent, and the roles of photochemistry, continental emissions, and transport from the stratosphere have been debated (Monks et al., 2000). Dry air conditions and weak radiation in early spring suppress HO_x production, which may lead to net ozone production in the low NO_x free troposphere (Yienger et al., 1999; Carpenter et al., 2000). The accumulation of NO_y and hydrocarbons due to longer lifetime in winter (e.g. PAN) has also been suggested as reasons for springtime O_3 peaks (e.g., Blake et al. (2003); Kramer et al. (2015)). Many studies have also reported impacts of downward transport from the stratosphere, e.g., Beekmann et al. (1994). Over the North Atlantic, Honrath et al. (1996) found that transport from the Arctic significantly increased ozone precursors during winter and spring.

In contrast to CO and O_3, NO_x and NO_y for NATL had maxima in summer at PMO. We speculated that the strengthened Azores-Bermuda anticyclone and downward transport during summer could be the reason for summertime maxima of NO_y. NO_y in the middle and upper troposphere over oceans was found to be higher than within the marine boundary layer due to less efficient removal processes, including wet and dry deposition (Hübler et al., 1992). Talbot et al. (1999) investigated the NO_y budget during the NASA SONEX aircraft mission in the region and reported a median value of 0.17 for PAN/NO_y in the middle and upper troposphere. The downward transport could have brought NO_y to the lower troposphere in summer, and the thermal dissociation of PAN during subsidence could lead to increased NO_x. The coverage of NO_x and NO_y measurements was not as good as that of CO and O_3 due to technical difficulties at the site (**Figure 1**). Further analysis and chemistry modeling work need to be done to identify the quantification of background NO_x and NO_y. The characteristics of NATL downward transport will be discussed further in Section 3.2.1.

Table 2: Statistics of carbon monoxide (CO), ozone (O_3), nitrogen oxides (NO_x and NO_y), and NMHCs (ethane, n-propane, and n-butane) for the determined transport patterns in three seasons. The statistics are given in the form of mean ± standard deviation (number of observations). Values in bold font indicate the highest concentrations across all the transport patterns (highest in each row). Results of transport patterns that occur at a lower frequency than 1% (see **Table 1**) are not provided. DOI: https://doi.org/10.1525/elementa.194.t2

Species	Season	NATL	NA-anthro	NA-anthro-aged	NA-anthro-young	NA-anthro-lifted	NA-anthro-low
CO (ppbv)	Spring[1]	113±14(2160)	128±17(782)	121±12(630)	**131±17(753)**	128±19(580)	127±13(271)
	Summer	83±15(5284)	97±18(4077)	92±19(1580)	100±19(1042)	101±18(841)	100±17(454)
	Fall	82±15(1002)	105±15(1659)	100±15(554)	**111±14(590)**	108±14(446)	N.A.
O_3 (ppbv)	Spring	39±11(2533)	49±10(2323)	47±10(996)	50±10(762)	49±10(672)	44±9(311)
	Summer	31±11(6359)	41±12(4565)	40±14(1818)	41±12(1149)	42±12(925)	34±17(543)
	Fall	33±8(1201)	42±8(1928)	41±8(690)	45±7(640)	44±7(466)	N.A.
NO_x (pptv)	Spring	22±15(299)	48±39(206)	40±41(70)	52±36(79)	50±36(64)	N.A.
	Summer	51±51(1584)	52±43(1232)	51±38(465)	56±54(297)	52±37(210)	50±36(147)
	Fall	36±30(269)	39±23(324)	33±13(160)	N.A.[2]	40±20(90)	N.A.
NO_y (pptv)	Spring	132±121(621)	344±343(470)	276±176(167)	394±381(187)	**401±463(142)**	153±180(56)
	Summer	201±157(1651)	290±313(1266)	264±172(488)	277±334(301)	**314±296(236)**	157±127(147)
	Fall	111±109(275)	205±127(352)	182±123(161)	219±137(62)	218±90(101)	N.A.
Ethane (pptv)	Spring	1109±226(15)	1460±452(62)	1329±362(11)	1393±412(23)	**1483±490(18)**	N.A.
	Summer	523±224(66)	710±203(66)	587±187(27)	**792±199(20)**	N.A.	724±251(20)
	Fall	549±185(13)	878±192(27)	N.A.	**887±215(20)**	868±127(12)	N.A.
n-Propane (pptv)	Spring	130±50(35)	281±143(62)	173±82(11)	312±139(23)	**321±165(18)**	N.A.
	Summer	31±26(67)	82±55(68)	50±42(28)	114±55(21)	92±66(11)	N.A.
	Fall	90±82(26)	146±76(26)	N.A.	**148±87(19)**	137±53(12)	N.A.
n-Butane (pptv)	Spring	26±15(36)	60±43(63)	38±19(12)	61±35(23)	**65±46(19)**	N.A.
	Summer	14±12(56)	18±14(63)	15±11(24)	**23±14(21)**	**23±17(11)**	N.A.
	Fall	20±28(26)	23±15(27)	N.A.	23±16(20)	19±11(12)	N.A.
i-Butane (pptv)	Spring	11±6(36)	28±20(63)	15±11(12)	30±18(23)	**31±21(19)**	N.A.
	Summer	5±4(39)	9±6(57)	7±5(17)	10±6(21)	N.A.	8±6(17)
	Fall	N.A.	N.A.	N.A.	N.A.	N.A.	N.A.
WVMR (g/kg)	Spring	5.54±1.91(2394)	4.67±2.03(4512)	4.19±2.02(960)	5.26±1.94(797)	**4.66±1.66(729)**	6.37±1.34(340)
	Summer	6.98±2.61(5506)	6.71±2.81(4521)	6.00±2.87(1654)	7.21±2.91(1238)	6.56±2.69(914)	8.48±1.93(547)
	Fall	8.24±2.40(1069)	6.23±2.75(1877)	6.87±2.72(613)	5.46±2.61(667)	5.66±2.66(529)	8.22±1.77(252)

contd.

Species	Season	Fire	Fire-aged	Fire-young	Upper	Upslope
CO (ppbv)	Spring	N.A.	N.A.	N.A.	123±11(206)	119±16(4141)
	Summer	**108±25(2296)**	107±20(885)	**108±32(538)**	92±16(655)	86±19(4642)
	Fall	N.A.	N.A.	N.A.	96±18(607)	86±17(2165)
O_3 (ppbv)	Spring	N.A.	N.A.	N.A.	**58±9(329)**	39±9(4787)
	Summer	47±13(2769)	45±11(968)	48±15(833)	**54±11(565)**	27±9(4650)
	Fall	42±12(63)	N.A.	N.A.	**49±8(635)**	32±7(2445)
NO_x (pptv)	Spring	N.A.	N.A.	N.A.	**63±47(82)**	37±38(591)
	Summer	59±35(542)	55±30(218)	60±44(137)	**66±65(129)**	48±52(1112)
	Fall	N.A.	N.A.	N.A.	**53±26(112)**	37±37(496)
NO_y (pptv)	Spring	N.A.	N.A.	N.A.	271±91(52)	150±183(1107)
	Summer	303±188(631)	265±152(259)	339±208(153)	234±69(127)	115±168(1106)
	Fall	N.A.	N.A.	N.A.	**229±85(113)**	73±79(492)
Ethane (pptv)	Spring	N.A.	N.A.	N.A.	N.A.	1256±320(174)
	Summer	773±225(19)	N.A.	N.A.	713±233(24)	626±229(74)
	Fall	N.A.	N.A.	N.A.	863±102(22)	672±288(37)
n-Propane (pptv)	Spring	N.A.	N.A.	N.A.	N.A.	196±102(175)
	Summer	**136±205(36)**	N.A.	N.A.	N.A.	58±46(75)
	Fall	N.A.	N.A.	N.A.	120±48(25)	90±58(59)
n-Butane (pptv)	Spring	N.A.	N.A.	N.A.	N.A.	42±28(170)
	Summer	22±25(34)	N.A.	N.A.	16±8(17)	16±11(68)
	Fall	N.A.	N.A.	N.A.	21±16(25)	28±21(58)
i-Butane (pptv)	Spring	N.A.	N.A.	N.A.	14±6(11)	19±14(170)
	Summer	**15±21(18)**	N.A.	N.A.	N.A.	11±15(67)
	Fall	N.A.	N.A.	N.A.	N.A.	12±8(58)
WVMR (g/kg)	Spring	5.78±2.38(80)	5.44±2.78(48)	6.29±2.48(32)	1.60±1.70(291)	**6.52±1.17(4141)**
	Summer	5.20±3.10(2214)	5.82±2.82(928)	4.70±3.13(801)	2.30±2.34(569)	**9.67±1.46(4642)**
	Fall	6.39±3.80(70)	3.02±3.62(27)	9.45±0.82(31)	3.06±2.83(713)	**9.48±1.50(2165)**

1. Spring: April and May; Summer: June, July and August; Fall: September, October.
2. N.A.: Results are not available because of lack of data and/or occurrence of the transport type is not frequent.

3.1.2. Transport affected by North American anthropogenic emissions

The transport affected by North American anthropogenic emission ("NA-anthro") had seasonal occurrence frequencies ranging from 13–16%. The integrated RT plot for NA-anthro (**Figure 3b**) shows a clear contribution from the northeastern states of the U.S. and a northwards curved transport pathway similarly to what has been found in previous North American transport events (Owen et al., 2006; Helmig et al., 2008, 2015). In the NA-anthro subsets, lifted transport ("NA-anthro-lifted") ranged from 2.5–4.0%, whereas the low altitude transport ("NA-anthro-low") had lower frequencies (1.0–3.1%). In terms of pollution age, transport older than 10 days ("NA- anthro-aged") occurred 4.5–6.2% of the time, whereas transport shorter than 7 days ("NA-anthro-young") ranged from 4.0 to 6.7%.

As a primary indicator of air pollution, CO enhancement (the amount that exceeds NATL) reflects the extent of pollution influence. North American anthropogenic emissions were estimated to add 15 ppbv (13%), 14 ppbv (17%), and 23 ppbv (28%) of CO to the background (NATL) in spring, summer, and fall, respectively (see **Table 2**). The larger percentiles in summer and fall were partly due to the low background of CO in the two seasons. In the NA-anthro sub-categories, CO for NA-anthro-low was statistically lower than the others; there was no significant difference among NA-anthro-lifted, -young, and -aged. The characteristic of NA-anthro-low was characterized by polluted air diluted by mixing with clean marine air within the MBL. Such findings suggest the mixing was more effective in reducing CO than photochemical aging in the transport from North America to PMO.

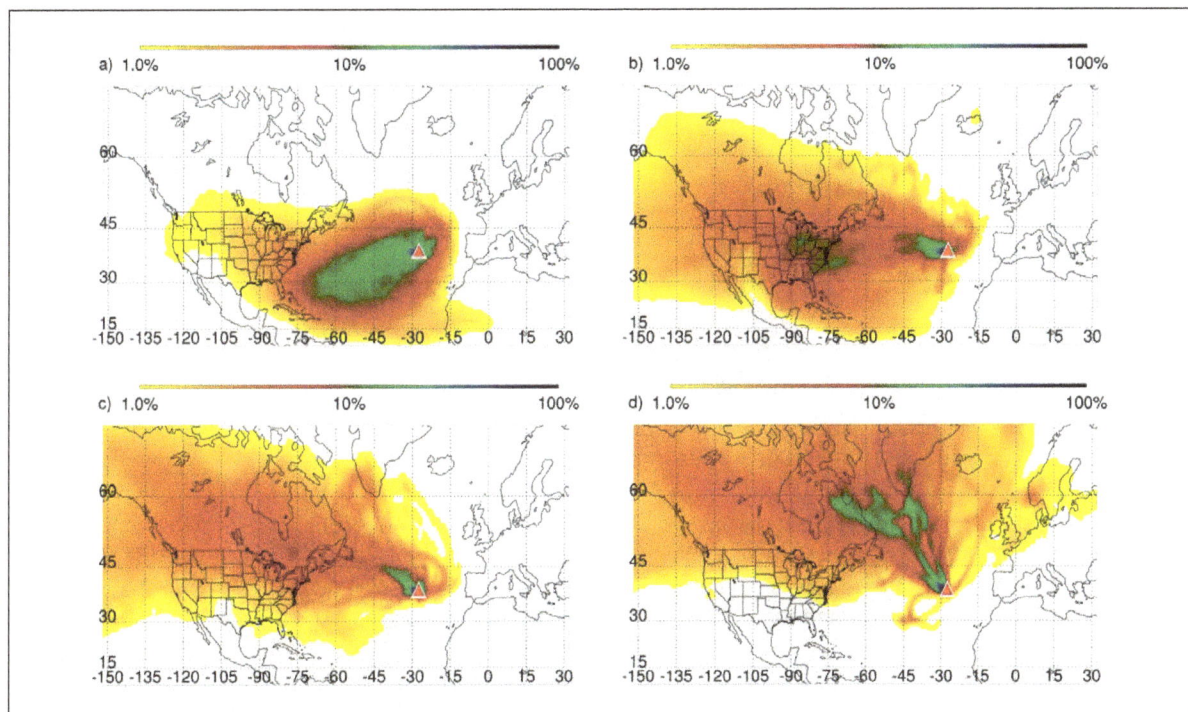

Figure 3: FLEXPART simulated summertime column integrated residence time for determined transport patterns: **(a)** NATL, **(b)** NA-anthro, **(c)** Fire, and **(d)** Upper in 2001–2010. The color coding indicates the relative abundance of residence time. Abbreviations of the transport patterns are given in Table 1. DOI: https://doi.org/10.1525/elementa.194.f3

Ozone enhancements for the NA-anthro were 10 ppbv (25%), 10 ppbv (32%), and ppbv (30%) in spring, summer, and fall, respectively. The enhanced percentiles are different because the background levels were lower in summer and fall. As shown in **Table 2**, the mean O_3 mixing ratios in the NA-anthro sub-categories were all significantly different from each other, possibly suggesting O_3 levels were sensitive to transport heights and distances. The mean O_3 level of NA-anthro-low was lower than NA-anthro-lifted, which is a result of more efficient O_3 removal in low altitude transport and potential O_3 production by PAN decomposition in the subsidence stage of the lifted transport (Hudman et al., 2004; Zhang et al., 2014). The lower O_3 of NA-anthro-aged was generally due to the longer transport time and lack of O_3 precursor sources over the North Atlantic.

NO_x and NO_y levels at PMO were found to be driven by transport of pollution emissions, which was also discussed for a shorter measurement period by Val Martín et al. (2008a). Val Martín et al. (2008a) pointed out that summertime peaks of NO_y were caused by efficient transport of boreal wildfire emissions and pollution from eastern North America. Here, we find that the enhancements in NO_y caused by anthropogenic emissions were the highest in spring when the analysis was extended to the ten-year observation period. The springtime enhancements could be the result of less efficient wet deposition of NO_y and faster zonal transport in mid-latitude spring. The colder spring conditions could lead to a large contribution of PAN to NO_y in export from continental regions (Bey et al., 2001; Koike et al., 2003; Li et al., 2004), which

reduces the conversion from NO_y to HNO_3, and the following loss through wet deposition (Moxim et al., 1996). For the same reasons, NO_y for NA-anthro-lifted was much higher than that for NA-anthro-low due to accumulation of PAN and less efficient loss of NO_y at higher altitude. This transformation of NO_y has been studied in detail for two transport events to PMO by Zhang et al. (2014). NO_y of NA-anthro-low was lower than the other transport patterns, reflecting efficient removal of NO_y in the moist marine boundary layer air.

Maxima of NMHC were found during NA-anthro transport and its subcategories in spring and fall. During NA-anthro-lifted and NA-anthro-young transport, NMHC were higher than that of aged and low transport, suggesting that NMHC loss during transport is due to a combination of mixing with the background and chemical sinks through oxidation reactions with OH.

3.1.3. Trend analyses of CO and O_3 enhancements associated with NA-anthro

In order to understand the impacts of North American anthropogenic emissions in light of recent reductions of emissions, we calculated the annual variability in trace gas enhancements as shown in **Figure 4**. This analysis was only conducted for summer months, because CO observations were not adequate in spring and fall for a couple of years. As shown in **Figure 4**, we find a decreasing trend of CO enhancements (-0.67 ± 0.60 ppbv/yr as mean $\pm$ one-sigma error, p-value = 0.15) and no clear change of enhancements in O_3 (-0.04 ± 0.44 ppbv/yr, p-value = 0.47). These findings add on recent studies about CO

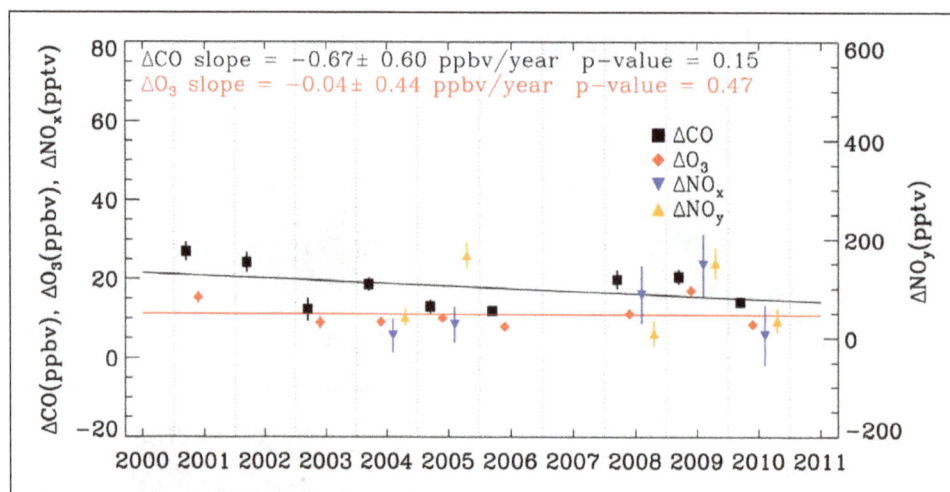

Figure 4: Trends of trace gas enhancements in transported North American anthropogenic emissions (NA-anthro minus NATL) in summer. The symbols represent the means of enhancements with vertical bars showing the 95% confidence intervals. The slopes were obtained from linear regression analyses for CO (black) and O_3 (red) enhancements. Slope ranges and p values are also shown in the figure. DOI: https://doi.org/10.1525/elementa.194.f4

trends in North American and globally. After the Environmental Protection Agency amended the Clean Air Act in 1990, air pollution emissions from the U.S. decreased (EPA, 2016). Globally-averaged CO has been found to have decreased at a rate of 1%/yr from 2001 to 2011 according to satellite observations (Worden et al., 2013). Trends of background O_3 in northern mid-latitudes have changed from positive to negative slopes around the year of 2000 (Parrish et al., 2012). In the eastern U.S., O_3 was found to have decreased at rates of –0.03 and –0.45 ppbv/year in spring and summer, respectively (Cooper et al., 2014). Previous work by Kumar et al. (2013) reported that both anthropogenic CO and NO_x emissions over the U.S. have decreased significantly since the beginning of the twenty-first century. The same study also found decreasing trends in CO (–0.31 ppbv/year) and O_3 (–0.21 ppbv/year) at PMO using a harmonic regression approach. The unclear trend in ozone enhancements in NA-anthro suggests that the decline in North American anthropogenic emissions was perhaps not significant enough to change ozone trends over the central North Atlantic. Instead, the decreasing trend found previously at PMO by Kumar et al. (2013) was probably a large regional global trend through North America to Europe (Cooper et al., 2014). Trend analyses for NO_x and NO_y are not shown here because the amount of qualified data for these species was not enough for an acceptable statistical test.

3.1.4. Transport affected by wildfire emissions
Transport affected by wildfire emissions ("Fire") occurred most frequently during summer when wildfires are more active. This transport pattern usually caused drastic changes in pollution levels at PMO due to the great amount of combustion emissions and fast transport in the free troposphere following the strong lifting created by large fires (Val Martín et al., 2006). The superimposed RT (**Figure 3c**) shows the main pathways from Canada to PMO, which are longer pathways from higher latitude than those of NA-anthro (**Figure 3b**). This region is fre-

quently dominated by persistent high pressure systems, which leads to dry climates in Canada and Alaska and increases the likelihood of wildfire occurrence in the regions (Macias Fauria and Johnson, 2006).

Boreal wildfires have been found to emit CO more intensively than anthropogenic activities and to cause significant O_3 production due to the large amount of CO emitted from large areas of vegetation burned in a short period of time (e.g., Parrington et al. (2013)). As shown in **Table 2**, the Fire transport pattern had an averaged CO mixing ratio of 108 ppbv, which was enhanced by 25 ppbv (30%) compared with the background (NATL). The enhancement was generally stronger than that caused by NA-anthro (14 ppbv, 17%) in summer. CO levels reported to represent fire signatures were different from those reported in previous studies (Val Martín et al., 2006, 2008a) because different methods were used in determining fire events. The average CO for fire events reported by Val Martín et al. (2006) was 139 ppbv, using a CO cutoff value of 110 ppbv in determining the events in 2004. Val Martín et al. (2008a) reported average CO of 124 ppbv and 105 ppbv, respectively, for fire affected periods in 2004 and 2005 at PMO using 75th percentiles of the FLEXPART_CO from fire emissions as the cutoff (compared to a fixed 15 ppbv of FLEXPART_CO used here). In spite of the differences, all studies indicated significant CO enhancements in transport of fire emissions. The enhancement of O_3 in fire plumes (16 ppbv, 50%) was also stronger than that in NA-anthro (10 ppbv, 32%) in summer. These facts suggest that fires act as a competing air pollution source to anthropogenic sources in the central North Atlantic, even though the frequency of Fire (7.3%) is about half of the frequency of NA-anthro in summer.

Vegetation fuel nitrogen can lead to high NO_x mixing ratios in fire plumes (Andreae and Merlet, 2001). The mixing ratios of NO_x and NO_y of Fire were higher than NATL by 9 pptv and 138 pptv, respectively, and these enhancements were in agreement with the values (9–30 pptv for NO_x, and 117–175 pptv for NO_y) reported by Val Martín et al. (2008a). The large enhancement of NO_y after long-rang transport could be due to chemical transformation of the nitrogen oxides in fire plumes. Through aircraft measurements, Alvarado et al. (2010) observed that 40% of the initial NO_x emissions were converted to PAN within a few hours in boreal biomass burning. Fast conversion of NO_x to NO_y may promote long-range transport of NO_y produced by wildfires.

3.1.5. Transport from the upper troposphere and lower stratosphere

Downward mixing from the UTLS ("Upper") had frequencies of 2.1–4.1%, with the highest frequency in fall. The fall maximum was likely caused by stratosphere-troposphere-exchange (STE) in the polar and middle latitude regions, which shifts southward from summer to winter (Holton et al., 1995). The transport pathways are shown in **Figure 3d**. A general transport direction from the northwest of PMO reflects the typical dynamics in anticyclones and tropopause folding in the mid-latitude.

The Upper transport pattern had CO means of 123, 92, and 96 ppbv for spring, summer, and fall, respectively (**Table 2**), which were ~10 ppbv higher than CO mixing ratios of NATL. CO has no large sources at high altitude and has been found to be mainly trapped in the lower troposphere (Liang et al., 2011). However, two potential reasons may lead to slight CO enhancement for the Upper transport pattern. In the superimposed RT plots of the two transport patterns, Upper and NATL, in **Figure 3a** and **3d**, air mass origins of Upper covered large sub-polar regions, while NATL covered the middle and low latitude regions. Air masses from the north should contain higher CO because of longer CO lifetime as a result of lower OH concentrations (Novelli et al., 1998). In addition, Upper transport also covered the whole boreal North American region (**Figure 3d**), so the Upper may also receive CO contributions from wildfires from that area.

The Upper transport had average O_3 mixing ratios of 58, 54, and 49 ppbv for spring, summer, and fall, respectively, which were the highest among all transport patterns. O_3 enhancement was ~20 ppbv, which is double that of NA-anthro. Such impact has been well recognized before and been associated with downward transport of naturally high O_3 in the UTLS (e.g., Logan, 1999; Neuman et al., 2012). The debate about stratospheric influence versus tropospheric production as the major source of tropospheric O_3 has existed for years (Monks et al., 2000). Subsidence of stratospheric ozone has been shown to have significa-tion contributions to episodic ozone enhancements in the lower troposphere. Parrish et al. (1993) reported ozone enhancements (up to 60 ppbv) due to injection of strato-spheric ozone at Cape Race, Canada. Springtime downward transport events were found to cause one-minute average ozone reaching 100 ppbv in Boulder, U.S., and a peak concentration of 215 ppbv in the Rocky Mountains (Langford et al., 2009). At an elevated site, Mount Washington, New Hampshire, U.S., a stratospheric influence was found to cause ozone enhancement in air flows associated with atmospheric anticyclones (Fischer et al., 2004). Although

we found downward transport caused the highest O_3 enhancements at PMO, the occurrence frequencies of this transport pattern (2.1%, 2.3% and 4.1%) were much lower than those of NA-anthro. Therefore, on an annual average basis, the total O_3 contribution from downward mixing was less significant than transport of anthropogenic pollution emissions. By comparing the products of the occurrence frequencies and O_3 enhancements associated to NA-anthro and Upper transport patterns, we estimated that the overall O_3 contribution by downward mixing is about 40% of the transport of anthropogenic emissions.

The Upper transport pattern also had the highest NO_x mixing ratios in all three seasons. The NO_x means were 63, 66, and 53 pptv in spring, summer, and fall, respectively, which were 15–41 pptv higher than those of the NATL. NO_x is produced in the UTLS with the source being the photochemical oxidation of N_2O. The NO_y means of Upper were 271, 234, and 229 pptv in spring, summer, and fall, respectively, which were all higher than NATL. The elevated NO_y during downward mixing can be attributed to accelerated hydrocarbon degradation followed by formation of PAN in stratosphere-troposphere air mass exchange (Liang et al., 2011). Hydrocarbon degradation accelerates because of increased OH production as a result of mixing of O_3-rich air from UTLS with moist tropospheric air in the downward transport. The higher NO_y could be due to downward transport of upper tropospheric air having a high ratio of PAN/NO_y, which, to some extent, reduces the rate of wet removal of NO_y. The subsequent PAN decomposition could have led to production of NO_x.

3.1.6. Transport affected by upslope flow

Upslope flow occurred 13–24% of the time, with the lowest frequency in summer. The seasonality found here is in agreement with the previous meteorological study at PMO, but has generally lower frequencies than that estimated for the period of 2004–2005 (4.2–37 % (Kleissl et al., 2007)). CO and O_3 in Upslope transport were statistically higher than NATL. This indicates upslope transport of local anthropogenic emissions to PMO. The Upslope transport had the lowest NO_y, which was likely due to efficient wet scavenging of NO_y during lifting (cloud condensation was predicted to take place for 30% of Upslope periods). The mixing ratios of n- and i-butane were found to be higher than the background in all seasons, which was likely due to butane being used as a primary domestic fuel on the island. Concentration enhancements of the butane isomers were higher in spring than the other seasons, possible due to increased butane usage as heating fuel in the cold spring.

3.2. Relationships of observed trace gases

Due to differences in chemical mechanisms and reaction rates, concentrations of trace gases change at different rates in long-range transport, which leads to recognizable relationships between observed trace gases at PMO. Such relationships are able to reflect chemical transformation history during long-range transport. For example, $d[O_3]/d[CO]$ values have been used to reflect O_3 production tendency based on the fact that CO has a much longer lifetime (a few months) compared with O_3 (lifetime of days to weeks) in the troposphere. During transport from North America to PMO, $d[O_3]/d[CO]$ changes mainly reflect O_3 chemistry in the absence of fresh emissions. Light NMHC (alkanes and alkynes) are detectable after transport from NA to PMO and are good tracers to study photochemical aging. Due to the differences in size and structure, different NMHC decay at varying reaction rates against OH. By examining the ratios between NMHC, the photochemical age of air masses can be estimated (McKeen and Liu, 1993). In the following sections, we examine these relationships in PMO observations with particular attention paid to two major transport patterns, NATL and NA-anthro.

3.2.1. $d[O_3]/d[CO]$

$d[O_3]/d[CO]$ was calculated as the reduced major axis (RMA) slope of the two species. RMA slope takes into account the variability in both x- and y-coordinates (Ayers, 2001); it is used as a proper tool for correlation analysis of observed trace gases. We present $d[O_3]/d[CO]$ grouped for the transport patterns, NATL, Fire, Upper, and NA-anthro, in spring, summer, and fall respectively (**Figure 5**). The data used in these analyses were uninterrupted observations for more than 12 hours. In this way, only prominent events were considered to fully include remarkable changes in CO and O_3, so that a clear and credible slope was more likely to be obtained. Yokelson et al. (2013) discussed how the ratio cannot be used to characterize source emissions and plume aging when mixing with air masses of different composition occurs (e.g., plume mixes with O_3-rich stratospheric air and then with clean marine boundary layer air). For this reason, we reported the statistical significance of the regression analyses. In cases of substantial changes in background air composition during transport, the significance of the regression will be low (high p-values).

$d[O_3]/d[CO]$ values for NA-anthro were 0.59 (spring, **Figure 5a**), 0.71 (summer, **Figure 5b**), and 0.45 (fall, **Figure 5c**) ppbv/ppbv, which were all lower than the values found in direct transport events to PMO, with slopes around 1.0 (Honrath et al., 2004; Zhang et al., 2014). The lower slopes in the current work were expected because events with long transport time were also included, which should be less intensive in O_3 production (lower positive $d[O_3]/d[CO]$). The values of $d[O_3]/d[CO]$ have been reported to be lower and relatively consistent at sites in the U.S. (0.20–0.37 ppbv/ppbv), which reflected a combination of aging processes of air pollutants and mixing with remote clean air mass during transport (Chin et al., 1994; Parrish et al., 1998; Mao and Talbot, 2004). Using the Tropospheric Emission Spectrometer retrievals on the Aura satellite, Hegarty et al. (2009) reported a spring slope value of 0.13 ppbv/ppbv in the free troposphere over the mid-latitude regions extending from North America to the North Atlantic. Interestingly, further into the North Atlantic ocean, higher slopes were found at PMO (Honrath et al. (2004) and this work) and Izana, Canary Islands (Cuevas et al., 2013).

The observations affected by wildfires had lower $d[O_3]/d[CO]$ values (slopes of 0.31 and 0.12 ppbv/ppbv for summer and fall respectively) than NA-anthro. The lower values were likely due to large amounts of CO produced in biomass burning and lower NO_x/CO emission ratios compared with anthropogenic sources (McKeen et al., 2002; Lapina et al., 2006). In addition, production of O_3 and NO_2 can be suppressed in wildfire plumes due to reduction of sunlight by heavy aerosol loadings (Real et al., 2007; Verma et al., 2009; Parrington et al., 2013). This variation in $d[O_3]/d[CO]$ in fire-affected patterns was consistent with a previous MOZART study at PMO (Pfister et al., 2006), in which lower $d[O_3]/d[CO]$ values were also found during periods associated with higher relative contribution of fire tracers. Val Martín et al. (2006) calculated the O_3 enhancement indicator for fire plumes captured at PMO in a different method, which is the $\Delta[O_3]/\Delta[CO]$ (Δ refers to the difference between observed concentrations and an estimated background), and reported a similar mean value of 0.20 ppbv/ppbv for summer 2004.

The slopes for NATL were higher than NA-anthro in spring and fall. In both seasons, a few high O_3 mixing ratios (greater than 50 ppbv) were observed for NATL. We investigated those periods specifically and found that these events received ~9% RT from above 5 km. This suggested that NATL receives non-negligible contributions from UTLS in general, even though strict criteria were applied to ensure that the majority RT originated from the lower troposphere. As a result, occasionally high O_3 mixing ratios combined with low CO should be expected in the background of PMO. Examples of such observations are marked by gold dotted rectangles in **Figure 5a** and 5b. These observations show high O_3 concentrations (up to 60 ppbv) in the lower end of CO ranges in spring and summer, pull the regression toward higher slopes, and lead to low correlation coefficients. Positive $d[O_3]/d[CO]$ caused by downward mixing was also observed in a previous study using GEOS-Chem to interpret global satellite observations of $d[O_3]/d[CO]$ (Kim et al., 2013). Negative slopes indicating anticorrelation between O_3 and CO were found in winter at a few ground sites (Parrish et al., 1998; Macdonald et al., 2011), which were attributed to titration of O_3 by NO in emissions. Strong vertical mixing carrying air mass from the upper troposphere was also found to cause negative slopes (Fishman and Seiler, 1983).

3.2.2. Oxidation of butane isomers

Figure 6 shows the observed ratios of [i-butane]/[n-butane] as a function of [n- butane] for all paired observations in the ten years. Observations for NA-anthro and NATL are shown in colors as indicated by the legends. The geometric means (red symbols) and standard deviations (bars) for the two transport patterns are also shown. Mixing ratios of n-butane on the x-axis are given in a logarithmic scale to better display the wide range. The [i-butane]/[n-butane] values towards the left show a more scattered distribution because of lower precision of measurements near the detection limit. The [n-butane] mean for NA-anthro was higher than NATL as a result of emission influence. The means of [i-butane]/[n-butane] were both 0.49 for NA-anthro and NATL, and uncertainties were similar, which indicated no significant difference

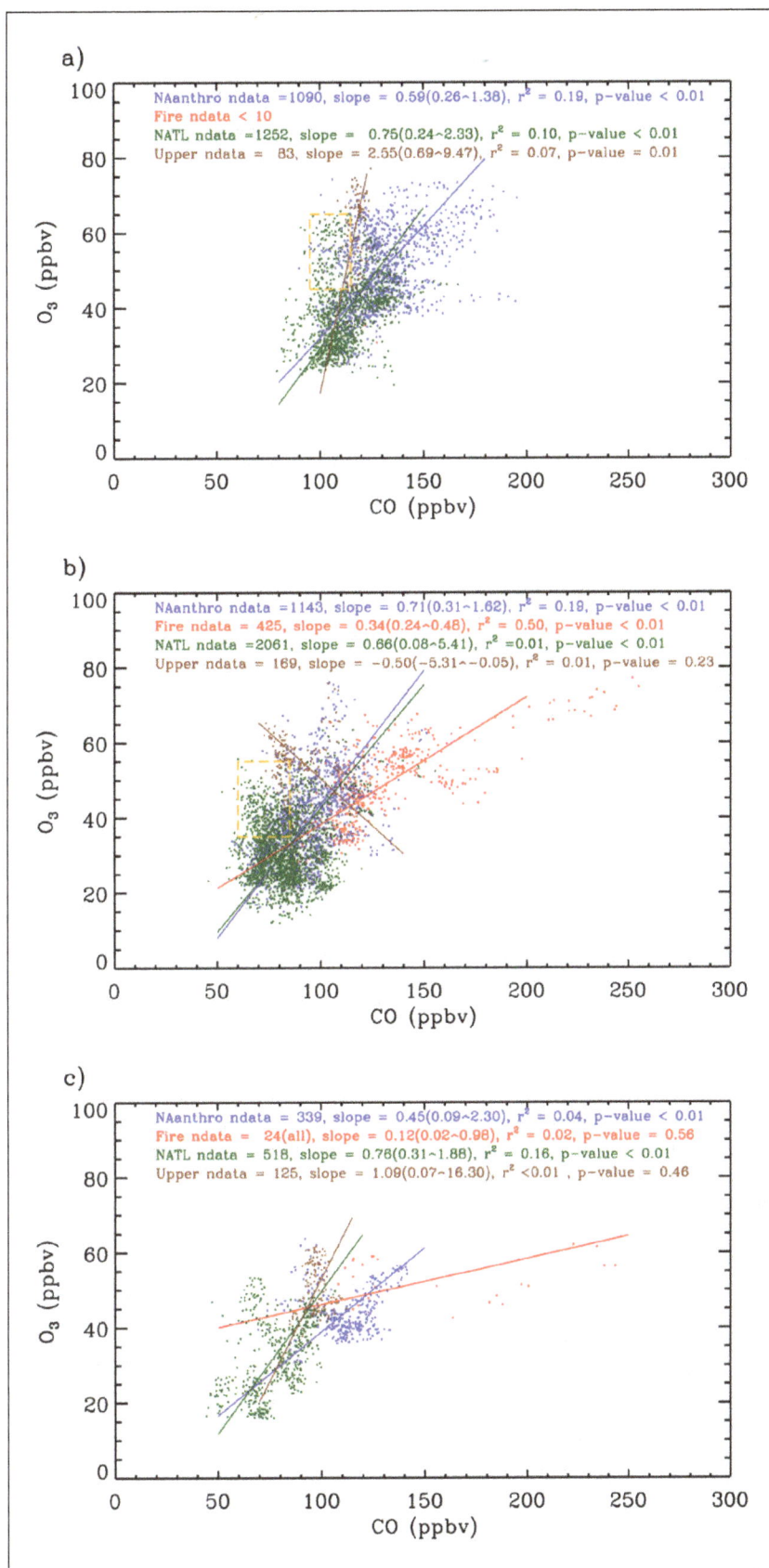

Figure 5: Regression analyses of observed O_3 and CO relationships for transport patterns NA-anthro, Fire, NATL, and Upper in spring **(a)**, summer **(b)**, and fall **(c)**. Only the observations of qualified continuous periods longer than 12 hour were used, except for the Fire transport pattern in fall, in which all observations were used. The gold dotted box marks observations likely influenced by subsidence air masses. See text for details. DOI: https://doi.org/10.1525/elementa.194.f5

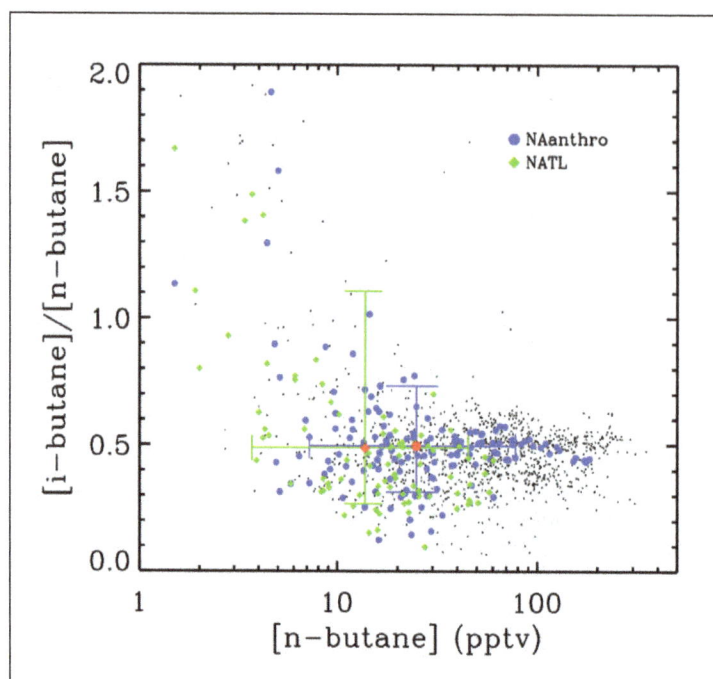

Figure 6: Relationships between [i-butane]/[n-butane] and [n-butane] observed in all seasons. The black dots show the data for all available paired measurements (also including data that do not belong to any determined transport patterns). Observations for two patterns, including NATL and NAanthro, are color coded as described in the legends. Red symbols indicate the geometric means for the two patterns; error bars show the 25 and 75 percentiles of [i-butane]/[n-butane]. DOI: https://doi.org/10.1525/elementa.194.f6

in aging rates between butane isomers. Chlorine atoms have been found to cause faster oxidation of n-isomers in Arctic marine air (Hopkins et al., 2002). Chlorine atoms from oceanic sources were also found to affect NMHC observations in coastal regions during time periods when wind blows from the Pacific (Gorham et al., 2010). However, similar values of [i-butane]/[n-butane] were found at PMO in well-aged background air and transport of pollution, suggesting that the difference in oxidation speed for the butane isomers was hardly detectable. This implies that aging in both species is driven by OH, while other oxidants, such as Cl/Br, play minor roles. These results are in agreement with earlier findings by Helmig et al. (2008).

3.2.3. NMHC photochemical clock
Observed relationships between NMHC can be used to indicate atmospheric processing during long-range transport (i.e., photochemical aging and mixing with background air). The logarithmic ratio of NMHC pairs (i.e., (ln [n-butane]/[ethane])/(ln [propane]/[ethane])) is a function of both OH, and NMHC concentrations in emission sources (Parrish et al., 2007). This approach works better than studying NMHC concentrations alone, since the ratio reduces the large variations of NMHC concentrations in emissions. The NMHC ratios are very effective tracers to study pollution transport to PMO, because NMHC have low concentrations and no significant emission over the ocean (McKeen and Liu, 1993).

The analyses of ln[n-butane]/[ethane] vs ln[propane]/[ethane] shown in **Figure 7** investigate photochemical aging and mixing extent during transport to PMO,

with specific attention paid to NATL and NA-anthro (**Figure 7b**). A kinetic scenario is an ideal condition, in which we assumed an isolated plume and no mixing with the marine air during transport. The kinetic scenario is indicated by the solid line with scales in both panels of **Figure 7**. The slope and aging scales were calculated using the same kinetic reaction rates as provided by Helmig et al. (2008) and an assumed constant [OH] of 1.0×10^6 molecule/cm^3. A global mean [OH] was chosen based on the estimate from Roelofs and Lelieveld (2000), but one should note that [OH] in the troposphere at northern mid-latitudes has been found higher in spring than in summer and fall (Spivakovsky et al., 1990). For this reason, the intervals between age scales in **Figure 7** would be longer in spring and shorter in fall. The end point on the upper right of the kinetic line was defined by the NMHC ratios in continental emissions derived by Helmig et al. (2008). Data for NATL and NA-anthro are shown in different colors in the legends in **Figure 7**. Most data locate between the kinetic line and the mixing line indicating transport involved both aging and mixing processes. The regression of all data (solid line without scales) gave a RMA slope of 1.44 with $r^2 = 0.55$ and $p < 0.01$. The ten-year observation gave a range of the NMHC age at PMO from 7 to more than 20 days, with a mean of 11 days and a standard deviation of 4 days. This estimated mean age was greater than that found in previous studies at PMO using FLEXPART (Honrath et al., 2004; Zhang et al., 2014). Again, the reason was mainly that previous studies focused on a few efficient transport events, whereas all transport event candidates from North America are included here. The

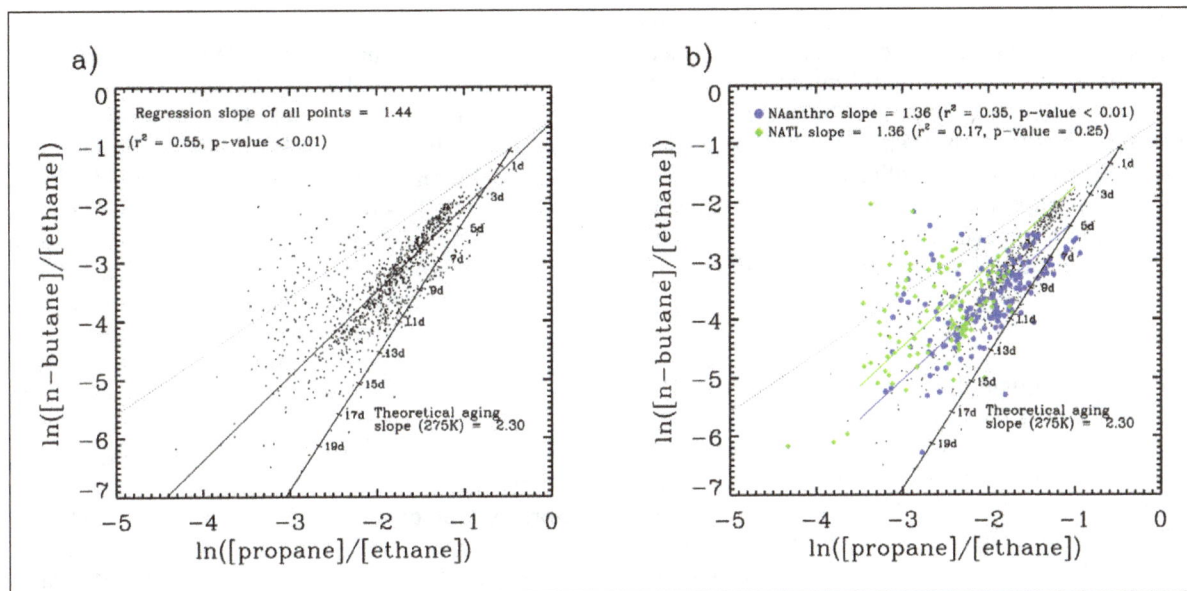

Figure 7: Regression analyses of ln([n-butane]/[ethane]) and ln([propane]/[ethane]) relationships for all observations **(a)** and two transport patterns **(b)**, (NATL and NA-anthro). The dotted lines in both figures indicate the mixing-only trend initiated from an assumed origin (upper right end) based on concentrations in emissions. The solid lines with scales show the photochemical decay slope from the same origin for a fixed OH concentration defined in the text. In both figures, the black dots show the data for all available paired measurements (including data that do not belong to any defined transport patterns). The method to compute the slopes is discussed in the text. The legend provides, for all observations and the two transport patterns, linear regression slopes, correlation coefficients, and p values. DOI: https://doi.org/10.1525/elementa.194.f7

regression slope for NA-anthro was 1.36 with $r^2 = 0.35$, which is consistent with observations in previous studies for long-range transport over the North Atlantic (Parrish et al., 2007; Helmig et al., 2008). The estimated mean for NA-anthro was 11. 5 days with a standard deviation of 3.5 days. The NA-anthro mean was similar to the mean estimated based on all observations, suggesting a dominating influence of NA-anthro. There was a much less credible regression for NATL ($p = 0.25$), which was likely due to information loss on the source after extended aging. The estimated mean of NATL age was 15 days with a standard deviation of 3.5 days.

4. Summary and Conclusions

We categorized the dominant transport patterns to Pico Mountain Observatory for the years 2001–2010 by considering factors that affect air mass composition including types of emission sources, atmospheric RT, transport height, and upslope flow. FLEXPART and simulated meteorological conditions were used to quantify these factors. Observations of trace gases for the transport patterns, viewed as chemical signatures, were compared and discussed to investigate the characteristics of long-range transport to the central North Atlantic.

Trace gas concentrations at PMO were driven by both pollution emissions and natural sources. Anthropogenic emissions contributed an additional 14–23 ppbv (13%–28%) CO and ~10 ppbv (25%–30%) O_3 to background levels at PMO depending on the seasons. Wildfires caused even higher pollutant levels (25 ppbv (30%) CO and 16 ppbv (45%) of O_3) in summer, and were comparable

sources of trace gases to anthropogenic emissions. The O_3 enhancements caused by UTLS downward mixing were about two times that caused by North American anthropogenic emissions. Downward mixing also caused slight CO enhancements at PMO, which were likely due to transport from high latitudes where CO has a longer lifetime. Higher NO_x concentrations were found in transport from North America. Variations of the enhancements in different transport patterns were attributed to temperature and height dependent processes, such as wet scavenging and chemical reactions. High NO_y and O_3 were found in lifted transport from North America, which supports previous findings that PAN preserves O_3 production tendency after long-range transport. Highest NO_x was found in downward transport, likely a result of thermal decomposition of PAN in subsidence.

The slope $d[O_3]/d[CO]$ has been used as an indicator of O_3 production tendency, and it reflected O_3 chemistry and composition of original air masses. The values of $d[O_3]/d[CO]$ for Fire (0.12–0.33) were lower than those for NA-anthro and NATL, which was likely due to the much larger amount of CO emission from wildfires. The mixing with O_3-rich air masses creates largely varying O_3 within a narrow CO range, which tends to result in the high variability in $d[O_3]/d[CO]$. Remarkable downward mixing showed highly positive or negative $[O_3]/d[CO]$ values, likely due to the opposite vertical gradients of O_3 and CO, but the strength of the correlation was relatively low.

Mixing ratios of light NMHC were used to study oxidation processing during transport to PMO. The ratios of butane isomers for NATL and NA-anthro were not significantly different, which implies that sources signatures and oxidation rates were similar. Enhanced levels of butane isomers were observed during periods when upslope flow occurred, likely due to butane being used in the fuel on the Island. The average NMHC photochemical ages for NA-anthro and the entire dataset were 11.5 ± 3.5 and 11.0 ± 4.0 days. This estimate reflected the general air mass age for long-range transport to PMO and confirmed the dominating role of North American emissions.

6. Acknowledgements
The authors thank Mike Dziobak (Michigan Technological University), Jacques Hueber (University of Colorado), and Paulo Fialho (University of the Azores) for helping with logistics and operation at the Pico Mountain Observatory. The authors also thank the management and employees of the Pico Mountain National Park for their help with the operation of the station.

7. Funding Information
The Pico Mountain Observatory is supported by the Azores Regional Secretariat for Science and Technology (project M1.2.1/I/006/2005, project M1.2.1/I/001/2008, and project M1.2.1/I/002/2008) and Program INTERREG IIIB, Azores, Madeira and Canarias (project CLIMARCOST FEDER-INTERREG IIIB-05/MAC/2.3/A1). This research was supported by U.S. National Science Foundation award NO. ATM- 0720955 (B. Zhang, R. C. Owen, J. A. Perlinger, and D. Helmig), AGS-1110059 (L. R. Mazzoleni, C. Mazzoleni, and B. Zhang), and Department of Energy award NO. DE-SC0006941 (C. Mazzoleni, L. R. Mazzoleni, and B. Zhang).

8. Competing Interests
Detlev Helmig is the Editor-in-Chief of the Elementa Atmospheric Science Domain. He was not involved in the peer review of the article.

9. Contributions
- Contributed to conception and design: B. Zhang, R. C. Owen, J. A. Perlinger, and D. Helmig.
- Contributed to acquisition of data: R. C. Owen, D. Helmig, M. Val Martín, L. Kramer, and B. Zhang.
- Contributed to analysis and interpretation of data: B. Zhang, R. C. Owen, J. A. Perlinger, and D. Helmig.
- Contributed to conducting FLEXPART simulations and model result analysis: B. Zhang and R. C. Owen.
- Drafted and/or revised the article: B. Zhang, R. C. Owen, J. A. Perlinger, D. Helmig, M. Val Martín, L. Kramer, L. R. Mazzoleni, and C. Mazzoleni.
- Approved the submitted version for publication: B. Zhang, R. C. Owen, J. A. Perlinger, D. Helmig, M. Val Martín, L. Kramer, L. R. Mazzoleni, and C. Mazzoleni.

References
Alvarado, MJ, Logan, JA, Mao, J, Apel, E, Riemer, D, Blake, D, Cohen, RC, Min, K-E, Perring, AE, Browne, EC, Wooldridge, PJ, Diskin, GS, Sachse, GW, Fuelberg, H, Sessions, WR, Harrigan, DL, Huey, G, Liao, J, Case-Hanks, A, Jimenez, JL, Cubison, MJ, Vay, SA, Weinheimer, AJ, Knapp, DJ, Montzka, DD, Flocke, FM, Pollack, IB, Wennberg, PO, Kurten, A, Crounse, J, Clair, JMS, Wisthaler, A, Mikoviny, T, Yantosca, RM, Carouge, CC and Le Sager, P 2010 Nitrogen oxides and PAN in plumes from boreal fires during ARC-TAS-B and their impact on ozone: an integrated analysis of aircraft and satellite observations, *Atmos. Chem. and Phys.*, **10**, 9739–9760. DOI: https://doi.org/ 10.5194/acp-10-9739-2010

Ambrose, J, Reidmiller, D and Jaffe, D 2011 Causes of high O 3 in the lower free troposphere over the Pacific Northwest as observed at the Mt. Bachelor Observatory, *Atmospheric Environment*, **45**, 5302–5315. DOI: https://doi.org/10.1016/j.atmosenv.2011.06.056

Andreae, MO and Merlet, P 2001 Emission of trace gases and aerosols from biomass burning, *Global biogeochemical cycles*, **15**, 955–966. DOI: https://doi.org/ 10.1029/2000GB001382

Ayers, G 2001 Comment on regression analysis of air quality data, Atmos. Environ., **35**, 2423–2425.

Beekmann, M, Ancellet, G and Mégie, G 1994 Climatology of tropospheric ozone in southern Europe and its relation to potential vorticity, *Journal of Geophysical Research: Atmospheres*, **99**, 12841–12853.

Bey, I, Jacob, DJ, Logan, JA and Yantosca, RM 2001 Asian chemical outflow to the Pacific in spring: Origins, pathways, and budgets.

Blake, NJ, Blake, DR, Sive, BC, Katzenstein, AS, Meinardi, S, Wingenter, OW, Atlas, EL, Flocke, F, Ridley, BA and Rowland, FS 2003 The seasonal evolution of NMHCs and light alkyl nitrates at middle to high northern latitudes during TOPSE, *Journal of Geophysical Research: Atmospheres*, **108**. DOI: https://doi.org/ 10.1029/2001jd001467

Brioude, J, Arnold, D, Stohl, A, Cassiani, M, Morton, D, Seibert, P, Angevine, W, Evan, S, Dingwell, A, and Fast, JD 2013 The Lagrangian particle dispersion model FLEXPART-WRF version 3.1, *Geoscientific Model Development*, **6**, 1889–1904. DOI: https://doi.org/10.5194/gmd-6-1889-2013

Carpenter, L, Green, T, Mills, G, Bauguitte, S, Penkett, S, Zanis, P, Schuepbach, E, Schmidbauer, N, Monks, P and Zellweger, C 2000 Oxidized nitrogen and ozone production efficiencies in the springtime free troposphere over the Alps, *Journal of Geophysical Research: Atmospheres*, **105**, 14547–14559.

Chin, M, Jacob, DJ, Munger, JW, Parrish, DD and Doddridge, BG 1994 Relationship of ozone and carbon monoxide over North America, *J. Geophys. Res.*, **99**, 14565–14573.

China, S, Scarnato, B, Owen, RC, Zhang, B, Ampadu, MT, Kumar, S, Dzepina, K, Dziobak, MP, Fialho, P, Perlinger, JA, et al. 2015 Morphology and mixing state of aged soot particles at a remote marine free troposphere site: Implications for optical properties, *Geophys. Res. Lett.*, **42**, 1243–1250. DOI: https://doi.org/10.1002/2014GL062404

Cooper, OR, Gao, R-S, Tarasick, D, Leblanc, T and Sweeney, C 2012 Long-term ozone trends at rural ozone monitoring sites across the United States, 1990–2010, *J. Geophys. Res.: Atmos. (1984–2012)*, 117.

Cooper, OR, Moody, JL, Parrish, DD, Trainer, M, Ryerson, TB, Holloway, JS, Hubler, G, Fehsenfeld, FC, Oltmans, SJ and Evans, MJ 2001 Trace gas signatures of the airstreams within North Atlantic cyclones: Case studies from the North Atlantic Regional Experiment (NARE '97) aircraft intensive, *J. Geophys. Res.*, **106**, 5437–5456. DOI: https://doi.org/10.1029/2000JD900574

Cooper, OR, Parrish, D, Ziemke, J, Balashov, N, Cupeiro, M, Galbally, I, Gilge, S, Horowitz, L, Jensen, N, Lamarque, J-F, et al. 2014 Global distribution and trends of tropospheric ozone: An observation-based review, Elementa: Science of the Anthropocene, **2**, 000–029.

Cuevas, E, González, Y, Rodríguez, S, Guerra, JC, Gómez-Peláez, AJ, Alonso-Pérez, S, Bustos, J, and Milford, C 2013 Assessment of atmospheric processes driving ozone variations in the subtropical North Atlantic free troposphere, *Atmos. Chem. and Phys.*, **13**, 1973–1998. DOI: https://doi.org/10.5194/acp-13-1973-2013

Dzepina, K, Mazzoleni, C, Fialho, P, China, S, Zhang, B, Owen, R, Helmig, D, Hueber, J, Kumar, S, Perlinger, J, et al. 2015 Molecular characterization of free tropospheric aerosol collected at the Pico Mountain Observatory: a case study with a long-range transported biomass burning plume, *Atmos. Chem. and Phys.*, **15**, 5047–5068. DOI: https://doi.org/10.5194/acp-15-5047-2015

Eckhardt, S, Stohl, A, Wernli, H, James, P, Forster, C, and Spichtinger, N 2004 A 15-year climatology of warm conveyor belts, *J. Climate*, **17**, 218–237. DOI: https://doi.org/10.1175/1520-0442(2004)017<0218:AYCOWC>2.0.CO;2

Emanuel, KA and Zivkovic-Rothman, M 1999 Development and evaluation of a convection scheme for use in climate models, *Journal of the Atmospheric Sciences*, **56**, 1766–1782. DOI: https://doi.org/10.1175/1520-0469(1999)056<1766:DAEOAC>2.0.CO;2

EPA 2016 Air Pollutant Emissions Trends Data, URL https://www.epa.gov/air-emissions-inventories/air-pollutant-emissions-trends-data, last accessed: November 2016.

Fehsenfeld, FC, Ancellet, G, Bates, TS, Goldstein, AH, Hardesty, RM, Honrath, R, Law, KS, Lewis, AC, Leaitch, R, McKeen, S, Meagher, J, Parrish, D, Pszenny, AAP, Russell, PB, Schlager, H, Seinfeld, J, Talbot, R and Zbinden, R 2006 International Consortium for Atmospheric Research on Transport and Transformation (ICARTT): North America to Europe – Overview of the 2004 summer field study, *J. Geophys. Res.*, **111**(D23), S01. DOI: https://doi.org/10.1029/2006JD007829

Fehsenfeld, FC, Trainer, M, Parrish, DD, VolzThomas, A and Penkett, S 1996 North Atlantic Regional Experiment 1993 summer intensive: Foreword, *J. Geophys. Res.*, **101**.

Fialho, P, Hansen, A and Honrath, R 2005 Absorption coefficients by aerosols in remote areas: a new approach to decouple dust and black carbon absorption coefficients using seven-wavelength Aethalometer data, *Journal of Aerosol Science*, **36**, 267–282. DOI: https://doi.org/10.1016/j.jaerosci.2004.09.004

Fischer, EV, Talbot, RW, Dibb, JE, Moody, JL and Murray, GL 2004 Summertime ozone at Mount Washington: Meteorological controls at the highest peak in the northeast, *J. Geophys. Res.*, **109**. DOI: https://doi.org/10.1029/2004jd004841

Fishman, J and Seiler, W 1983 Correlative nature of ozone and carbon monoxide in the troposphere: Implications for the tropospheric ozone budget, *J. Geophys. Res.*, **88**, 3662–3670, DOI: https://doi.org/10.1029/JC088iC06p03662

GDAS: NCEP Global Data Assimilation System, URL http://ready.arl.noaa.gov/gdas1.php, accessed: Nov., 2014.

Gorham, KA, Blake, NJ, VanCuren, RA, Fuelberg, HE, Meinardi, S and Blake, DR 2010 Seasonal and diurnal measurements of carbon monoxide and nonmethane hydrocarbons at Mt. Wilson, California: Indirect evidence of atomic Cl in the Los Angeles basin, *Atmos. Environ.*, **44**, 2271–2279. DOI: https://doi.org/10.1016/j.atmosenv.2010.04.019

Hegarty, J, Mao, H and Talbot, R 2009 Synoptic influences on springtime tropospheric O_3 and CO over the North American export region observed by TES, *Atmos. Chem. Phys.*, **9**, 3755–3776. DOI: https://doi.org/10.5194/acp-9-3755-2009

Helmig, D, Muñoz, M, Hueber, J, Mazzoleni, C, Mazzoleni, L, Owen, RC, Val-Martin, M, Fialho, P, Plass-Duelmer, C, Palmer, PI, et al. 2015 Climatology and atmospheric chemistry of the non-methane hydrocarbons ethane and propane over the North Atlantic, *Elementa: Science of the Anthropocene*, **3**, 000–054.

Helmig, D, Tanner, DM, Honrath, RE, Owen, RC and Parrish, DD 2008 Nonmethane hydrocarbons at Pico Mountain, Azores: 1. Oxidation chemistry in the North Atlantic region, *J. Geophys. Res.*, **113**. DOI: https://doi.org/10.1029/2007jd008930

Holton, JR, Haynes, PH, McIntyre, ME, Douglass, AR, Rood, RB and Pfister, L 1995 Stratosphere-troposphere exchange, *Rev. Geophys.*, **33**, 403–439. DOI: https://doi.org/10.1029/95RG02097

Honrath, RE, Hamlin, AJ and Merrill, JT 1996 Transport of ozone precursors from the Arctic troposphere to the North Atlantic region, *Journal of Geophysical Research: Atmospheres,* **101**, 29335–29351.

Honrath, RE, Helmig, D, Owen, RC, Parrish, DD and Tanner, DM 2008 Nonmethane hydrocarbons at Pico Mountain, Azores: 2. Event-specific analyses of the impacts of mixing and photochemistry on hydrocarbon ratios, *J. Geophys. Res.,* **113**(D20), S92. DOI: https://doi.org/10.1029/2008JD009832

Honrath, RE, Owen, RC, Val Martin, M, Reid, JS, Lapina, K, Fialho, P, Dziobak, MP, Kleissl, J and Westphal, DL 2004 Regional and hemispheric impacts of anthropogenic and biomass burning emissions on summertime CO and O_3 in the North Atlantic lower free troposphere, *J. Geophys. Res.,* **109**(D24), 310. DOI: https://doi.org/10.1029/2004JD005147

Hopkins, J, Jones, I, Lewis, A, McQuaid, J and Seakins, P 2002 Non-methane hydrocarbons in the Arctic boundary layer, *Atmos. Environ.,* **36**, 3217–3229. DOI: https://doi.org/10.1016/S1352-2310(02)00324-2

Hübler, G, Fahey, D, Ridley, B, Gregory, G and Fehsenfeld, F 1992 Airborne measurements of total reactive odd nitrogen (NO_y), *Journal of Geophysical Research: Atmospheres,* **97**, 9833–9850. DOI: https://doi.org/10.1029/91JD02326

Hudman, RC, Jacob, DJ, Cooper, OR, Evans, MJ, Heald, CL, Park, RJ, Fehsenfeld, F, Flocke, F, Holloway, J, Hubler, G, Kita, K, Koike, M, Kondo, Y, Neuman, A, Nowak, J, Oltmans, S, Parrish, D, Roberts, JM and Ryerson, T 2004 Ozone production in transpacific Asian pollution plumes and implications for ozone air quality in California, *J. Geophys. Res.,* **109**. DOI: https://doi.org/10.1029/2004jd004974

Kim, P, Jacob, D, Liu, X, Warner, J, Yang, K, Chance, K, Thouret, V and Nedelec, P 2013 Global ozone-CO correlations from OMI and AIRS: constraints on tropospheric ozone sources, *Atmos. Chem. and Phys.,* **13**, 9321–9335. DOI: https://doi.org/10.5194/acp-13-9321-2013

Kleissl, J, Honrath, RE, Dziobak, MP, Tanner, D, Martin, MV, Owen, RC and Helmig, D 2007 Occurrence of upslope flows at the Pico mountaintop observatory: A case study of orographic flows on a small, volcanic island, *J. Geophys. Res.,* **112**(D10), S35. DOI: https://doi.org/10.1029/2006JD007565

Koike, M, Kondo, Y, Kita, K, Takegawa, N, Masui, Y, Miyazaki, Y, Ko, M, Weinheimer, A, Flocke, F, Weber, R, et al. 2003 Export of anthropogenic reactive nitrogen and sulfur compounds from the East Asia region in spring, *Journal of Geophysical Research: Atmospheres,* **108**. DOI: https://doi.org/10.1029/2002jd003284

Kramer, L, Helmig, D, Burkhart, J, Stohl, A, Oltmans, S and Honrath, RE 2015 Seasonal variability of atmospheric nitrogen oxides and non-methane hydrocarbons at the GEOSummit station, Greenland, *Atmospheric Chemistry and Physics,* **15**, 6827–6849. DOI: https://doi.org/10.5194/acp-15-6827-2015

Kumar, A, Wu, S, Weise, M, Honrath, R, Owen, R, Helmig, D, Kramer, L, Val Martin, M and Li, Q 2013 Free-troposphere ozone and carbon monoxide over the North Atlantic for 2001–2011, *Atmos. Chem. and Phys.,* **13**, 12537–12547.

Langford, A, Aikin, K, Eubank, C and Williams, E 2009 Stratospheric contribution to high surface ozone in Colorado during springtime, *Geophys. Res. Lett.,* **36**. DOI: https://doi.org/10.1029/2009gl038367

Lapina, K, Honrath, RE, Owen, RC, Martin, MV and Pfister, G 2006 Evidence of significant large-scale impacts of boreal fires on ozone levels in the mid-latitude Northern Hemisphere free troposphere, *Geophys. Res. Lett.,* **33**(L10), 815. DOI: https://doi.org/10.1029/2006GL025878

Liang, Q, Rodriguez, JM, Douglass, AR, Crawford, JH, Olson, JR, Apel, E, Bian, H, Blake, DR, Brune, W, Chin, M, Colarco, PR, da Silva, A, Diskin, GS, Duncan, BN, Huey, LG, Knapp, DJ, Montzka, DD, Nielsen, JE, Pawson, S, Riemer, DD, Weinheimer, AJ and Wisthaler, A 2011 Reactive nitrogen, ozone and ozone production in the Arctic troposphere and the impact of stratosphere-troposphere exchange, *Atmos. Chem. and Phys.,* **11**, 13181–13199. DOI: https://doi.org/10.5194/acp-11-13181-2011

Li, Q, Jacob, DJ, Munger, JW, Yantosca, RM and Parrish, DD 2004 Export of NOy from the North American boundary layer: Reconciling aircraft observations and global model budgets, *Journal of Geophysical Research: Atmospheres,* **109**.

Li, QB, Jacob, DJ, Bey, I, Palmer, PI, Duncan, BN, Field, BD, Martin, RV, Fiore, AM, Yantosca, RM, Parrish, DD, Simmonds, PG and Oltmans, SJ 2002 Transatlantic transport of pollution and its effects on surface ozone in Europe and North America, *J. Geophys. Res.,* **107**. DOI: https://doi.org/10.1029/2001jd001422

Logan, JA 1999 An analysis of ozonesonde data for the lower stratosphere: Recommendations for testing models, *J. Geophys. Res.,* **104**, 16151–16170. DOI: https://doi.org/10.1029/1999JD900216

Logan, JA, Prather, MJ, Wofsy, SC and McElroy, MB 1981 Tropospheric chemistry: A global perspective, *J. Geophys. Res.,* **86**, 7210–7254. DOI: https://doi.org/10.1029/JC086iC08p07210

Macdonald, A, Anlauf, K, Leaitch, W, Chan, E and Tarasick, D 2011 Interannual variability of ozone and carbon monoxide at the Whistler high elevation site: 20022006, *Atmos. Chem. Phys,* **11**, 11431–11446. DOI: https://doi.org/10.5194/acp-11-11431-2011

Macias Fauria, M and Johnson, EA 2006 Large-scale climatic patterns control large lightning fire occurrence in Canada and Alaska forest regions, *Journal of Geophysical Research: Biogeosciences,* **111**. DOI: https://doi.org/10.1029/2006jg000181

Mao, HT and Talbot, R 2004 O-3 and CO in New England: Temporal variations and relationships, *J. Geophys. Res.,* **109**.

McKeen, S and Liu, S 1993 Hydrocarbon ratios and photochemical history of air masses, *Geophys. Res. Lett.*, **20**, 2363–2366. DOI: https://doi.org/10.1029/93GL02527

McKeen, S, Wotawa, G, Parrish, D, Holloway, J, Buhr, M, Hübler, G, Fehsenfeld, F and Meagher, J 2002 Ozone production from Canadian wildfires during June and July of 1995, *J. Geophys. Res.*, **107**, ACH-7. DOI: https://doi.org/10.1029/2001JD000697

Monks, PS, Salisbury, G, Holland, G, Penkett, SA and Ayers, GP 2000 A seasonal comparison of ozone photochemistry in the remote marine boundary layer, *Atmospheric Environment*, **34**, 2547–2561. DOI: https://doi.org/10.1016/S1352-2310(99)00504-X

Moxim, W, Levy, H and Kasibhatla, P 1996 Simulated global tropospheric PAN: Its transport and impact on NO$_x$, *Journal of Geophysical Research: Atmospheres*, **101**, 12621–12638. DOI: https://doi.org/10.1029/96JD00338

Mu, M, Randerson, JT, van der Werf, GR, Giglio, L, Kasibhatla, P, Morton, D, Collatz, GJ, DeFries, RS, Hyer, EJ, Prins, EM, Griffith, DWT, Wunch, D, Toon, GC, Sherlock, V and Wennberg, PO 2011 Daily and 3-hourly variability in global fire emissions and consequences for atmospheric model predictions of carbon monoxide, *J. Geophys. Res.*, **116**(D24), 303. DOI: https://doi.org/10.1029/2011JD016245

Neuman, JA, Trainer, M, Aikin, KC, Angevine, WM, Brioude, J, Brown, SS, de Gouw, JA, Dube, WP, Flynn, JH, Graus, M, Holloway, JS, Lefer, BL, Nedelec, P, Nowak, JB, Parrish, DD, Pollack, IB, Roberts, JM, Ryerson, TB, Smit, H, Thouret, V and Wagner, NL 2012 Observations of ozone transport from the free troposphere to the Los Angeles basin, *J. Geophys. Res.*, **117**. DOI: https://doi.org/10.1029/2011jd016919

Novelli, P, Masarie, K and Lang, P 1998 Distributions and recent changes of carbon monoxide in the lower troposphere, *J. Geophys. Res.*, **103**, 19015–19033.

Olivier, J and Berdowski, J 2001 The Climate System, A.A. Balkema, Swets and Zeitlinger.

Owen, RC, Cooper, OR, Stohl, A and Honrath, RE 2006 An analysis of the mechanisms of North American pollutant transport to the central North Atlantic lower free troposphere, *J. Geophys. Res.*, **111**(D23), S58. DOI: https://doi.org/10.1029/2006JD007062

Parrington, M, Palmer, PI, Henze, D, Tarasick, D, Hyer, E, Owen, RC, Helmig, D, Clerbaux, C, Bowman, K, Deeter, M, et al. 2012 The influence of boreal biomass burning emissions on the distribution of tropospheric ozone over North America and the North Atlantic during 2010, *Atmos. Chem. and Phys.*, **12**, 2077–2098. DOI: https://doi.org/10.5194/acp-12-2077-2012

Parrington, M, Palmer, PI, Lewis, AC, Lee, JD, Rickard, AR, Di Carlo, P, Taylor, JW, Hopkins, JR, Punjabi, S, Oram, DE, Forster, G, Aruffo, E, Moller, SJ, Bauguitte, SJ-B, Allan, JD, Coe, H and Leigh, RJ 2013 Ozone photochemistry in boreal biomass burning plumes, *Atmos. Chem. and Phys.*, **13**, 7321–7341. DOI: https://doi.org/10.5194/acp-13-7321-2013

Parrish, DD, Hahn, CJ, Williams, EJ, Norton, RB, Fehsenfeld, FC, Singh, HB, Shetter, JD, Gandrud, BW and Ridley, BA 1992 Indications of photochemical histories of Pacific air masses from measurements of atmospheric trace species at Point Arena, California, *J. Geophys. Res.*, **97**, 15883–15901.

Parrish, DD, Holloway, JS, Trainer, M, Murphy, PC, Forbes, GL and Fehsenfeld, FC 1993 Export of North American ozone pollution to the north Atlantic Ocean, *Science*, **259**, 1436–1439. DOI: https://doi.org/10.1126/science.259.5100.1436

Parrish, DD, Law, KS, Staehelin, J, Derwent, R, Cooper, OR, Tanimoto, H, Volz-Thomas, A, Gilge, S, Scheel, H-E, Steinbacher, M and Chan, E 2012 Long-term changes in lower tropospheric baseline ozone concentrations at northern mid-latitudes, *Atmos. Chem. and Phys.*, **12**, 11485–11504. DOI: https://doi.org/10.5194/acp-12-11485-2012

Parrish, DD, Stohl, A, Forster, C, Atlas, EL, Blake, DR, Goldan, PD, Kuster, WC and de Gouw, JA 2007 Effects of mixing on evolution of hydrocarbon ratios in the troposphere, *J. Geophys. Res.*, **112**. DOI: https://doi.org/10.1029/2006jd007583

Parrish, DD, Trainer, M, Holloway, JS, Yee, JE, Warshawsky, MS, Fehsenfeld, FC, Forbes, GL and Moody, JL 1998 Relationships between ozone and carbon monoxide at surface sites in the North Atlantic region, *J. Geophys. Res.*, **103**, 13357–13376. DOI: https://doi.org/10.1029/98JD00376

Pfister, GG, Emmons, LK, Hess, PG, Honrath, R, Lamarque, J-F, Val Martin, M, Owen, RC, Avery, MA, Browell, EV, Holloway, JS, Nedelec, P, Purvis, R, Ryerson, TB, Sachse, GW and Schlager, H 2006 Ozone production from the 2004 North American boreal fires, *J. Geophys. Res.*, **111**. DOI: https://doi.org/10.1029/2006jd007695

Randerson, J, Chen, Y, Werf, G, Rogers, B and Morton, D 2012 Global burned area and biomass burning emissions from small fires, *Journal of Geophysical Research: Biogeosciences*, **117**. DOI: https://doi.org/10.1029/2012jg002128

Real, E, Law, KS, Weinzierl, B, Fiebig, M, Petzold, A, Wild, O, Methven, J, Arnold, S, Stohl, A, Huntrieser, H, Roiger, A, Schlager, H, Stewart, D, Avery, M, Sachse, G, Browell, E, Ferrare, R and Blake, D 2007 Processes influencing ozone levels in Alaskan forest fire plumes during long-range transport over the North Atlantic, *J. Geophys. Res.*, **112**, 19. DOI: https://doi.org/10.1029/2006JD007576

Roelofs, G-J and Lelieveld, J 2000 Tropospheric ozone simulation with a chemistry- general circulation model: Influence of higher hydrocarbon chemistry, *Journal of Geophysical Research: Atmos.*, **105**, 22697–22712. DOI: https://doi.org/10.1002/qj.49708235418

Sheppard, P 1956 Airflow over mountains, Quart. J. Roy. Meteor. Soc., **82**, 528–529.

Spivakovsky, C, Yevich, R, Logan, J, Wofsy, S, McElroy, M and **Prather, M** 1990 Tropospheric OH in a three-dimensional chemical tracer model: An assessment based on observations of CH3CCl3, *J. Geophys. Res.,* **95**, 18441–18471.

Stohl, A, Hittenberger, M and **Wotawa, G** 1998 Validation of the Lagrangian particle dispersion model FLEXPART against large-scale tracer experiment data, *Atmos. Environ.,* **32**, 4245–4264. DOI: https://doi.org/10.1016/S1352-2310(98)00184-8

Stohl, A, Sodemann, H, Eckhardt, S, Frank, A, Seibert, P and **Wotawa, G** 2011 The Lagrangian particle dispersion model FLEXPART version 8.2, FLEXPART user guide, url: http://zardoz.nilu.no/flexpart/flexpart/flexpart82.pdf (March 14, 2012).

Talbot, R, Dibb, JE, Scheuer, E, Kondo, Y, Koike, M, Singh, H, Salas, L, Fukui, F, Ballenthin, J, Meads, R, et al. 1999 Reactive nitrogen budget during the NASA SONEX mission, *Geophysical Research Letters.* DOI: https://doi.org/10.1029/1999GL900589

Tanner, D, Helmig, D, Hueber, J and **Goldan, P** 2006 Gas chromatography system for the automated, unattended, and cryogen-free monitoring of C2 to C6 nonmethane hydrocarbons in the remote troposphere, *Journal of Chromatography A,* **1111**, 76–88. DOI: https://doi.org/10.1016/j.chroma.2006.01.100

Val Martín, M, Honrath, RE, Owen, RC and **Lapina, K** 2008a Large-scale impacts of anthropogenic pollution and boreal wildfires on the nitrogen oxides over the central North Atlantic region, *J. Geophys. Res.,* **113**. DOI: https://doi.org/10.1029/2007jd009689

Val Martín, M, Honrath, RE, Owen, RC and **Li, QB** 2008b Seasonal variation of nitrogen oxides in the central North Atlantic lower free troposphere, *J. Geophys. Res.,* **113**(D17), 307. DOI: https://doi.org/10.1029/2007JD009688

Val Martín, M, Honrath, RE, Owen, RC, Pfister, G, Fialho, P and **Barata, F** 2006 Significant enhancements of nitrogen oxides, black carbon, and ozone in the North Atlantic lower free troposphere resulting from North American boreal wildfires, *J. Geophys. Res.,* **111**(D23), S60. DOI: https://doi.org/10.1029/2006JD007530

Verma, S, Worden, J, Pierce, B, Jones, DBA, Al-Saadi, J, Boersma, F, Bowman, K, Eldering, A, Fisher, B, Jourdain, L, Kulawik, S and **Worden, H** 2009 Ozone production in boreal fire smoke plumes using observations from the Tropospheric Emission Spectrometer and the Ozone Monitoring Instrument, *J. Geophys. Res.,* **114**. DOI: https://doi.org/10.1029/2008jd010108

Vogelezang, D and **Holtslag, A** 1996 Evaluation and model impacts of alternative boundary-layer height formulations, *Boundary-Layer Meteorology,* **81**, 245–269. DOI: https://doi.org/10.1007/BF02430331

Worden, HM, Deeter, MN, Frankenberg, C, George, M, Nichitiu, F, Worden, J, Aben, I, Bowman, KW, Clerbaux, C, Coheur, PF, de Laat, ATJ, Detweiler, R, Drummond, JR, Edwards, DP, Gille, JC, Hurtmans, D, Luo, M, Martinez-Alonso, S, Massie, S, Pfister, G, and **Warner, JX** 2013 Decadal record of satellite carbon monoxide observations, *Atmos. Chem. and Phys.,* **13**, 837–850. DOI: https://doi.org/10.5194/acp-13-837-2013

Yienger, J, Carmichael, G and **Klonecki, A** 1999 An evaluation of chemistry's role in the winter-spring ozone maximum found in the northern mid-latitude free troposphere, *Journal of geophysical research,* **104**, 3655–3667. DOI: https://doi.org/10.1029/1998JD100043

Yokelson, RJ, Burling, IR, Gilman, J, Warneke, C, Stockwell, CE, Gouw, Jd, Akagi, S, Urbanski, S, Veres, P, Roberts, JM, et al. 2013 Coupling field and laboratory measurements to estimate the emission factors of identified and unidentified trace gases for prescribed fires, *Atmospheric Chemistry and Physics,* **13**, 89–116. DOI: https://doi.org/10.5194/acp-13-89-2013

Zhang, B, Owen, RC, Perlinger, J, Kumar, A, Wu, S, Val Martin, M, Kramer, L, Helmig, D and **Honrath, R** 2014 A semi-Lagrangian view of ozone production tendency in North American outflow in the summers of 2009 and 2010, *Atmos. Chem. and Phys.,* **14**, 2267–2287. DOI: https://doi.org/10.5194/acp-14-2267-2014

9

Hydrological controls on the tropospheric ozone greenhouse gas effect

Le Kuai[*], Kevin W. Bowman[†], Helen M. Worden[‡], Robert L. Herman[†] and
Susan S. Kulawik[§,‖]

The influence of the hydrological cycle in the greenhouse gas (GHG) effect of tropospheric ozone (O_3) is quantified in terms of the O_3 longwave radiative effect (LWRE), which is defined as the net reduction of top-of-atmosphere flux due to total tropospheric O_3 absorption. The O_3 LWRE derived from the infrared spectral measurements by Aura's Tropospheric Emission Spectrometer (TES) show that the spatiotemporal variation of LWRE is relevant to relative humidity, surface temperature, and tropospheric O_3 column. The zonally averaged subtropical LWRE is ~0.2 W m^{-2} higher than the zonally averaged tropical LWRE, generally due to lower water vapor concentrations and less cloud coverage at the downward branch of the Hadley cell in the subtropics. The largest values of O_3 LWRE over the Middle East (>1 W/m²) are further due to large thermal contrasts and tropospheric ozone enhancements from atmospheric circulation and pollution. Conversely, the low O_3 LWRE over the Inter-Tropical Convergence Zone (on average 0.4 W m^{-2}) is due to strong water vapor absorption and cloudiness, both of which reduce the tropospheric O_3 absorption in the longwave radiation. These results show that changes in the hydrological cycle due to climate change could affect the magnitude and distribution of ozone radiative forcing.

Keywords: ozone greenhouse gas effect; longwave radiative effect; instantaneous Radiative Kernels

1. Introduction

Tropospheric ozone (O_3) has increased substantially since the preindustrial era due to anthropogenic emissions (Lamarque et al., 2010). In contrast to long-lived greenhouse gases (GHG) such as carbon dioxide, the lifetime of tropospheric O_3 ranges from hours to weeks leading to a significantly more variable spatial distribution. In the latest Intergovernmental Panel for Climate Change (IPCC) assessment (AR5) (Myhre et al., 2013), the estimated radiative forcing (RF) of tropospheric O_3 computed using chemistry-climate models ranged widely from +0.2 to +0.6 W m^{-2} but the cause of this large inter-model spread remains unexplained. Most of the longwave O_3 RF is in the 9.6-μm O_3 band, which accounts for more than 97% of total longwave O_3 absorption (Rothman et al., 1987). Shortwave forcing, which accounts for ~1/4 of total tropo-

spheric O_3 RF (Myhre et al., 2013), is not considered here, and we therefore restrict our analysis and conclusions to longwave GHG effects. Spectrally resolved satellite measurements of this O_3 band allow evaluation of the sensitivity of top-of-atmosphere (TOA) flux to tropospheric O_3 change by means of instantaneous Radiative Kernels (IRK) and longwave radiative effect (LWRE). The IRK represents the TOA flux sensitivity to the vertical distribution of O_3 and LWRE is the net reduction in TOA flux due to the tropospheric O_3 column, respectively (Worden et al., 2011). The tropospheric O_3 GHG effect can be quantified with IRK or LWRE.

O_3 LWRE estimates, inferred from the satellite observational based IRK, for example from Tropospheric Emission Spectrometer (TES) or Infrared Atmospheric Sounding Interferometer (IASI) data (Worden et al., 2011; Doniki et al., 2015), are useful for further diagnosing the inter-model spread of O_3 RF in the IPCC AR5 assessment. For example, Bowman et al. (2013) used TES retrieved O_3 and IRKs to reduce the inter-model uncertainty of the preindustrial-to-present O_3 RF of the chemistry-climate models participating in the Atmospheric Chemistry and Climate Model Intercomparison Project (ACCMIP) (Lamarque et al., 2013) by 30%. Their study also suggested that the vertical TOA flux sensitivity to tropospheric O_3 is highly variable but is structurally consistent with atmospheric opacity, which determines on the amount of upwelling longwave radiation available for O_3 absorption (Lacis and Hansen, 1974; Berntsen et al., 1997; Worden et al., 2008; Worden

[*] Joint Institute for Regional Earth System Science and Engineering, University of California, Los Angeles, California, US

[†] Jet Propulsion Laboratory, California Institute of Technology, Pasadena, California, US

[‡] National Center for Atmospheric Research, Boulder, Colorado, US

[§] Bay Area Environmental Research Institute, Mountain View, California, US

[‖] NASA's Ames Research Center, Mountain View, California, US

Corresponding author: Le Kuai (lkuai@g.ucla.edu)

et al., 2011). The atmospheric opacity is highly correlated with the three key variables in the hydrological cycle: water vapor, clouds and temperature. Water vapor has significant absorption over a wide spectral range in the longwave; as a result, the outgoing longwave radiation is strongly reduced over high water vapor regions and hence the TOA flux sensitivity to tropospheric O_3 changes is low. Similarly, clouds (liquid or solid state of water vapor) absorb more OLR than water vapor and reduce atmospheric transparency by attenuating upwelling radiation below the cloud layer, also leading to a low TOA flux sensitivity to tropospheric O_3 changes. Lastly, besides its control of the phase transition between water vapor and cloud through the Clausius-Clapeyron relation, air temperature also determines the amount of longwave emission: the warmer the troposphere is, the higher of the TOA flux sensitivity to tropospheric O_3 changes will be.

In this paper, we will extend the work of Bowman et al. (2013) and further elucidate the influence of water vapor, clouds, temperature and tropospheric O_3 column to the O_3 LWRE. For consistency, we aim to be able to use data from a single instrument. Aura TES provides observations of water vapor, temperature, relative humidity (RH) and tropospheric O_3. Aura TES does not provide direct cloud measurements but we shall assume that RH is a good proxy for cloud coverage. RH is also a good variable that represents simultaneously the coupled effects of water vapor, temperature and cloud in the O_3 LWRE (Bowman et al., 2013). Following Doniki et al, (2015), we update the calculation of tropospheric O_3 TOA flux sensitivity (IRK) using a more accurate, 5-angle Gaussian Quadrature approximation, which is described in section 2. In section 3, we examine the relationship of TES TOA flux sensitivity to tropospheric O_3 with TES RH, water vapor, tropospheric O_3 column, and surface temperature in January and July of 2006 in order to investigate the primary drivers that determine the variation of tropical and sub-tropical O_3 GHG effect. Section 4 summarizes our main results. At the end, we discusses some implications from our study.

2. Method and data
2.1. O_3 IRK
The TOA flux in 9.6-μm O_3 band (F_{TOA}) can be computed as

$$F_{TOA} = \int_{v_1}^{v_2} \int_0^{2\pi} \int_0^{\frac{\pi}{2}} L_v\left(\theta, O_3(z), q(z)\right) \cos\theta \sin\theta \, d\theta \, d\phi \, dv, \quad (1)$$

where $L_v(\theta, O_3(z), q(z))$ is the TOA spectral radiance in units of W m^{-2} sr^{-1} cm^{-1}. θ is the zenith angle from 0 to $\frac{\pi}{2}$, ϕ is the azimuthal angle from 0 to 2π, and v is the spectral frequency from $v_1 = 980$cm^{-1} to $v_2 = 1080$cm^{-1} (or, equivalently, from 10.20 μm to 9.26 μm). L_v depends on the vertical distribution of O_3 as well as other atmospheric states, such as water vapor, temperature, cloud optical depth and emissivity; these atmospheric states are collectively denoted by q.

The O_3 instantaneous radiative kernel (IRK; in the unit of W m^{-2} ppb^{-1}) is defined as the sensitivity of the TOA radiative flux to increase in the vertical distribution of O_3 (see Worden et al., 2011):

$$\frac{\partial F_{TOA}}{\partial O_3(z_l)} = 2\pi \int_{v_1}^{v_2} \int_0^{\frac{\pi}{2}} \frac{\partial L_v\left(\theta, O_3(z), q(z)\right)}{\partial O_3(z_l)} \cos\theta \sin\theta \, d\theta \, dv, \quad (2)$$

where $O_3(z_l)$ is O_3 concentration at level z_l. $\frac{\partial L_v}{\partial O_3(z_l)}$ is the spectral radiance Jacobians, which is analytically calculated in the TES retrieval algorithm. Note that O_3 IRK also depends on geophysical quantities $q(z)$ other than O_3 itself. In this study, we are particularly interested in understanding how the key players in the hydrological cycle, including water vapor, temperature, and clouds, impact the O_3 IRK variation; therefore, this is to study the second derivatives $\frac{\partial^2 F_{TOA}}{\partial O_3(z_l)\partial q(z_l)}$.

The zenith integral in Eq. (2) can be rewritten as the first moment of $\frac{\partial L_v}{\partial O_3(z_l)}$ using the change of variable $x = \cos\theta$:

$$\int_0^{\frac{\pi}{2}} \frac{\partial L_v(\theta, ...)}{\partial O_3(z_l)} \cos\theta \sin\theta \, d\theta = \int_0^1 \frac{\partial L_v\left(\cos^{-1} x, ...\right)}{\partial O_3(z_l)} x \, dx. \quad (3)$$

In general, the first moment of any function can be a Gaussian quadrature (GQ): $\int_0^1 xf(x)dx \approx \sum_{k=1}^N w_k f(x_k)$, where w_k are the weights at selected abscissas x_k, which are the roots of the Jacobi polynomial (Abramowitz and Stegun, 1964). w_k and x_k can be obtained using the methods discussed in the Appendix of Li (2000). For the application to Eq. (3), Doniki et al. (2015) show that a 5-point ($N = 5$) GQ is more accurate than the previous single-angle anisotropy estimate (Worden et al., 2011). With this quadrature, Eq. (2) can be evaluated as

$$\frac{\partial F_{TOA}}{\partial O_3(z_l)} = 2\pi \sum_{k=1}^5 w_k \int_{v_1}^{v_2} \frac{\partial L_v(\theta_k, ...)}{\partial O_3(z_l)} dv \quad (4)$$

where $\theta_k = \cos^{-1} x_k$. The values of w_k and θ_k are shown in **Table 1**.

For convenience, we also define a logarithmic instantaneous radiative kernel (LIRK) with respect to $\ln O_3(z_l)$ to represent the change in TOA flux due to fractional change of O_3 abundance at each level:

$$\frac{\partial F_{TOA}}{\partial \ln O_3(z_l)} = 2\pi \sum_{i=1}^5 w_i \int_{v_1}^{v_2} \frac{\partial L_v(\theta_i, ...)}{\partial \ln O_3(z_l)} dv. \quad (5)$$

LIRK is in units of W m^{-2}.

Table 1: The weights and viewing angles to nadir for the 5-point Gaussian quadrature approximation. DOI: https://doi.org/10.1525/elementa.208.t1

Weight, w_i	TOA Nadir Angle, θ_i (°)	Surface Zenith Angle (°)
0.015748	63.6765	84.3452
0.073909	59.0983	72.2698
0.146387	48.1689	55.8040
0.167175	32.5555	36.6798
0.096782	14.5752	16.2213

2.2 Long wave radiative effect (LWRE)

The instantaneous radiative kernel, at a given location i and sampling time j, can be used to calculate the TOA longwave radiative effect (LWRE; in W m^{-2}), defined as the reduction in TOA longwave flux due to increase of O_3 abundance $\left(\Delta O_3^i(z_l)\right)$:

$$\Delta_{LWRE}^{i,j}(z_l) = \frac{\partial F_{TOA}^{i,j}}{\partial O_3^{i,j}(z_l)} \Delta O_3^{i,j}(z_l), \qquad (6)$$

which can also be obtained in terms of fractional changes using LIRK:

$$\Delta_{LWRE}^{i,j}(z_l) = O_3^{i,j}(z_l) \frac{\partial F_{TOA}^{i,j}}{\partial O_3^{i,j}(z_l)} \frac{\Delta O_3^{i,j}(z_l)}{O_3^{i,j}(z_l)} = \frac{\partial F_{TOA}^{i,j}}{\partial \ln O_3^{i,j}(z_l)} \frac{\Delta O_3^{i,j}(z_l)}{O_3^{i,j}(z_l)}. \qquad (7)$$

We then define a tropospheric LWRE, $\Delta_{LWRE}^{Troposphere}(i,j)$, as the net change of TOA flux due to the total tropospheric column absorption or 100% change of the species in the entire troposphere. $\Delta_{LWRE}^{Troposphere}(i,j)$ can be obtained by integrating IRK vertically from surface to the tropopause (TP) and letting $\Delta O_3^{i,j}(z_l)$ to be the change from O_3 abundance (e.g. measured by TES) to zero in that layer:

$$\Delta_{LWRE}^{Troposphere}(i,j) = \sum_{z_l=0}^{z_{TP}} \frac{\partial F_{TOA}^{i,j}}{\partial O_3^{i,j}(z_l)} O_3^{i,j}(z_l) = \sum_{z_l=0}^{z_{TP}} \frac{\partial F_{TOA}^{i,j}}{\partial \ln O_3^{i,j}(z_l)}. \qquad (8)$$

Following Bowman et al. (2013) to avoid different results due to different tropopause definitions [Stevenson et al., 2004], we define a chemical tropopause as the 150 ppb ozone isopleth, also used by Young et al., (2013) and Rap et al., (2015). Therefore, Z_{TP} in Eq. (8) corresponds to the altitude of the 150 ppb isopleth over a particular location.

The new version of monthly averaged O_3 IRK, LIRK and LWRE for 2006 has been run using about 16 global surveys (GS) TES data in each month. Each GS takes approximately every two days (~26 hours) to finish 16 orbits. One GS has about 3000 soundings to make the global coverage. Therefore, there are about 48000 soundings to grid bin to a 2° by 2.5° monthly mean map.

Note that the LWRE data we have processed (January, April, July, and October of 2006) with the new 5-point GQ is only 10% of the complete 5-year record. But the spatial patterns and the climatology of the LWRE obtained from these data are consistent with the previous 5-year record using the 1-point GQ LWRE (**Figure 1** of Bowman et al., 2013). All the figures in Bowman et al. (2013) are well replicated with the 5-point GQ data.

2.3 Relative humidity

The relative humidity (RH) represents the amount of water vapor in air in a percentage of the amount of water vapor needed for saturation at the same temperature. RH is estimated as the ratio of the partial pressure e_w of water vapor to the equilibrium water vapor pressure e_w^* at a given temperature

$$RH(z) = \frac{e_w\left(H_2O(z), P(z)\right)}{e_w^*\left(T(z), P(z)\right)}, \qquad (9)$$

where P is pressure. Regions with RH close to 100% typically have clouds; consequently, RH is often used as a proxy for the presence of clouds. In contrast, RH much less than 100% means water vapor is far from saturation at the given temperature and therefore is indicative of clear sky.

Since individual variations of temperature, water vapor, and clouds may change the O_3 GHG effect in different directions, RH, as a function of $H_2O(z)$ and $T(z)$, and as a proxy of cloud coverage, may be a better variable to describe the aggregated effect of the hydrological cycle on the O_3 GHG effect.

3. Results

3.1 Signatures of temperature, water vapor, clouds, and tropospheric O_3 on O_3 LWRE

Figure 1 shows the global distribution of the TOA flux in 9.6-μm O_3 band, tropospheric O_3 LWRE, RH data at 500 hPa, water vapor total column amount, tropospheric column O_3, and surface temperature in January and July 2006 all observed by TES. Since the averaging kernels of the TES retrievals peak at 500 hPa, we choose RH at 500 hPa. The spatial pattern of O_3 LWRE is quite similar to that of TOA flux, which is denoted as outgoing longwave radiation (OLR) (top two panels). At high latitudes, tropospheric O_3 LWRE is generally weak (< 0.3 W m^{-2} poleward of 45°) and decreases rapidly with latitude as a consequence of the negative gradient in surface temperature and O_3 variation (Doniki et al., 2015). The lower temperature results in less OLR and consequently less sensitivity to O_3. The processes controlled O_3 variation include ozone hole in the Polar regions and stratospheric intrusion into the troposphere.

The spatial pattern of O_3 LWRE is high (> 0.8 W m^{-2}) for values of RH less than 20%. Conversely, low LWRE regions (< 0.3 W m^{-2}) occur when RH is greater than 80%. Low RH generally occurs in subtropical dry regions (near 30°) such as the Sahara and Middle East in July, and subtropical Atlantic in both January and July. The zonal average of O_3 LWRE at subtropics is about 0.6 W m^{-2}. In contrast, the zonal average O_3 LWRE is 0.4 W m^{-2} in the ITCZ, especially at the convective continental regions, e.g. tropical rainforests, where RH is usually high.

In the tropics, LWRE values < 0.3 W m^{-2} are associated with RH greater than 80% and low TOA flux less than 15 W m^{-2}. The lower values of tropical O_3 LWRE also follow the meridional transitions of ITCZ between January and July. For example, in January, four deep convection zones, with tropical maxima in RH close to 100% and low TOA flux, ~13 W m^{-2}, for the 9.6-μm O_3 band are found in the central Pacific and tropical continents (Amazon basin, Congo basin, and Indonesia). These regions of maximum RH all correspond to minima in LWRE (< 0.3 W m^{-2}) and are mostly south of the equator. When the ITCZ moves to north of the equator in July, the convection zones with RH > 80% and low TOA flux near 13 W m^{-2} shift to

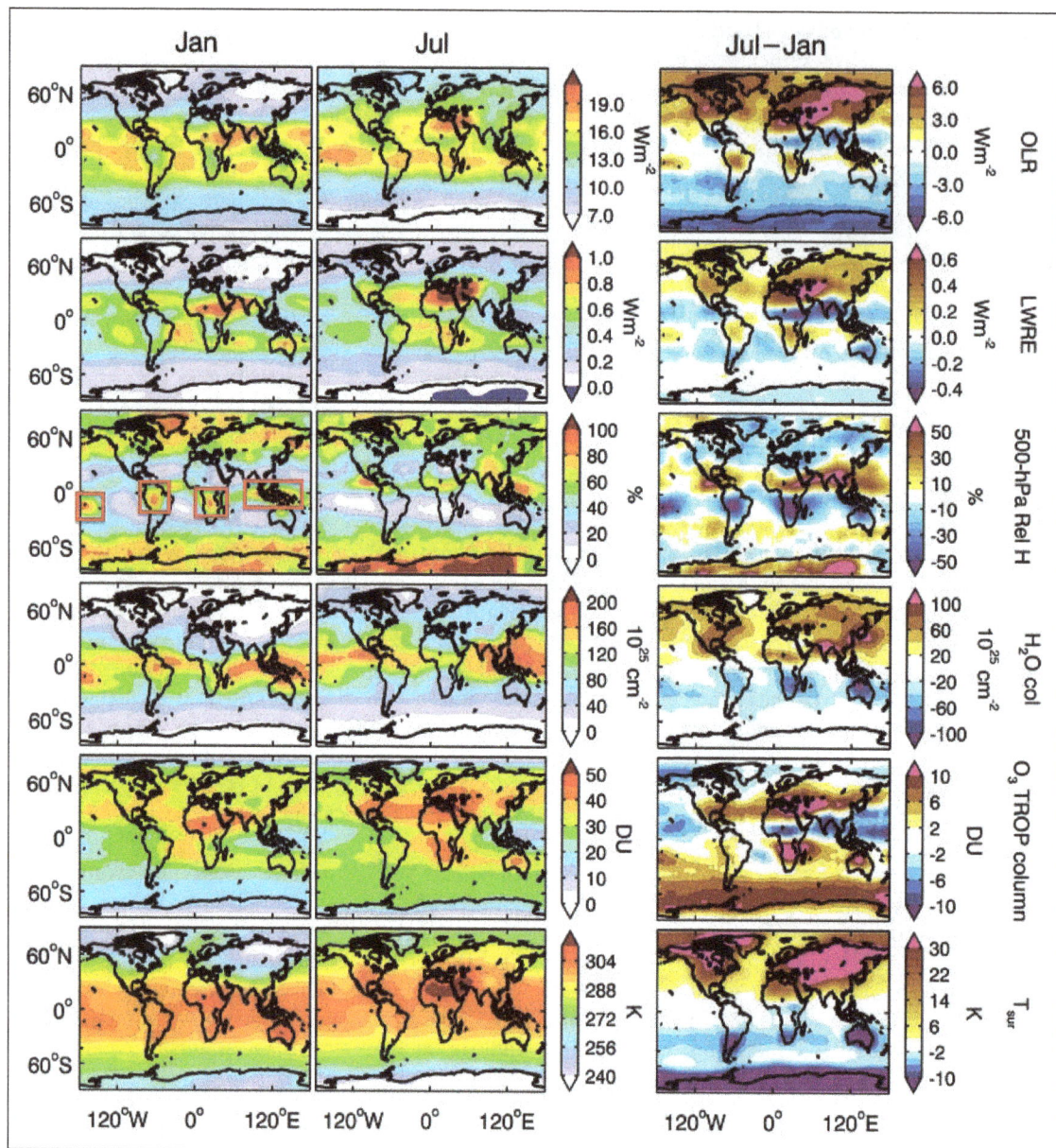

Figure 1: The comparison of outgoing longwave radiation, longwave radiative effect, relative humidity, water vapor, tropospheric O_3 column, and surface temperature. From top to bottom are the TES observations of outgoing longwave radiation (OLR) in 9.6-μm O_3 band, longwave radiative effect (LWRE), 500-hPa relative humidity (RH), and water vapor (H_2O) column, tropospheric O_3 column, and surface temperature January (left column), July (middle column) in 2006, and the difference between July and January (right column). DOI: https://doi.org/10.1525/elementa.208.f1

the East Tropical Pacific near the coast of Mexico, Africa Savanna, extending east across India and Southeast Asia to the north of Indonesia, as shown in **Figure 1**. The LWRE are consistently low over high RH regions in the tropical deep convection zones, where cloud layer shields the OLR flux originating from the troposphere and the OLR flux observed at TOA is mostly from O_3 above the cloud. This cloud effect has been demonstrated in **Figure 3** of Worden et al. (2011), suggesting that a strong sensitive layer in the mid-troposphere for clear-sky cases becomes a thinner layer that is shifted upwards in the all-sky cases in O_3 IRK data. In contrast, **Figure 1** also shows that the areas between these deep convection zones at the same

latitude are dominated by downwelling and are drier and cloud free with values of LWRE values that are higher due to less atmospheric attenuation.

The seasonal differences shown in the right column of **Figure 1** also suggest the opposing temporal changes in LWRE and RH. For example, when ITCZ moves to north in July, the four deep convection zones (Central Tropical Pacific, Amazon basin, Congo basin, and Indonesia) in January become dryer with low RH resulting in enhanced LWRE. The same relationship holds for most of the other regions except Australia, where effects other than water and cloud coverage, such as surface temperature dominate, described below.

Over Australia, the high LWRE values in January (> 0.7 W m^{-2}) are mainly due to high surface temperature and larger thermal contrast over the desert. A larger thermal contrast will amplify the sensitivity of the TOA flux to the tropospheric O$_3$ (Bowman et. al., 2013). In July, the lower RH and higher tropospheric O$_3$ column comparing with those in January, both of which should have led to a larger LWRE, however is compensated by the reduction of surface thermal contrast during the austral winter. Therefore, despite the Australian O$_3$ enhancement in July, the LWRE did not increase.

However, there are many regions where LWRE variations correspond to changes in tropospheric O$_3$ column. For example, a strong O$_3$ LWRE in the Africa savanna in January and Congo basin in July is related to O$_3$ enhancement due to biomass burning during their respective winters (Bowman et al., 2013). Since the ITCZ migrates from the Congo to the African savanna from January to July, the enhanced LWRE in both regions is also associated with lower RH.

In July, there is a large olive-shaped area with the global highest O$_3$ LWRE (> 1 W m^{-2}) over Sahara deserts and the Middle East, which is a result of multiple effects. The surface temperature is highest (> 310 K), which causes the highest values of TOA flux (> 18 W m^{-2}). The local minimum of RH ($< 20\%$) suggests clear sky. Cloud formation is less likely since tropopause descent and a dominant sinking flow of the Hadley circulation further prohibit upward movement of upper tropospheric air and adiabatic cooling. The Hadley cell is a tropical atmospheric circulation

in vertical and meridional directions and is characterized as a rising motion near the equator up to the tropopause, where the air separates into northward and southward branches reaching across to the subtropics (Holton and Hakim, 2012). The air parcel then descends and both branches return to the equator near the surface where subsequent convection closes the circulation.

Therefore, the atmosphere in the subtropical region is more transparent during boreal summer. In addition, an O$_3$ enhancement (tropospheric column > 45 DU) that is partly produced locally as well as transported both westward from an extended Asian monsoon anticyclone system and eastward by an Arabian anticyclone (Li et al., 2001; Liu et al., 2009) contributes to this O$_3$ LWRE maximum. Consequently, the maximum Middle East LWRE in July is a result of high surface temperatures and O$_3$ combined with low water vapor and cloud-free conditions.

In summary, we found that the combined impact of water vapor, cloud and temperature on LWRE is mostly explained by RH, especially in tropics and subtropics. Tropospheric O$_3$ variation may control the LWRE when the local atmosphere is transparent and TOA flux sensitivity is non-zero.

3.2 Vertical distribution of O$_3$ LIRK
Figure 2 shows the zonal pattern of RH and O$_3$ LIRK averaged from 0° to 25°S in January and from 0° to 5°N in July to account for the seasonal shift in the ITCZ belt. The ITCZ belt in July is much narrower so we chose 0° to 5°N instead of 0° to 25°N. An anti-correlation between RH and O$_3$ LIRK

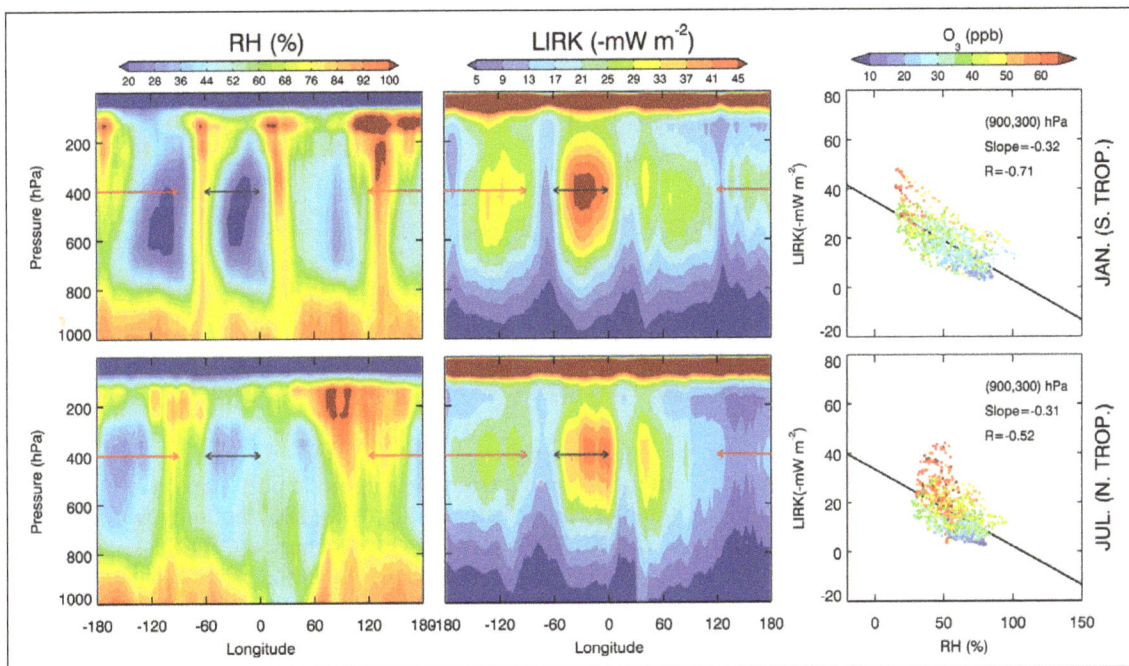

Figure 2: Vertical distribution of RH and O$_3$ LIRK in zonal pattern in ITCZ. TES RH (left), O$_3$ LIRK (middle), and their correlation between 300 and 900 hPa (right). Top row is data averaged from 0° to 25°S for S. Trop. in January. Bottom row is data averaged from 0° to 5°N for N. Trop. in July. The red arrows indicate the longitude range of Pacific and the blue arrows suggest the longitude range of Atlantic Ocean. The color of the scatter points on the correlation plots represents the O$_3$ amount in the color bar above. DOI: https://doi.org/10.1525/elementa.208.f2

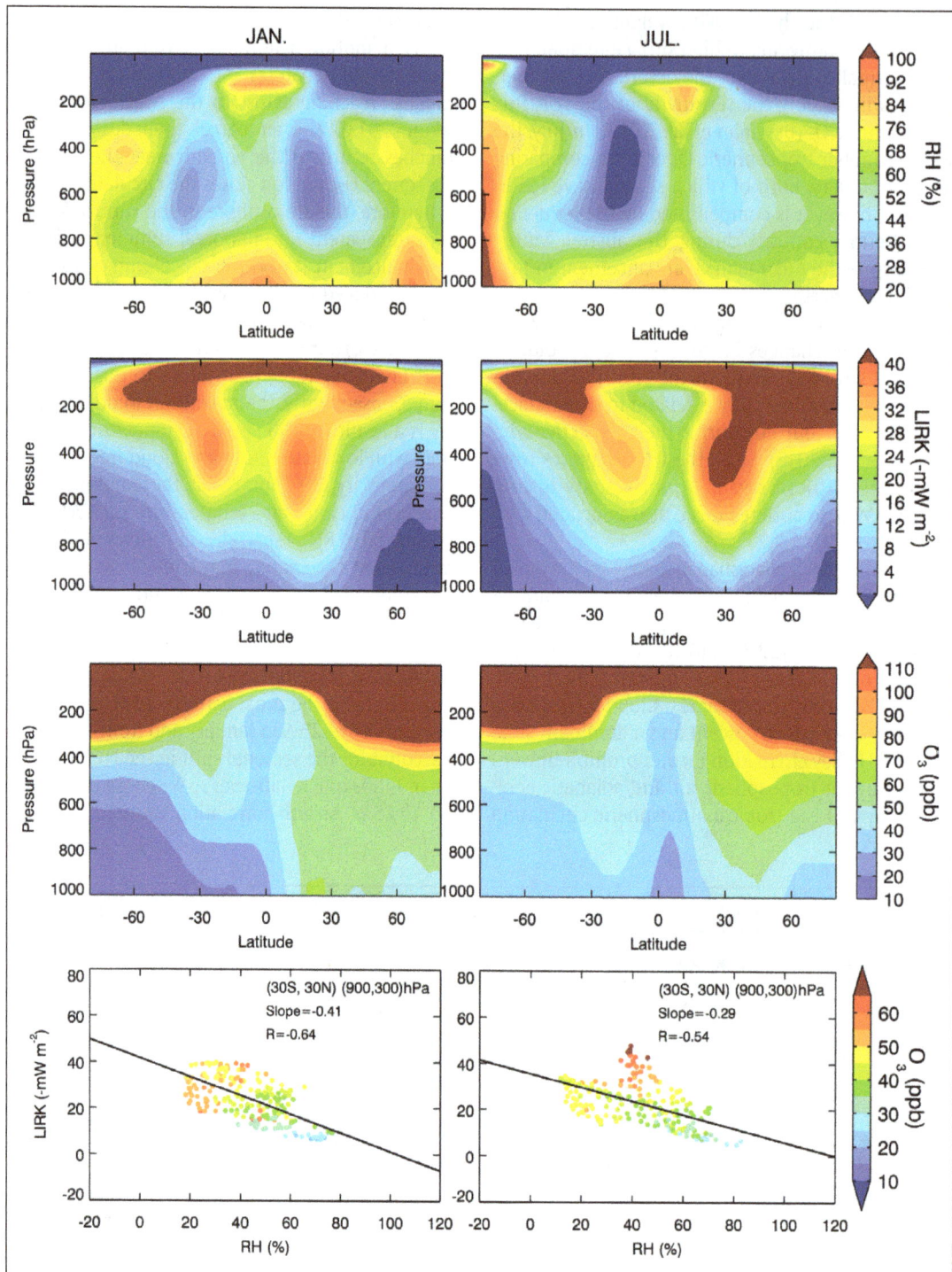

Figure 3: The comparison of RH, LIRK, and O$_3$ in meridional pattern from pole to pole. Zonally averaged TES RH (top row), O$_3$ LIRK (2nd row), and O$_3$ (3rd row) and the correlation between RH and O$_3$ LIRK (bottom row) for the region between 300 and 900 hPa and within in latitude of ±30°. The right column for January and the left column for July. The color of the scatter points on the correlation plots represents the O$_3$ amount in the color bar on the right. DOI: https://doi.org/10.1525/elementa.208.f3

is found in both S. Tropics and N. Tropics in the free troposphere (300–900 hPa) with the magnitude of correlation coefficients (R) 0.71 and 0.52, respectively and with similar slopes of –0.32 and –0.31 mW m^{-2} per % RH. Convection zones (e.g. tropical rainforests) characterized by high RH greater than 80% in the free troposphere between

300 and 900 hPa are associated with O$_3$ LIRK magnitudes less than 14 mW m^{-2}. The opposite holds for dry regions between the convection zones. The regions where RH is less than 40% correspond to LIRK magnitude greater than 20 mW m^{-2}. In July, the correlation plot suggests when RH is below 50%, LIRK tend to be high. Further increase

of LIRK are found due to O_3 enhancement (red points in **Figure 2** for July, > 60 ppb). These O_3 enhancement are found at N. tropical Atlantic (see **Figure 1**).

The Walker circulation characterized by high RH in January is a primary driver for the Central Pacific deep convection zones, near 170°W (Lau and Yang, 2003). The Walker circulation describes the motion of airflow in the zonal and vertical direction in the tropical troposphere and explains the formation of monsoons in the Indian and Central Pacific Ocean (Walker and Bliss, 1932; Lau and Yang, 2003). These effects dominate the tropospheric features in the LIRK in the Central Pacific, where the peak values are reduced significantly.

In addition to the Central Pacific, other deep convection zones near 60°W, 20°E, and 130°E respectively in **Figure 2** coincide with the local maxima in 500 hPa RH over the Amazon basin, Congo basin, and Indonesia, south of equator in **Figure 1**. All these regions also correspond to LWRE minima due to the reduction of mid tropospheric sensitivity. Areas between the convection zones are affected by the downward flow in the circulation resulting in regions that stay cool, dry, and cloud free. Consequently, the atmosphere is more transparent leading to higher TOA flux sensitivity to mid tropospheric O_3 and higher values of LIRK.

In July, the region with RH greater than 80% and O_3 LIRK less than 14 mW m^{-2} is over the Southeast Asia (90°E–150°E) as a consequence of the Asian monsoon, which transports water vapor from the Pacific and brings more precipitation to the southeast coast of Asia. Other low LIRK regions such as the eastern tropical Pacific near the coast of Mexico (~110°E) and African Savanna (~10°E) are also characterized by relatively high RH between 60–70%. In contrast, the regions with RH less than 40% and O_3 LIRK greater than 30 mW m^{-2} are found over the tropical Atlantic (–60°E to 0°) and Indian Ocean (50°E to 70°E). The LIRK in the eastern tropical Pacific, (–180°E to –120°E), where RH is also less than 40%, has a moderately high LIRK at ~25 mW m^{-2}. **Figure 1** suggests that tropospheric O_3 columns are much higher in the tropical Atlantic (more than 30 DU) than in the tropical Pacific (less than 20 DU). When RH and cloud optical depth are both low, attenuation by water vapor absorption or clouds on the TOA flux sensitivity to tropospheric O_3 is negligible resulting in higher tropospheric O_3 LWRE that can vary with O_3 amount. Under these conditions, more tropospheric O_3

over the tropical Atlantic than tropical Pacific will further strengthen the greenhouse gas effect. **Table 2** summarizes what are discussed above based on **Figure 2**.

Figure 3 shows the meridional distribution of RH and O_3 LIRK globally zonal averaged profiles. The correlation coefficients between RH and O_3 LIRK between 300 and 900 hPa at tropics and subtropics (±30°) are approximately –0.6. O_3 LIRK is characterized by a strong tropical stratospheric component (not the focus of this paper), with two legs extending into subtropical mid-troposphere, coinciding with subtropical low RH at the descending branches of the Hadley cell. The subtropical high O_3 LIRK in middle and upper troposphere is also partly due to free tropospheric O_3 enhancements in the upper troposphere from stratosphere-troposphere exchange (especially in the northern hemisphere). In these area, O_3 (shown in **Figure 3**, the 3rd row) is greater than 50 ppb due to stratosphere-troposphere exchange (Hitchman and Rogal, 2010; Rogal et al., 2010; Tilmes et al., 2010; Neu et al., 2014), lightning (Christian et al., 2003; Sauvage et al., 2007; Murray et al., 2012), or biomass burning (Bowman et al., 2009; van der Werf et al., 2010). The correlation scatter plot between RH and LIRK for July again show some of the cases, RH about 40%, the O_3 enhancement at Middle East further increase the LIRK, the same as in **Figure 2** correlation plot for July. The effect of O_3 enhancement in July under hot dry condition also explains why the correlation coefficient is always slightly lower in July than in January.

The tropical low O_3 LIRK is associated with the tower of high RH, which is partly due to the upwelling motion in Hadley cell causing deep convection in the ITCZ. **Figure 3** also show the O_3 LIRK tropical minimum shifts around the equator seasonally following the ITCZ.

4. Conclusions

The large-scale atmospheric circulation has a strong influence on the hydrological cycle, which impacts the ozone GHG effect. From a latitudinal perspective, the Hadley cell leads to high amounts of water vapor and clouds in the tropics and low water vapor and clear sky in the subtropics. From a meridional perspective, the Walker circulation drives the formation of the deep convection for upwelling and clear sky over subsidence regions at tropical Pacific and tropical Indian Ocean. These in turn modulate the large-scale pattern of the ozone GHG effect.

Table 2: The summary of specific regions in ITCZ highlighted by Figure 2. DOI: https://doi.org/10.1525/elementa.208.t2

	RH	O_3 LIRK (mW m^{-2})	Regions
Jan.	>70%	< 14	Central Pacific deep convection zones (–170°E) Amazon basin near (–60°E)
Southern tropics			Congo basin near (20°E) Indonesia near (130°E)
Jul.	>70%	< 14	Southeast Asia (90°E–150°E) Tropical Pacific near the coast of Mexico (–110°E) African Savanna (10°E)
Northern tropics	<40%	> 25	the tropical Atlantic (–50°E to 0°) Indian Ocean (50°E to 70°E) the eastern tropical Pacific, (–180°E to –120°E)

In the tropics, the low O_3 GHG effect, represented by LWRE and LIRK, is primarily driven by high concentration of water vapor and more clouds, both of which have the effect to attenuate the LWRE and LIRK (e.g. in the central tropical Pacific, Amazon basin, Congo basin, and Indonesia in January; the west tropical Pacific, Africa Savanna, and Southeast Asia in July).

Conversely, the subtropical high O_3 GHG effect is dominated by high surface temperature and elevated tropospheric O_3 amount, especially over desert regions where the effect by water vapor and clouds are much smaller than the tropics as a result of the downwelling branch of Hadley cell. Consequently, the high surface temperature and large thermal contrast amplifies the TOA flux sensitivity to tropospheric O_3. The mid and upper tropospheric O_3 enhancement further increases the O_3 GHG effect over the region with the transparent atmosphere. The confluence of low RH, high tropospheric O_3, and high surface temperature lead to a global maximum of the O_3 GHG effect over the Middle East in summer.

The TES observations of O_3 and RH are used to identify the primary large-scale circulation that determines the typical climate condition in the tropical and subtropical regions. TES RH observations link water vapor, temperature, and clouds and allow us to assess the primary drivers to the variations of O_3 GHG effect in the tropics and subtropics. Although the dependence on O_3 amount is not controlled exclusively by RH, we showed that the O_3 GHG effect is only large when water vapor and cloud attenuation is reduced significantly and thermal contrast is large enough for the sensitivity of O_3 GHG effect to tropospheric O_3 to become significant.

5. Implications

Exploration how the hydrological cycles controls the patterns and magnitude of O_3 TOA flux and flux sensitivity variation can help improve chemistry-climate model simulations and contribute to better estimates of present-day and future radiative forcing. For example, the downwelling of the Hadley circulation is the primary driver for the subtropical maximum of O_3 GHG effect. The width of the Hadley cell has been expanding (Seidel and Randel, 2007) and is expected to continue expanding under current climate change scenarios, with increases in global mean temperature and pole-to-equator temperature gradient (Frierson et al., 2007). The poleward shift of the downward branch of the Hadley cell means a poleward expansion of subtropical dry zones, which have the strongest O_3 GHG effect. The causes of this poleward shift include the increase of static stability in subtropics, tropopause height increase near the subtropics, and a shift in the ITCZ farther away from the equator due to the response to CO_2 forcing (Held, 2000; Lu et al., 2007; Kang and Lu, 2012). These changes would all affect the global distribution of O_3 GHG effect, especially for the subtropical maximum, and would have a positive feedback on global warming.

Furthermore, Pal et al. (2016) find that under business-as-usual emissions scenarios, climate extremes in some region like the Middle East, may hit wet-bulb temperatures (a combined measure of temperature and humidity) of 35°C before the end of this century. This wet-bulb temperature is at the limit of the human habitability where humidity prevents sweat from effectively cooling down the human body. People can survive in such heat, but only for a few hours. Their study shows that such heat waves will likely occur in places like Dubai or other Arabian Gulf regions. We show here that the Middle East currently has the global maximum O_3 GHG effect due to high surface temperatures, high tropospheric O_3 abundance and low RH. Without changes to the earth's energy budget, increasing surface temperatures and widening of the Hadley circulation will add additional O_3 radiative forcing to this region.

Finally, the Asian monsoon also tends to be strengthened by global warming (Li et al., 2010; Singh et al., 2014). The O_3 transport from Southeast Asia is an important driver for summer-time O_3 enhancements in the Middle East, which add another positive feedback to the Middle East O_3 GHG effect. These feedbacks would combine to accelerate the increase in surface temperature and subsequent climate extremes in the Middle East.

List of abbreviations

ACCMIP   Atmospheric Chemistry and Climate Model Intercomparison Project
GHG   Green House Gas
IRK   Instantaneous Radiative Kernel (mW/m²/ppb); Eq. (2) or (4)
LIRK   Logarithm Instantaneous Radiative Kernel (mW/m²); Eq. (5)
LWRE   Long Wave Radiative Effect (W/m²); Eq. (8)
OLR   Outgoing Longwave Radiation
RF   Radiative Forcing (W/m²)
RH   Relative Humidity (%); Eq. (9)
TOA   Top of atmosphere
F_{TOA} The TOA flux in 9.6-μm O_3 band (W/m²); Eq. (1)
L_v The TOA spectral radiance (W m⁻² sr⁻¹ cm⁻¹)

Acknowledgements
We thank the TES team at the Jet Propulsion Laboratory for their contribution to provide the data. L.K. also acknowledges Thomas Walker and King-Fai Li for helpful discussions. The TES products can be downloaded at http://tes.jpl.nasa.gov/data/.

Funding information
This work was funded by NASA Grant NNX14AE84G and carried out at the Jet Propulsion Laboratory, California Institute of Technology.

Competing interests
The authors have no competing interests to declare.

Author contributions
· Contributed to conception and design: LK, KWB, HMW.
· Acquisition of data: LK, RH, SSK.

- Analysis and interpretation of data: LK, KWB.
- Review of issues and summary of findings: All authors.
- Wrote the manuscript: LK, KWB.
- Contributed revisions to the manuscript: HMW, RH, SSK.
- Approved the submitted version for publication: All authors.

References

Abramowitz, M and Stegun, IA 1964 *Handbook of Mathematical Functions with Formulas, Graphs, and Mathematical Tables*. National Bureau of Standards. Applied Mathematics Series, Vol. **55**, 1045 pp.

Berntsen, TK, Isaksen, ISA, Myhre, G, Fuglestvedt, JS, Stordal, F, et al. 1997 Effects of anthropogenic emissions on tropospheric ozone and its radiative forcing. *J Geophys Res* **102**: 28101–28126. DOI: https://doi.org/10.1029/97JD02226

Bowman, KW, Jones, DBA, Logan, JA, Worden, HM, Boersma, F, et al. 2009 The zonal structure of tropical O_3 and CO as observed by the Tropospheric Emission Spectrometer in November 2004 — Part 2: Impact of surface emissions on O_3 and its precursors. *Atmos Chem Phys* **9**: 3563–3582. DOI: https://doi.org/10.5194/acp-9-3563-2009

Bowman, KW, Shindell, DT, Worden, HM, Lamarque, JF, Young, PJ, et al. 2013 Evaluation of ACCMIP outgoing longwave radiation from tropospheric ozone using TES satellite observations. *Atmos Chem Phys* **13**: 4057–4072. DOI: https://doi.org/10.5194/acp-13-4057-2013

Christian, HJ, Blakeslee, RJ, Boccippio, DJ, Boeck, WL, Buechler, DE, et al. 2003 Global frequency and distribution of lightning as observed from space by the Optical Transient Detector. *J Geophys Res* **108**: 4005. DOI: https://doi.org/10.1029/2002JD002347

Doniki, S, Hurtmans, D, Clarisse, L, Clerbaux, C, Worden, H, et al. 2015 Instantaneous longwave radiative impact of ozone: an application on IASI/MetOp observations. *Atmos Chem Phys* **15**: 12971–12987. DOI: https://doi.org/10.5194/acp-15-12971-2015

Frierson, DMW, Lu, J and Chen, G 2007 Width of the Hadley cell in simple and comprehensive general circulation models. *Geophys Res Lett* **34**(L18): 804. DOI: https://doi.org/10.1029/2007GL031115

Held, IM 2000 The general circulation of the atmosphere. 2000 Program in Geophysical Fluid Dynamics. Woods Hole, MA: Woods Hole Oceanographic Institute.

Hitchman, MH and Rogal, MJ 2010 Influence of tropical convection on the Southern Hemisphere ozone maximum during the winter to spring transition. *J Geophys Res* **115**(D14): 118. DOI: https://doi.org/10.1029/2009JD012883

Holton, JR and Hakim, GJ 2012 *An Introduction to Dynamic Meteorology*. 5th ed. Waltham, MA: Academic Press, 552 pp.

Kang, SM and Lu, J 2012 Expansion of the Hadley cell under global warming: Winter versus summer. *J Climate* **25**: 8387–8393. DOI: https://doi.org/10.1175/JCLI-D-12-00323.1

Lacis, AA and Hansen, J 1974 A parameterization for the absorption of solar radiation in the earth's atmosphere. *Journal of the Atmospheric Sciences* **31**: 118–133. DOI: https://doi.org/10.1175/1520-0469(1974)031<0118:APFTAO>2.0.CO;2

Lamarque, JF, Shindell, DT, Josse, B, Young, PJ, Cionni, I, et al. 2013 The Atmospheric Chemistry and Climate Model Intercomparison Project (ACCMIP): overview and description of models, simulations and climate diagnostics. *Geosci Model Dev* **6**: 179–206. DOI: https://doi.org/10.5194/gmd-6-179-2013

Lau, KM and Yang, S 2003 Walker circulation. *Encycl Atmos Sci* **6**: 2505–2510. DOI: https://doi.org/10.1016/B0-12-227090-8/00450-4

Li, J 2000 Gaussian quadrature and its application to infrared radiation. *J Atmos Sci* **57**: 753–765. DOI: https://doi.org/10.1175/1520-0469(2000)057<0753:GQAIAT>2.0.CO;2

Li, J, Wu, Z, Jiang, Z and He, J 2010 Can global warming strengthen the east Asian summer monsoon? *J Climate* **23**: 6696–6705. DOI: https://doi.org/10.1175/2010JCLI3434.1

Lu, J, Vecchi, GA and Reichler, T 2007 Expansion of the Hadley cell under global warming. *Geophys Res Lett* **34**(L06): 805. DOI: https://doi.org/10.1029/2006GL028443

Murray, LT, Jacob, DJ, Logan, JA, Hudman, RC and Koshak ,WJ 2012 Optimized regional and interannual variability of lightning in a global chemical transport model constrained by LIS/OTD satellite data. *J Geophys Res* **117**(D20): 307. DOI: https://doi.org/10.1029/2012JD017934

Myhre, G, Shindell, D, Bréon, F-M, Collins, W, Fuglestvedt, J, et al. 2013 Anthropogenic and Natural Radiative Forcing. In: Stocker, TF, Qin, D, Plattner, G-K, Tignor, M, Allen, SK et al. , eds. Climate Change: The Physical Science Basis Contribution of Working Group I to the Fifth Assessment Report of the Intergovernmental Panel on Climate Change. Cambridge University Press, Cambridge, United Kingdom and New York, NY, USA.

Pal, JS and Eltahir, EAB 2016 Future temperature in southwest Asia projected to exceed a threshold for human adaptability. *Nat Clim Change* **6**: 197–200. DOI: https://doi.org/10.1038/nclimate2833

Rap, A, Richards, NAD, Forster, PM, Monks, SA, Arnold, SR and Chipperfield, MP 2015 Satellite constraint on the tropospheric ozone radiative effect. *Geophys Res Lett* **40**(12): DOI: https://doi.org/10.1002/2015GL064037

Rogal, M, Hitchman, MH, Buker, ML, Tripoli, GJ, Stajner, I, et al. 2010 Modeling the effects of Southeast Asian monsoon outflow on subtropical anticyclones and midlatitude ozone over the Southern Indian Ocean. *J Geophys Res* **115**(D20): 101. DOI: https://doi.org/10.1029/2009JD012979

Rothman, LS, Gamache, RR, Goldman, A, Brown, LR, Toth, RA, et al. 1987 The HITRAN database: 1986 edition. *Appl Opt* **26**: 4058–4097. DOI: https://doi.org/10.1364/AO.26.004058

Sauvage, B, Martin, RV, van Donkelaar, A and Ziemke, JR 2007 Quantification of the factors controlling tropical tropospheric ozone and the South Atlantic maximum. *J Geophys Res* **112**(D11): 309. DOI: https://doi.org/10.1029/2006JD008008

Seidel, DJ and Randel, WJ 2007 Recent widening of the tropical belt: Evidence from tropopause observations. *J Geophys Res* **112**(D20): 113. DOI: https://doi.org/10.1029/2007JD008861

Singh, D, Tsiang, M, Rajaratnam, B and Diffenbaugh, NS 2014 Observed changes in extreme wet and dry spells during the South Asian summer monsoon season. *Nat Clim Change* **4**: 456–461. DOI: https://doi.org/10.1038/nclimate2208

Stevenson, DS, Doherty, RM, Sanderson, MG, Collins, WJ, Johnson, CE, and Derwent, RG 2004 Radiative forcing from aircraft NO$_x$ emissions: Mechanisms and seasonal dependence, *J. Geophys Res* **109**(D17): 307. DOI: https://doi.org/10.1029/2004JD004759

Tilmes, S, Pan, L, Hoor, P, Atlas, E, Avery, MA, et al. 2010 An aircraft-based upper troposphere lower stratosphere O$_3$, CO, and H$_2$O climatology for the Northern Hemisphere. *J Geophys Res* **115**(D14): 303. DOI: https://doi.org/10.1029/2009JD012731

van der Werf, GR, Randerson, JT, Giglio, L, Collatz, GJ, Mu M, et al. 2010 Global fire emissions and the contribution of deforestation, savanna, forest, agricultural, and peat fires (1997–2009). *Atmos Chem Phys* **10**: 11707–11735. DOI: https://doi.org/10.5194/acp-10-11707-2010

Worden, HM, Bowman, KW, Kulawik, SS and Aghedo, AM 2011 Sensitivity of outgoing longwave radiative flux to the global vertical distribution of ozone characterized by instantaneous radiative kernels from Aura-TES. *J Geophys Res* **116**(D14): 115. DOI: https://doi.org/10.1029/2010JD015101

Worden, HM, Bowman, KW, Worden, JR, Eldering, A and Beer, R 2008 Satellite measurements of the clear-sky greenhouse effect from tropospheric ozone. *Nature Geosci* **1**: 305–308. DOI: https://doi.org/10.1038/ngeo182

Young, PJ, Archibald, AT, Bowman, KW ,Lamarque, JF, Naik, V, Stevenson, DS, Tilmes, S, Voulgarakis, A, Wild, O, Bergmann, D and Cameron-Smith, P 2013 Pre-industrial to end 21st century projections of tropospheric ozone from the Atmospheric Chemistry and Climate Model Intercomparison Project (ACCMIP), *Atmos. Chem. Phys.*, **13**(4): 2063–2090. DOI: https://doi.org/10.5194/acp-13-2063-2013

10

Effective inundation of continental United States communities with 21st century sea level rise

Kristina A. Dahl*, Erika Spanger-Siegfried[†], Astrid Caldas[‡] and Shana Udvardy[‡]

Recurrent, tidally driven coastal flooding is one of the most visible signs of sea level rise. Recent studies have shown that such flooding will become more frequent and extensive as sea level continues to rise, potentially altering the landscape and livability of coastal communities decades before sea level rise causes coastal land to be permanently inundated. In this study, we identify US communities that will face effective inundation—defined as having 10% or more of livable land area flooded at least 26 times per year—with three localized sea level rise scenarios based on projections for the 3rd US National Climate Assessment. We present these results in a new, online interactive tool that allows users to explore when and how effective inundation will impact their communities. In addition, we identify communities facing effective inundation within the next 30 years that contain areas of high socioeconomic vulnerability today using a previously published vulnerability index. With the Intermediate-High and Highest sea level rise scenarios, 489 and 668 communities, respectively, would face effective inundation by the year 2100. With these two scenarios, more than half of communities facing effective inundation by 2045 contain areas of current high socioeconomic vulnerability. These results highlight the timeframes that US coastal communities have to respond to disruptive future inundation. The results also underscore the importance of limiting future warming and sea level rise: under the Intermediate-Low scenario, used as a proxy for sea level rise under the Paris Climate Agreement, 199 fewer communities would be effectively inundated by 2100.

Keywords: climate change; sea level rise; coastal resilience; socioeconomic vulnerability; United States

1. Introduction

Sea level rise as a consequence of ongoing climate change poses a threat to millions of people worldwide (Hinkel et al. 2014). In the United States alone, the combination of global sea level rise, population growth and land use change is projected to expose between 4 and 13 million people to inundation by the year 2100 (Hauer et al. 2016). Left unabated, rising seas could affect upwards of 20 million US residents through the end of this century and beyond (Strauss et al. 2015).

While the number of people and communities affected by future inundation depends on the pace and magnitude of sea level rise, recurrent tidal flooding is already emerging as one of the most visible and quantifiable present-day signs of climate change. The East and Gulf Coasts of the US experienced some of the world's fastest rates of sea level rise during the 20th century (National Oceanic and Atmospheric Administration 2013a; Dangendorf et al. 2017). These rising seas have caused tidal flooding–coastal flooding that is driven in large part by routine tidal

fluctuations rather than precipitation or storm surge–to become an increasingly frequent occurrence in US coastal communities. Whereas minor coastal flooding along the East, Gulf, and West coasts of the US occurred just once every one to five years in the 1950s, it was occurring about once every three months by 2012 (Sweet et al. 2014).

Sea level rise is expected to make recurrent tidal flooding both more frequent and more extensive (Sweet & Park 2014; Moftakhari et al. 2015; Dahl et al. 2017; Kulp & Strauss 2017). While the tidal datums associated with the mean higher high water (MHHW) mark could be revised upward as sea level rises, the water level at which a community begins to flood will not change, thus leading to an increase in flood frequency. With this increase, many areas will flood with such frequency–potentially facing dozens to hundreds of minor coastal floods per year by mid-century–that, in the absence of protective measures, they could be rendered unusable before they actually fall at or below the present day MHHW level.

The definition of an area permanently inundated by the ocean is conceptually and functionally straightforward: Any area that is under water at high tide would be considered permanently inundated. Sea level rise is projected to permanently inundate many coastal areas in the US this century [e.g. NOAA, 2017]. Many communities, however, are already facing disruptive, even transformative flooding

* Dahl Scientific, San Francisco, CA, US

[†] Union of Concerned Scientists, Cambridge, MA, US

[‡] Union of Concerned Scientists, Washington, DC, US

Corresponding author: Kristina A. Dahl (kdahl@alum.mit.edu)

long before they will be rendered permanently inundated (Spanger-Siegfried et al. 2014). In places such as Annapolis, Maryland, Norfolk, Virginia, and Miami Beach, Florida, substantial investments of time and money are being made to cope with frequent tidal flooding that disrupts daily life and business operations (City of Annapolis 2011; Applegate 2014; Weiss 2016).

As sea level rises, more coastal communities will begin to see increasingly frequent tidal flooding that is both expansive enough to preclude normal daily life in certain areas (hindering work and school transportation, impeding commerce, damaging property, etc.) and frequent enough to make adjusting to this disruption costly—in some cases prohibitively so—or untenable (Spanger-Siegfried et al. 2014; Sweet & Park 2014; Moftakhari et al. 2015; Moftakhari et al. 2017). Investments into protective measures such as bulkheads or pump systems can make a substantial difference to community-level flood severity (Allen 2016). While not specifically addressed in the present study, such measures have the potential to forestall the onset of disruptive flooding.

The consequences of frequent flooding for communities that already face socioeconomic challenges are likely to be even more disruptive than for those with greater resources. While the causes of socioeconomic vulnerability are complex and can encompass a wide range of variables—including income, race, education, and health insurance coverage—communities with high socioeconomic vulnerability are traditionally more impacted when faced with environmental hazards such as flooding and have fewer resources to cope and adapt (Adger et al. 2009; Lane et al. 2013; Dilling et al. 2015).

In this study, we examine what we call "effective inundation." We consider an effectively inundated area to be one in which flooding is so frequent that it renders the area's current use no longer feasible. In this sense, effective inundation is the point at which a community is forced to make changes to ensure its residents are safe and its infrastructure and services are functional.

Effective inundation exists along an inundation trajectory that begins with no tidal flooding, then shifts as sea level rises to infrequent tidal flooding, then advances further into frequent tidal flooding, which becomes effective inundation, and eventually, permanent inundation (**Figure 1**).

Despite a general understanding that sea level rise will bring more frequent flooding to many areas (Dahl et al., 2017 and others) and that permanent inundation is a long-term risk, there are few tools available to communities that assess the growing land area likely to be affected by frequent, disruptive flooding within timeframes associated with community planning horizons.

Previously published tools and studies have focused on either a) the frequency and intensity of coastal flooding, either tidally-driven or from storm surge, at defined time points in the future (e.g. Sweet & Park 2014; Moftakhari et al. 2015; Dahl et al. 2017); b) national-scale visualization and analysis of inundation with defined increments of sea level rise (e.g. 0.5 m, 1.0 m) relative to MHHW irrespective of the fact that sea level is not expected to rise uniformly along our coasts (Marcy et al. 2011; Climate Central 2014; Hauer et al. 2016); or c) local-scale visualizations and analyses reflecting the amount of sea level rise projected locally for a given year (e.g. TMAC, 2015).

The first approach (a) is useful in communicating the magnitude and extent of projected future flooding; however a frequency alone (e.g. 180 floods per year) without a tie to the area affected by such flooding limits how much a community can do with the information. The second approach (b) relies on users to have some *a priori* knowledge about the local pace of sea level rise as well as different sea level rise scenarios. For users whose expertise lies outside of sea level rise science, having to do additional research to link the mapped increments of sea level rise to timeframes could be an impediment to well-informed decision-making. The third approach (c) has its greatest utility for the communities that have undertaken such efforts, but is not universally available to all communities.

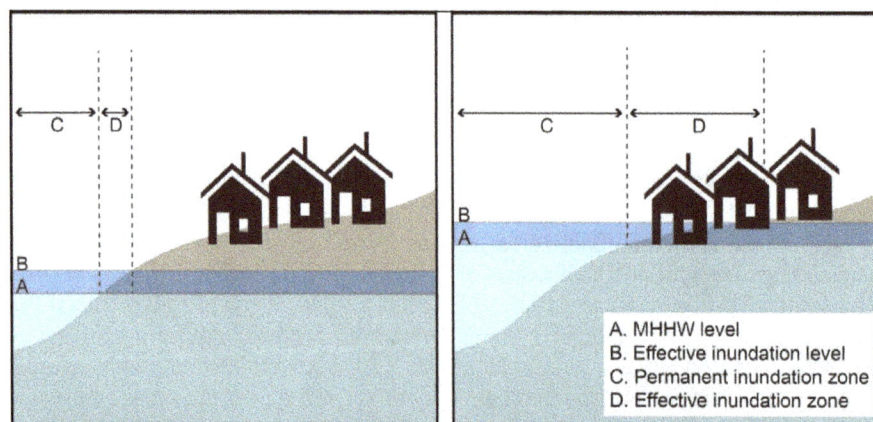

Figure 1: Sea level rise expands the zone of effective inundation. Areas that fall below the mean higher high water (MHHW; light blue) level today are permanently inundated, as infrastructure below the MHHW level would be inundated, on average, once daily. Areas that lie just above the MHHW level (darker blue) flood regularly enough that their use is limited. These areas are effectively inundated. Compared to today (left panel), sea level rise will expand both the permanent inundation zone and the effective inundation zone (right panel). DOI: https://doi.org/10.1525/elementa.234.f1

With all of these approaches, the link between potential reductions in greenhouse gas emissions and community-level coastal impacts is not explicit.

With these gaps in the existing decision-making tools in mind, we have undertaken a novel analysis that identifies where and when sea level rise effectively inundates coastal communities in the continental US through the end of this century. We also evaluated the intersection of this physical exposure and socioeconomic vulnerability as measured by the Social Vulnerability Index, recognizing that compounding risk factors will create additional challenges for many communities (SoVI; Cutter, Boruff, & Shirley, 2003; Martinich, Neumann, Ludwig, & Jantarasami, 2013).

We do this by:

1. Developing a method to quantify "effective inundation";
2. Mapping the extent of effective inundation within the 23 coastal states of the continental US at a series of time steps between now and 2100 using tide gauge-specific sea level rise projections based on three global sea level rise scenarios published for the Third US National Climate Assessment (NCA hereafter) (Parris et al. 2012; Walsh et al. 2014);
3. Explicitly connecting the concept of reductions in greenhouse gas emissions to effective inundation by using the NCA Intermediate-Low projection as a proxy for sea level rise under a scenario in which future warming is capped at 2°C;
4. Identifying cohorts of communities at the Census county subdivision level that meet the effectively inundated threshold for each future time horizon;
5. Evaluating the proportion of exposed communities that contain at least one Census tract with high socioeconomic vulnerability as defined by the SoVI;
6. Developing a practical online interactive planning tool that allows users to explore the extent of effective inundation at any location with different sea level rise projections at specific years in the future.

2. Methodology

2.1 Determining a frequency threshold

Flood risk tolerance will vary from community to community. In order to conduct a nationally-consistent spatial and temporal analysis, however, we defined a single flooding frequency associated with effective inundation and a land area threshold above which a community would be considered effectively inundated. In addition to reviewing the literature on tipping points in the flood frequencies that communities can cope with (Sweet & Park 2014), we conducted interviews with community experts in East and Gulf Coast communities including Annapolis, Maryland, Charleston, South Carolina, Broad Channel, New York, and consulted publicly available sources such as National Weather Service alerts to determine this frequency. These interviews are summarized in Table S1.

The current frequency of minor coastal flooding—often called nuisance flooding—in these communities ranges from approximately 24 in Charleston to 50 floods per year in Annapolis, on average (Dahl et al. 2017). Despite this large range, and speaking to the issue of different tolerance levels to flooding, each community was already developing or implementing a response to frequent flooding. A city official from Annapolis noted that the city initiated a response to flooding long before reaching the level of 50 floods per year (L. Craig, pers. comm.), while in the 1980s Charleston developed a comprehensive drainage master plan in response to flooding—when flooding was not as frequent as it is today (City of Charleston 2015). In Broad Channel, flooding on certain streets around each full and new moon (about 2 times per month) had driven the neighborhood association to lobby for and secure $28 million for sea walls and road elevation (Katz 2016).

These conversations conveyed that communities were responding to flooding, typically of limited areas, out of necessity and long before it reached the level of 50 events per year. Two of the four communities we spoke with—Charleston and Broad Channel—had taken action by the time they were coping with about 25 flood events per year.

This research suggests that 26 floods or more per year has, for affected communities, required substantial planning and investment. We therefore settled on this frequency as a threshold for defining effectively inundated areas.

In addition to defining a frequency threshold, we defined a threshold of affected land area above which we consider a community to be chronically inundated. Based on an evaluation of our results of present day effectively inundated communities and conversations with experts within those communities, we posit that if 10% or more of a community's usable land area is flooded 26 times per year or more, major municipal challenges will ensue in many cases. These challenges could include, for example, the need for: significant investments in shoreline protection structures; reallocation of land to open space to allow floodwaters to ebb and flow; raising homes, streets, and other infrastructure; or relocating coastal residents to inland areas.

In reality, the impact of flooding on a community has as much or more to do with *what* is being flooded as with *the area* being flooded. Based on our initial results of the effectively inundated area today, the communities of Annapolis, Maryland, and Miami Beach, Florida, do not experience flooding of 10% or more of their area 26 times per year or more. And yet frequent flooding of critical areas of those communities has prompted major investments of time and money (City of Annapolis 2011; Weiss 2016). In contrast, there are low-lying coastal communities where much more than 10% of the land area floods 26 or more times per year, but the flooded area is largely rural and uninhabited and thus does not affect the local people.

Our interviews with local experts revealed that there is no one land area threshold that applies universally to all communities. However, 80% of the 91 communities that meet both the frequency and 10% land area thresholds for flooding today largely fall within two regions with well-documented flooding problems: Louisiana and the Eastern Shore of Maryland. Frequent flooding in Isle de Jean Charles, Louisiana, for example, led residents there to seek and receive federal assistance for relocation (Maldonado et al. 2014). And the population on Smith Island, Maryland, has declined by more than one-third since 2010, in part due to frequent flooding (Holland 2016; TownCharts 2017). In speaking with local experts representing the majority of communities that met the effective inundation threshold today, both the area we mapped as effectively inundated and the frequency of inundation within that area were confirmed as consistent with their current observations and experience (Table S1). In one case (West Wildwood, New Jersey), very recent upgrades to bulkheads had reduced flooding below the extent and frequency indicated by our analysis, suggesting the importance of continuing to update local digital elevation models as protective measures are put in place.

Tidal events that exceed the effective inundation threshold could be affected by a number of factors in addition to tidal variability. These factors include storminess, long-term changes in regional climate patterns–such as the prevailing wind direction–or the Pacific Decadal Oscillation. This analysis does not attempt to separate out the differing causes of flood events. Tidally driven flood events tend to cluster around times when a new or full moon coincides with lunar perigee–the point at which the moon is closest to the Earth–because these conditions amplify the normal tidal range. These events tend to occur more in the spring and fall rather than being spaced evenly throughout the year.

2.2 Tide gauge data to identify the water level associated with 26 exceedances per year

In order to determine the physical areas of the US that are inundated at least 26 times per year, we utilized a set of 93 tide gauges (66 on the East and Gulf Coasts, 27 on the West Coast) maintained by the National Ocean Service. Using 20 years (1996–2015) of hourly, verified water level data for each gauge, we determined the threshold water level relative to the present MHHW level that was exceeded 26 ± 1 times annually (Table S2).

The water level at each gauge associated with the 26 floods per year threshold—hereafter referred to as the effective inundation threshold–could be influenced by a number of factors, both natural and anthropogenic in cause. On interannual and interdecadal timescales, the 18.6 year nodal tidal cycle and the 8.85 year cycle of lunar perigee are both known to influence mean sea level and MHHW along the US East and Gulf Coasts and elsewhere (Flick et al. 2003; Haigh et al. 2011; Wadey et al. 2014). The El Niño Southern Oscillation also affects sea level and extreme water levels on both the East and West Coasts of the US (Sweet & Zervas 2011; Hamlington et al. 2015). On shorter timescales, extreme sea levels such as occurred along the US East Coast in 2009–2010 have the potential to influence flood frequency and the water level associated with effective inundation threshold (Sweet et al. 2009; Goddard et al. 2015). By using 20 years of tide gauge data, our results encompass a full nodal tidal cycle, more than two cycles of lunar perigee, and several El Niño events.

While 30 years is typically considered the modern climate epoch, sea level has also risen substantially in that time period, which has caused an increase in the frequency of tidal flooding at the nuisance level and would likely affect the water level exceeded 26 times per year (Church & White 2011; Ezer & Atkinson 2014; Sweet et al. 2014; Hay et al. 2015; Moftakhari et al. 2015). In using 20 years of data, we aimed to use enough data to encompass the 18.6 and 8.85 year cycles mentioned above while also reasonably capturing modern sea level conditions. This is consistent with our previously published research (Dahl et al. 2017) and considerably longer than the tide gauge reference period used for projections of future flood frequency by previous studies (Sweet & Park 2014). Our projections assume that future tidal ranges will not differ substantially from those during the reference period, though there is evidence that sea level rise may increase tidal range (Flick et al. 2003; Passeri et al. 2016). While the sea level rise projections we use incorporate local rates of vertical land movement, we do not model any changes in coastal morphology, although such changes are likely to occur as sea level rises (FitzGerald et al. 2008; Lentz et al. 2016).

The effective inundation threshold was determined for each year (January–December) of the tide gauge record. Years for which 10% or more of the hourly observations were missing were excluded from the analysis (Table S2). The threshold water level for each gauge was determined recursively using a script that counted the number of exceedances of a specified water level starting with the MHHW level. The script then adjusted the water level in increments of 0.30 mm (0.0010 ft) and recounted the number of exceedances until the number of exceedances was 26 ± 1. This, and all other scripts used developed for this analysis, are available in a public GitHub repository at https://github.com/kristydahl/permanent_inundation.

We then used the mean threshold water level for all of the years to define the effective inundation threshold relative to MHHW for each gauge. We used the standard deviation about the mean effective inundation threshold for the full set of gauges as a component in the combined linear error used to define the time steps we analyzed from now through 2100 (see section 2.4).

It is important to note that tide gauges record variations in water levels relative to local benchmarks that are ideally situated on bedrock. This is not always the case, however. In Louisiana, where benchmarks are typically located tens of meters below the land surface, gauges are recording water level variations relative to those subsurface benchmarks (Jankowski et al. 2017). While this study does not attempt to correct for this phenomenon, it is important to note that in places like Louisiana, the determination of inundation thresholds could be affected by long-term deep subsidence.

2.3 Elevation data and inundated areas

Mapping the extent of the effectively inundated area based on water levels from tide gauges requires a digital elevation model (DEM). We obtained DEMs for the continental US from the National Oceanic and Atmospheric

Administration (Marcy et al. 2011). The resolution of the DEMs varies between ⅓ arc second (~10 meters) and 1/9 arc second (~3 m), though much of the East Coast is at the latter, higher resolution. The DEMs, which were used in the creation of NOAA's Sea Level Rise Viewer, are lidar-based and were conditioned and created specifically for sea level rise mapping (Marcy et al. 2011; NOAA 2017). Because the original data sources vary, so does the vertical uncertainty (root mean square error, or RMSE) of the DEMs. All of the DEMs meet or exceed the 18.5 cm RMSE standard for the National Flood Insurance Program (NOAA 2017). Investigation of the DEM metadata showed that RMSEs were less than 10 cm for most of the East Coast and higher for some parts of the Gulf Coast. Vertical accuracy data was not reported within the metadata of the West Coast DEMs. We assume an average RMSE of 9.25 cm, which we use in our calculations of combined linear error and minimum sea level rise interval below.

2.4 Sea level rise projections

To determine the height of the effective inundation threshold over time, we used local projections based on three global sea level rise projections originally developed for the 3rd National Climate Assessment (NCA; Parris et al. 2012; Walsh et al. 2014). The NCA Highest scenario, which projects 2 m of sea level rise globally by 2100, assumes ocean warming in accordance with IPCC AR4 projections and an estimate of maximum possible ice loss (Pfeffer et al. 2008). The NCA Intermediate-High scenario projects 1.2 m of sea level rise globally by 2100 and assumes warming associated with the upper end of the IPCC AR4 projections while ice loss is modeled using a semi-empirical approach (e.g. Horton et al., 2008; Vermeer & Rahmstorf, 2009; Jevrejeva, Moore, & Grinsted, 2010). The NCA Intermediate-Low scenario assumes that sea level rise is driven primarily by ocean warming with very little contribution of ice loss. This scenario is associated with an average global temperature increase of 1.8°C and a 0.5 m rise in sea level by 2100 (Parris et al. 2012).

There is no published sea level rise projection developed specifically with the goals of the Paris Climate Agreement–namely limiting warming to less than 1.5 or 2°C above pre-industrial levels–as a basis. One recent study projects a 0.8 m rise above 2000 levels by 2100 with 2°C of warming, for example (Schaeffer et al. 2012). Another states that limiting warming to below 2°C is associated with sea level rise near or below 1 m by 2100 (Strauss et al. 2015). Because the warming associated with the NCA Intermediate-Low scenario is in line with the Paris goals and the scenario can be easily localized using USACE guidelines, we determined it to be the most useful proxy for a Paris Agreement sea level rise scenario.

The NCA scenarios described above represent globally averaged sea level rise. Sea level is not expected to rise uniformly, however, due to regional factors such as land subsidence, tectonics, changes in ocean circulation, gravitational fingerprinting, groundwater pumping, and dredging, which together account for local vertical land movement (Milliken et al. 2008; Moucha et al. 2008; Mitrovica et al. 2009; Konikow 2011; Ezer et al. 2013). We calculated local

sea level rise projections (E) at each tide gauge and at each future time using the equation described by the US Army Corps of Engineers (Huber & White 2015):

$$E(t) = Mt + bt^2$$

Where:

- t is years since 1992
- M is the eustatic sea level rise rate (0.0017 mm/yr) plus the local vertical land movement rate (Huber & White 2015; Zervas et al. 2013)
- b is a variable that determines the pace of sea level rise. This variable is set to 1.56E-04, 8.71E-05, and 2.71E-05 for the NCA Highest, Intermediate-High, and Intermediate-Low scenarios, respectively (Huber & White 2015).

Estimates of vertical land movement (VLM) come directly from the tide gauge records. These estimates were derived by decomposing the records into a number of components, including seasonal variability and global sea level trends, to calculate average VLM rates over the length of each record (Zervas et al. 2013). In places like Louisiana, where subsidence rates are closely linked to rates of fluid extraction and have varied considerably over the last century, the average VLM rates used here may mask any accelerations or decelerations in subsidence rates over the last 20 years (Kolker et al. 2011). Because this average VLM is held constant when calculating future local sea level rise, this calculation could underestimate or overestimate future sea level rise in locations where VLM is highly variable, such as Louisiana.

2.5 Determining the minimum sea level rise increment

In order for the inundation zones for each time interval to be meaningfully different from each other, they must be spaced far enough apart to be outside of the range of statistical uncertainty associated with the underlying datasets (Gesch 2013). There are several sources of statistical uncertainty in this analysis:

1. The vertical accuracy of the DEMs (9.25 cm)
2. Tide gauge measurement errors (3.0 cm; National Oceanic and Atmospheric Administration, 2013a)
3. Datum uncertainty (1.5 cm; National Oceanic and Atmospheric Administration, 2013b)
4. Standard deviation about the mean effective inundation threshold (5.3 cm; this study)

Using the average values (reported above) for (1), (2), and (3), we calculated a cumulative vertical error of 11.2 cm using a sum of squares approach. We then calculate a combined linear error by multiplying the cumulative vertical error by 1.28 for the 80% confidence level. This confidence level is lower than that suggested by Gesch (2013), but consistent with the level employed by NOAA using the same underlying DEMs (NOAA 2017). We then calculate the minimum sea level by multiplying the combined linear error by two (Gesch 2013). These calculations result in an average minimum sea level rise interval of 28.6 cm that we apply across all tide gauges.

For each sea level rise scenario, we used the minimum sea level rise interval and the average of the projected sea level rise values for each year for all of the tide gauges to determine the years for future analysis. Because sea level is projected to rise quickly with the Highest scenario, we analyzed seven future years in addition to the present-day: 2030, 2045, 2060, 2070, 2080, 2090, and 2100. For the Intermediate-High scenario, we analyzed 2035, 2060, 2080, and 2100. For the Intermediate-Low scenario, we analyzed just 2060 and 2100.

2.6 Spatial analysis of inundated areas
Our spatial analysis largely follows the methods outlined by the National Oceanic and Atmospheric Administration (NOAA Office for Coastal Management 2012). We determined the effectively inundated areas by creating a transect of points perpendicular to the coast at each gauge and assigning the gauges and their associated transect points the height of the effective inundation threshold above MHHW. Analyzing the West and contiguous East and Gulf Coast regions separately, we then interpolated between those points using the natural neighbor method. This yielded a spatially variable water level surface above MHHW, which we then added to a MHHW surface developed and published by NOAA (NOAA 2016) and referenced to the NAVD88 vertical datum. This total water level surface represented the height of the effective inundation threshold above NAVD88. For future time steps, we added the corresponding projected amount of sea level rise for that gauge to the effective inundation threshold value and interpolated again to create a future water level surface.

We then subtracted the DEMs from the total water level surface to create an inundation surface for each time step. To ensure that the inundated areas were hydrologically connected to the ocean, not just low-lying areas that might, in actuality, be disconnected from the ocean by higher elevation barriers, we performed a region grouping and extracted only hydrologically connected areas (Figure S1).

2.7 Community-level area analysis
For the purposes of this study, we defined communities using US Census county subdivision areas. The Census defines county subdivisions as "the primary divisions of counties and equivalent entities" (US Census Bureau 2010). County subdivisions vary both in size and population. Unlike Census tracts or counties, county subdivisions tend to represent recognizable towns and their boundaries. Examples include Miami Beach, FL, Atlantic City, NJ, and Galveston, TX.

Using standard spatial analysis tools, we determined the area of each county subdivision above MHHW that was inundated at each time step. In order to assess how much of the inundated area was developed or developable land, we first did the area analysis including all county subdivision land areas above MHHW. We then removed wetland and areas protected by federal levees from each county

subdivision and inundation surface and calculated the non-wetland area above MHHW that was inundated at each time step (US Fish & Wildlife Service 2016; USACE 2017). Leveed areas were removed because any errors in levee height or representation within the DEMs could results in false inundation. Additional protective structures such as bulkheads and seawalls were included to the degree with which they were represented within the DEMs.

For each time step, we define a cohort of effectively inundated communities (EICs) based on the percentage of usable land area inundated, excluding wetlands and leveed areas. The impact of coastal flooding on a community will depend highly on *what* is being inundated, not just the frequency, as discussed above. Given the variable levels of resilience to the percentage of land area exposed to flooding, we explored using a higher percentage threshold than the 10% discussed above. Using a higher threshold, e.g. 25% or 50% yields fewer EICs; however the trend—an increasing number of EICs as sea level rises—remains (see section 3.4).

2.8 Analysis of socioeconomically vulnerable communities
Our analysis of socioeconomically vulnerable communities relies on the social vulnerability index (SoVI; Cutter et al., 2003). SoVI provides a relative measure of vulnerability to environmental hazards based on 29 socioeconomic variables. These variables are collected primarily by the US Census Bureau and include economic measures (e.g. per capita income and median household value) as well as demographic measures (e.g. median age and race/ethnicity; see Hazards and Vulnerability Research Institute, 2013 for a full list of underlying variables). Census tract level data were developed and provided by Martinich et al. 2013. For each Census tract, the variables were normalized to z-scores with a mean of zero and a standard deviation of 1, then reduced to an overall SoVI score using a principal components analysis. The overall SoVI score helps to identify places that are significantly above or below mean levels of vulnerability.

Because socioeconomic vulnerability and its causes vary greatly, it does not lend itself to straightforward comparisons across regions. Therefore, the SoVI data were broken into four regions: North Atlantic (ME through VA), South Atlantic (NC through Monroe County, FL), Gulf Coast (Collier County, FL through TX), and Pacific (CA through WA) (Martinich et al. 2013). The overall SoVI scores were normalized within each region such that the mean SoVI score for a region is zero and the standard deviation is 1. We defined tracts with high vulnerability as those with SoVI scores greater than 0.5 standard deviations above the mean, as previous studies have done (Martinich et al. 2013).

We chose to use SoVI over other environmental justice indices (such as EJ Screen) or a smaller subset of variables because of its extensive use by previous studies (Dunning

& Durden 2013; Martinich et al. 2013) and because it encompasses a wide range of social, economic, and demographic variables, all of which can contribute to an overall level of vulnerability (Cutter et al. 2003).

After defining the cohort of EICs for each time horizon and sea level rise projection, we used GIS intersection tools to determine which EICs contain at least one Census tract with high socioeconomic vulnerability. Demographics change over time, and rising seas could force large-scale changes in coastal populations (Hauer et al. 2016). For this reason, we limit our primary analysis of the intersection between tracts with high vulnerability and inundated areas to time steps within the next 30 years (2030 and 2045 for the Highest scenario; 2035 for the Intermediate-High scenario).

2.9 Uncertainty

We assess sources of uncertainty, but do not conduct an explicit error analysis in this study. The primary source of uncertainty in projecting the impact of future sea level rise on coastal communities is likely the future pace and magnitude of the sea level rise itself, which will be a product of both past and future greenhouse gas emissions as well as the Earth system response to those emissions. Because future emissions trajectories are highly uncertain, we do not assign any probability or likelihood to the three sea level rise scenarios analyzed here, but rather see them as a range that brackets uncertainty in future emissions choices, the global ice sheet response to those emissions, and the associated magnitude of sea level rise over the course of this century. Future changes in coastal demographics are an additional source of uncertainty. Because SoVI is a static assessment of socioeconomic vulnerability, we cannot exclude the possibility that time and exposure to flooding will substantially change patterns of socioeconomic vulnerability along the coast.

Other sources of uncertainty derive from the data underlying our analyses–the vertical error in the DEMs, interannual variation in the effective inundation threshold, and tide gauge measurement and datum transformation errors. We have implicitly incorporated these combined errors into our analyses by following conventions for defining the minimum sea level rise interval for mapping (Titus et al. 2009; Gesch 2013). When using the same underlying elevation datasets, previous studies have mapped sea level rise intervals of 12 inches, on par with the 28.6 cm minimum sea level rise interval for this study (Marcy et al. 2011; Climate Central 2014). While recent attempts have been made to quantify errors in spatial sea level rise assessments of limited geographic scope (e.g. Leon, Heuvelink, & Phinn, 2014), there is also precedent for relying on best estimates of uncertainty in studies with a national or regional geographic scope (Weiss et al. 2011; Strauss et al. 2012). Because the underlying DEMs have varying degrees of accuracy, it is likely that the level of uncertainty in our results will also vary along the coasts.

A full, strict uncertainty analysis may also be of limited value because future conditions rely on unknowns, such as the magnitude of sea level rise at a given location in a given year (Schmid et al. 2014). Having used the 80% confidence level to calculate the minimum sea level rise interval, we assume that differences in the results we present for sequential years are statistically significant at the 80% confidence level.

3. Results and Discussion

3.1 Tide gauge analysis

For the 93 gauges in our set, the mean height of the effective inundation threshold was 0.33 m above MHHW, and the mean standard deviation about that height was 5.27 cm (see supplementary online material). The threshold at most gauges falls between MHHW and the minor coastal flooding threshold set by the National Weather Service, which averages 0.56 m above MHHW for East and Gulf Coast gauges (Table S2).

3.2 Verification of present-day conditions

We identified EICs in which at least 10% of the non-wetland, non-leveed area above MHHW falls below the effective inundation threshold in the present day (**Figure 2**). Nationally, there are 91 EICs today that cluster into just 29 counties (**Figure 3**). Nearly half of the EICs (59) are in Louisiana, where high rates of land subsidence have exacerbated sea level rise to date (Kolker et al. 2011; Zervas et al. 2013). This present-day cohort includes widely reported coastal flooding hot spots such as Somerset and Dorchester Counties in Maryland (Gertner 2016), the Florida Keys (Union of Concerned Scientists 2015), and Terrebonne and St. Mary Parishes in Louisiana (Marshall et al. 2014). For each of the counties that contain EICs, we contacted local experts in an effort to ground truth our present day results. We spoke with representatives from local National Flood Insurance, sustainability, and environmental planning offices, as well as citizens, who confirmed that the extent of effective inundation we had mapped for the present day was representative of the frequency and extent of flooding observed locally (Table S1). It is important to note that additional shoreline protection measures (e.g. bulkheads, seawalls, etc.) would likely change the frequency and extent of flooding a community experiences, as was the case for one community expert with whom we spoke.

3.3 A flooded future

The number of EICs on a national basis increases steadily as sea level rises (**Figure 3**). By 2035, the number of EICs nearly doubles (to 167) compared to today with the Intermediate-High scenario. That number rises to 272, 365, and 489 in the years 2060, 2080, and 2100, respectively. In addition to the simple rise in the number of EICs, the land area inundated within the EICs increases over the course of the century. Whereas 47 of today's 91 EICs have 25% or more of their land area effectively inundated, by

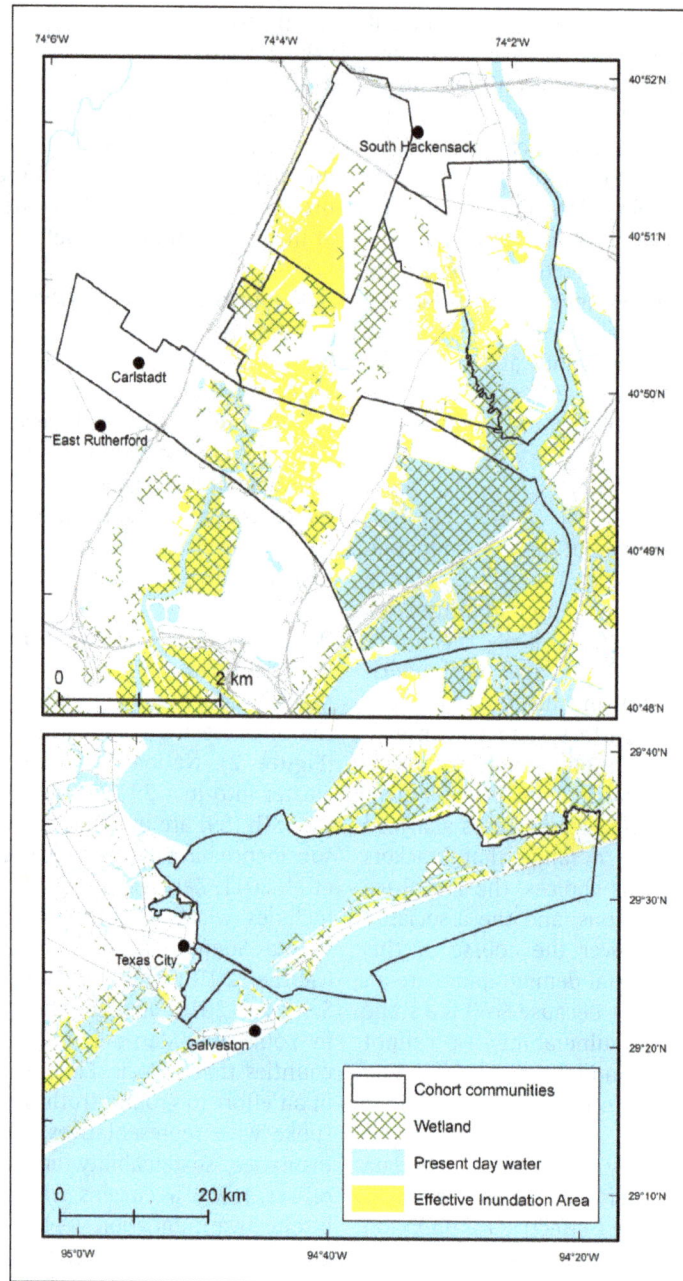

Figure 2: Present day effectively inundated areas. Areas below mean higher high water (blue) and below the effective inundation threshold (yellow) for two example regions within the national analysis: northern New Jersey **(a)**; and the Galveston, Texas, region **(b)**. Wetland areas are shown with cross hatching and present day cohort communities are outlined in black. DOI: https://doi.org/10.1525/elementa.234.f2

2100, 60% of the EICS (or 294 in all) are inundated at that level with the Intermediate-High scenario (**Table 1**). Full lists of communities inundated in year and scenario can be found in Table S3.

3.4 Early EICs: 2030 through 2045

More than 70 new communities face effective inundation by 2035 with the Intermediate-High scenario (**Figure 4**). These early EICs cluster in several regions: the eastern shore of Maryland, the mainland side of North Carolina's Pamlico Sound, the New Jersey shore, South Carolina's

Lowcountry, Louisiana west of New Orleans, and the northern coast of Texas between the Louisiana border and Brazosport.

The Highest scenario projects a similar number of newly inundated communities in 2030 as the Intermediate-High scenario in 2035—178 for the Highest scenario compared with 167 for the Intermediate-High. The clusters of affected communities with the two scenarios are also similar. By 2045, with the Highest scenario, however, the number of EICs expands to 265—a rise of nearly 100 in the 15 years since 2030. And by 2045, with the Highest

Figure 3: Effectively inundated communities for present and future time horizons. Effectively inundated communities with the NCA Intermediate-High sea level rise scenario for the present (yellow), and in 2035 (blue), 2060 (green), and 2100 (pink). DOI: https://doi.org/10.1525/elementa.234.f3

scenario, the Atlantic Coast of Florida goes from having just one EIC 15 years earlier to having 8. New Jersey also experiences a large increase in EICs between 2030 and 2045—from 26 to 55.

3.5 Mid-century EICs

Between 2035 and 2060, an additional 105 communities face effective inundation with the Intermediate-High scenario. Whereas the clusters of EICs in 2035 tend to simply expand areas with clusters of EICs today, by 2060 entirely new stretches of the coastline are exposed to effective inundation (**Figure 3**). South Carolina, for example, goes from just 2 EICs in 2035 to 12 in 2060, spanning most of the state's coastline. Likewise, Florida's Atlantic coast goes from just one EIC in 2035 to 8 in 2060, including Miami Beach. The greater Boston area,

northern New Jersey, and the Atlantic coast of the Delmarva Peninsula, including Lewes, Delaware, and Ocean City, Maryland, all face effective inundation in 2060. Regional inundation patterns—as well as total numbers of EICs—with the Intermediate-High scenario in 2060 are similar to those of the Highest scenario in 2045.

The Highest scenario would expose an additional 88 communities to effective inundation by 2060 compared to the Intermediate-High scenario (Figure S2). Fifty of these communities—more than half—are concentrated in just three states: Florida (14 communities), New Jersey (18 communities), and North Carolina (18 communities). Additional regions with clusters of communities that would face effective inundation with the Highest scenario but not the Intermediate-High scenario in 2060 include

Table 1: Number of effectively inundated communities with each sea level rise scenario. Total number of effectively inundated communities for the present, Intermediate-Low, Intermediate-High, and High scenarios. Number of communities affected to different degrees of inundation reported for four classes of inundation: 10 – 25%, 25 – 50%, 50 – 75%, and >75%. DOI: https://doi.org/10.1525/elementa.234.t1

% inundation	Present	Int-Low Scenario		Int-High Scenario				Highest Scenario						
		2060	2100	2035	2060	2080	2100	2030	2045	2060	2070	2080	2090	2100
10–25%	44	63	112	64	103	133	195	75	109	132	165	208	226	240
25–50%	31	53	61	49	71	76	102	55	78	89	89	110	123	155
50–75%	12	42	58	37	50	74	59	30	44	71	69	71	81	76
>75%	4	25	59	17	48	82	133	18	34	68	104	134	170	197
Total	91	183	290	167	272	365	489	178	265	360	427	523	600	668

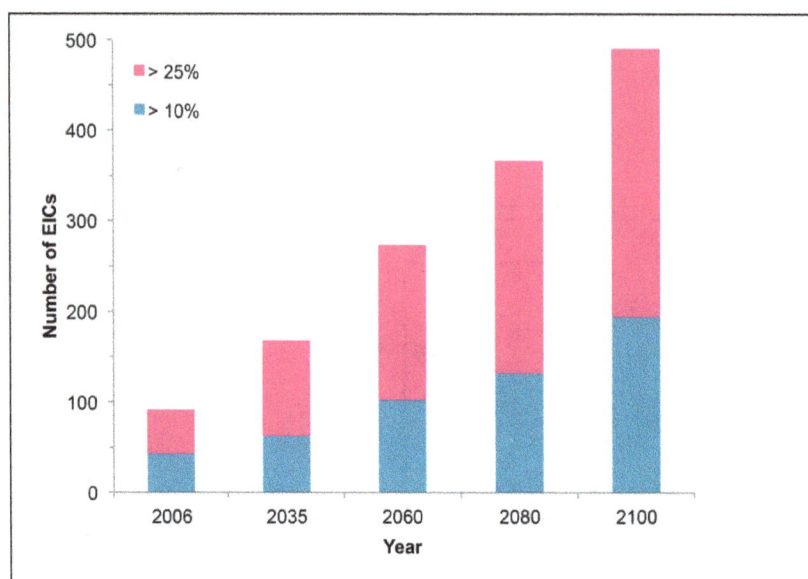

Figure 4: Number of effectively inundated communities nationwide. The number of effectively inundated communities (EICs) increases as sea level rises. Results shown here are for the NCA Intermediate-High scenario. Bar height is inclusive of all communities with 10% or more effective inundation; colors indicate the number of communities at 10 to 25% (blue) and >25% (pink) effective inundation. DOI: https://doi.org/10.1525/elementa.234.f4

the greater Charleston, SC, area, Iberville and St. Martin parishes in Louisiana, and Alameda, CA.

3.6 End of century EICs

By 2100, 489 communities–including nearly all of the immediate coastal communities in New Jersey, Maryland, northern North Carolina, South Carolina, Georgia, Louisiana, and northern Texas–face effective inundation with the Intermediate-High scenario (**Figure 3**). The 2100 cohort includes previous unaffected communities in the San Francisco region (San Mateo and Alameda) as well as the greater Los Angeles region (North Coast). Notably, there are 29 EICs with present day populations over 100,000, including Boston, MA, Newark, NJ, and St. Petersburg, FL (US Census Bureau, 2010).

An additional 179 communities would face effective inundation with the Highest scenario that would not be affected with the Intermediate-High scenario (**Figure 5**). Three-quarters (75%) of these communities fall into eight

states—Florida, Louisiana, Maryland, Massachusetts, New Jersey, New York, North Carolina, and Virginia—all of which have 10 or more communities that would be effectively inundated with the Highest scenario but not the Intermediate-High scenario. Significant clusters of communities that fall into this category include the San Francisco Bay Area, much of the Georgia coast and the Florida Panhandle, Hancock County, MS, southern Texas, and Long Island, NY. With the Highest scenario, the number of EICs with present day populations over 100,000 rises to 52, including four of the five boroughs of New York City.

3.7 State trends

Today, the 23 states (including the District of Columbia) included in our analysis have a mean of four EICs. Whereas Louisiana has the most EICs (59), the majority (15) of those states have no EICs, and the mean number of EICs per state today is zero. By 2100, the mean number of

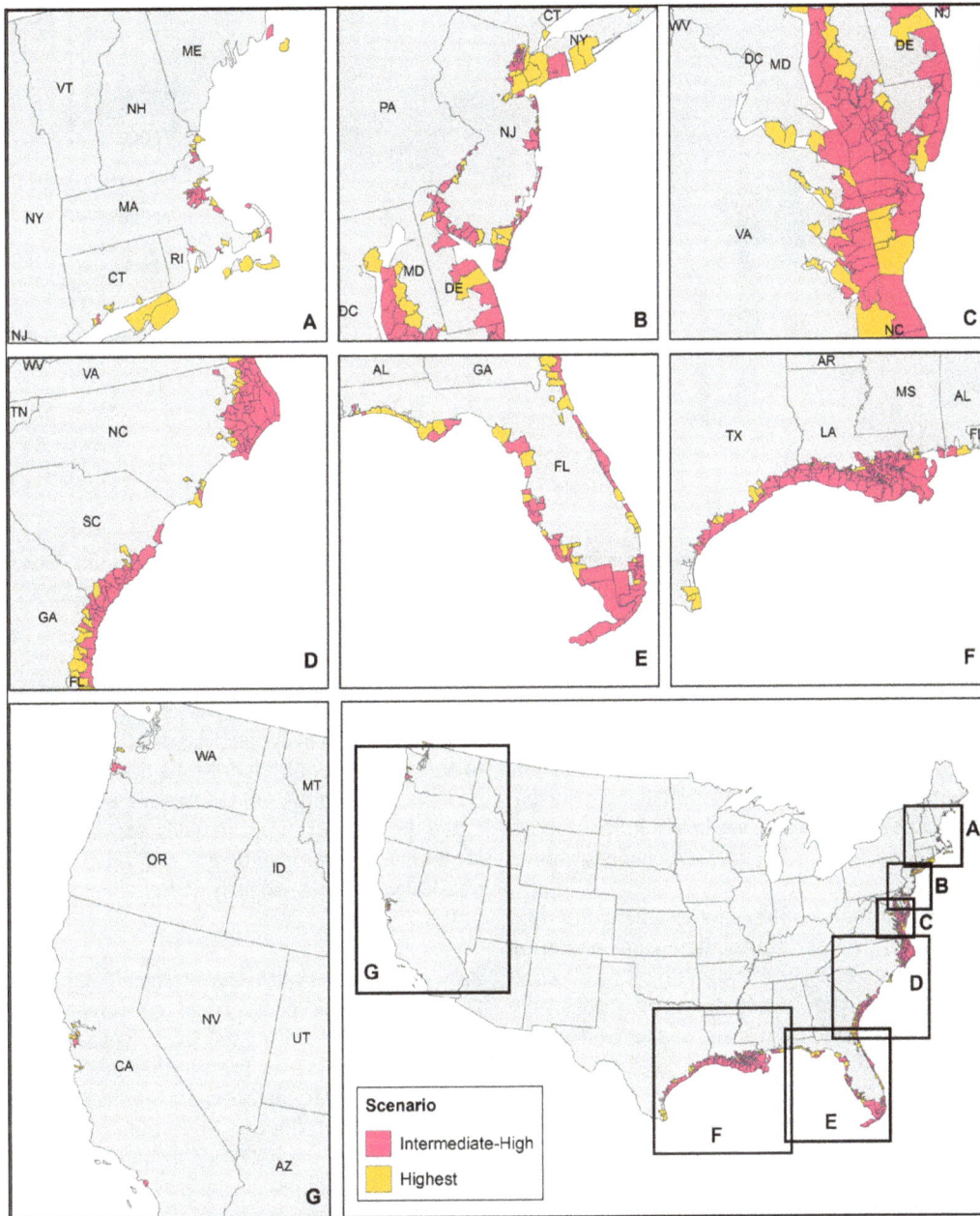

Figure 5: Effectively inundated communities in 2100 with the Intermediate-High and Highest scenarios. Effectively inundated communities with the Intermediate-High scenario in 2100 are shown in pink. Additional communities that would face effective inundation with the Highest scenario are shown in yellow. DOI: https://doi.org/10.1525/elementa.234.f5

EICs per state with the Intermediate-High scenario is 21: roughly a five-fold increase.

Averages, while useful, obscure stark differences in the number of EICs in each state as well as the pace of growth as sea level rises. For example, while the number of EICs in Louisiana grows rapidly—from 59 today to 131 in 2100 with the Intermediate-High scenario—the rate of the increase in EICs in New Jersey is faster (**Figure 6**). By 2100, there are 103 EICs in New Jersey compared to just 7 today—an increase of more than one order of magnitude. Several states—South Carolina, Massachusetts, Texas, and Georgia—go from two or fewer EICs today to 10 or more in 2100. By the end of the century, more than 40% (10) of the 23 coastal states are projected to have 10 or more EICs (**Table 2**).

3.8 Physically exposed and socially vulnerable

Hurricane Katrina and other natural disasters have highlighted the fact that socially vulnerable communities often bear the brunt of disasters and, in the aftermath, face additional challenges to restoring their living situations (Kuhl et al. 2014; Cleetus et al. 2015). Lack of transportation to evacuate a flooded area, living in older, less flood-resistant housing, or working minimum wage

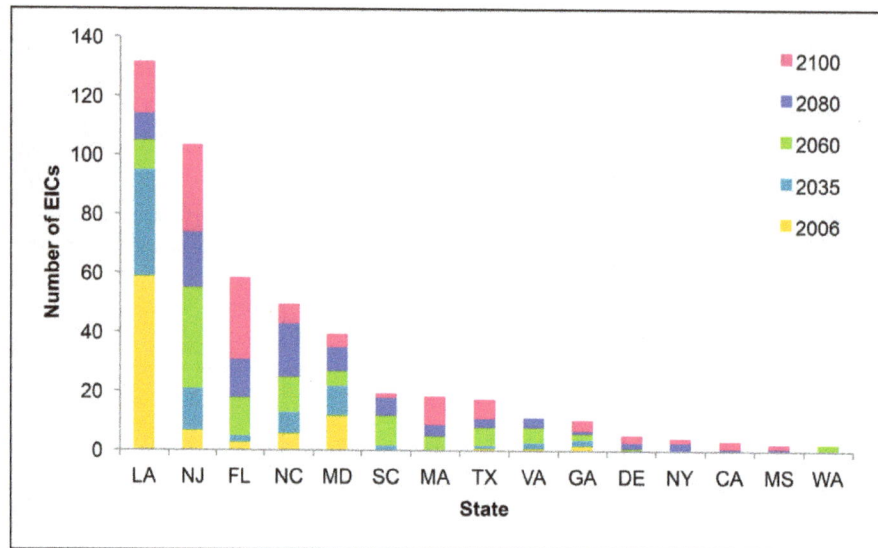

Figure 6: Effectively inundated communities by state. Effectively inundated communities (EICs) for each state with the Intermediate-High scenario. Total bar height represents the total number of EICs for each state by 2100. Note that states with one or zero EICs by 2100 are not shown. DOI: https://doi.org/10.1525/elementa.234.f6

service jobs in a flood-prone coastal region are just a few examples of how socioeconomic vulnerability contributes to heightened environmental risk. While extreme events provide a window into the additional challenges facing socially vulnerable communities, sea level's more gradual rise also has the potential to bring these challenges into closer view.

We find that, nationally, 55% of the 2035 EICs (92 out of 167 total under the Intermediate-High scenario) contain at least one Census tract with a high SoVI score (**Figure 7**). Similar to previous findings, over 40% (39) of these socially vulnerable EICs are in the Gulf Coast region (Martinich et al. 2013). Of those, the vast majority are in the state of Louisiana. Despite the Gulf Coast's concentration of socially vulnerable EICs, there are a number of clusters of EICs in other regions that stand out as well. These include: The Eastern Shore/Chesapeake Coast of Maryland; the mainland side of Pamlico Sound in North Carolina; the New Jersey Shore; Kiawah and Edisto Islands in South Carolina's Lowcountry; and the Florida Keys. At 54%, the percentage or EICs containing a tract with a high SoVI score is similar for both the 2030 and 2045 Highest cohorts. Our results suggest that these regions and communities will require particular attention, and potentially additional resources, as coastal communities begin to build resilience to coastal flooding.

The demographic variables driving high SoVI scores vary from place to place. Within the Gulf Coast region, for example, which has a large African-American population, high SoVI scores tend to be driven by poverty and race. Along Maryland's Eastern Shore, a large elderly population, likely with reduced mobility, contributes to high social vulnerability. The varying suite of factors contributing to social vulnerability within our cohort of EICs suggests that resilience building and/or coastal retreat strategies will need to vary in accordance with the

specific social vulnerability challenges each community faces. A comprehensive analysis of the factors contributing to social vulnerability is beyond the scope of this work. However, the range of causes of social vulnerability noted here contributes to calls for tailored initiatives for enhancing preparedness and adaptive capacity in physically exposed, socially vulnerable areas (Emrich & Cutter 2011).

3.9 Comparisons with the Intermediate-Low scenario

The pace at which sea level rises has a great bearing on the number of communities that face effective inundation this century. Differences in the number of EICs between the three scenarios we analyzed are significant by 2060 and dramatic by 2100 (**Figure 8**). In 2060, the Intermediate-High scenario projects 272 EICs. That figure is 32% higher (360) with the Highest scenario and 33% lower (183) with the Intermediate-Low scenario. The percentage differences are similar for 2100, with the Highest scenario projecting 37% more EICs (668 in total) than then Intermediate-High, and the Intermediate-Low projecting 41% fewer EICs (290 in total).

With all three scenarios, and for all years, between 52 and 64% of EICs have 25% or more of their land area subject to effective inundation. While these percentages are relatively unvarying, there are large differences in the total numbers of EICs with 25% or more inundation. The Intermediate-High scenario projects 169 EICs with 25% or more inundation by 2060, and 294 by 2100. With the Highest scenario, those numbers rise to 228 and 428, respectively. With the Intermediate-Low scenario, they fall to 120 and 178.

In 2060, there are several clusters of communities that could be spared effective inundation with the Intermediate-Low scenario relative to the Intermediate-High (Figure S3). These clusters include the greater Boston area, northern New Jersey and 13 communities along the

Table 2: Effective inundated communities by state. Number of effectively inundated communities per state for the present, Intermediate-Low, Intermediate-High, and Highest scenarios for each year analyzed. DOI: https://doi.org/10.1525/elementa.234.t2

State	Present	Intermediate-Low		Intermediate-High				Highest						
		2060	2100	2035	2060	2080	2100	2030	2045	2060	2070	2080	2090	2100
AL	0	0	0	0	0	1	1	0	0	1	1	1	1	2
CT	0	0	0	0	0	0	1	0	0	0	1	1	3	5
DE	0	0	1	0	1	3	5	0	1	3	5	5	6	7
DC	0	0	0	0	0	0	0	0	0	0	0	0	0	0
FL	3	5	19	5	18	31	58	5	18	32	48	69	85	90
GA	2	4	6	4	6	7	10	4	6	7	7	14	17	18
LA	59	97	112	95	105	114	131	89	101	110	116	129	139	146
ME	0	0	0	0	0	1	1	0	0	1	1	2	3	4
MD	12	23	30	22	27	35	39	23	27	35	37	41	44	51
MA	0	0	5	0	5	9	18	1	5	9	13	18	20	28
MS	0	0	0	0	0	1	2	0	0	1	2	2	4	5
NH	0	0	0	0	0	0	1	0	0	0	0	2	2	4
NJ	7	27	58	21	55	74	103	26	55	73	87	110	120	131
NY	0	0	1	0	0	3	4	0	0	3	4	6	9	14
NC	6	15	26	13	25	43	49	20	26	43	47	53	61	63
PA	0	0	0	0	0	0	0	0	0	0	0	0	0	0
RI	0	0	0	0	0	0	1	0	0	0	0	2	3	3
SC	0	3	12	2	12	18	19	3	10	18	19	20	20	22
TX	1	5	10	2	8	11	17	3	7	10	15	18	20	26
VA	1	4	8	3	8	11	24	4	7	11	20	25	34	38
CA	0	0	0	0	0	1	3	0	0	1	2	3	6	7
OR	0	0	0	0	0	0	0	0	0	0	0	0	0	0
WA	0	0	2	0	2	2	2	0	2	2	2	2	3	4

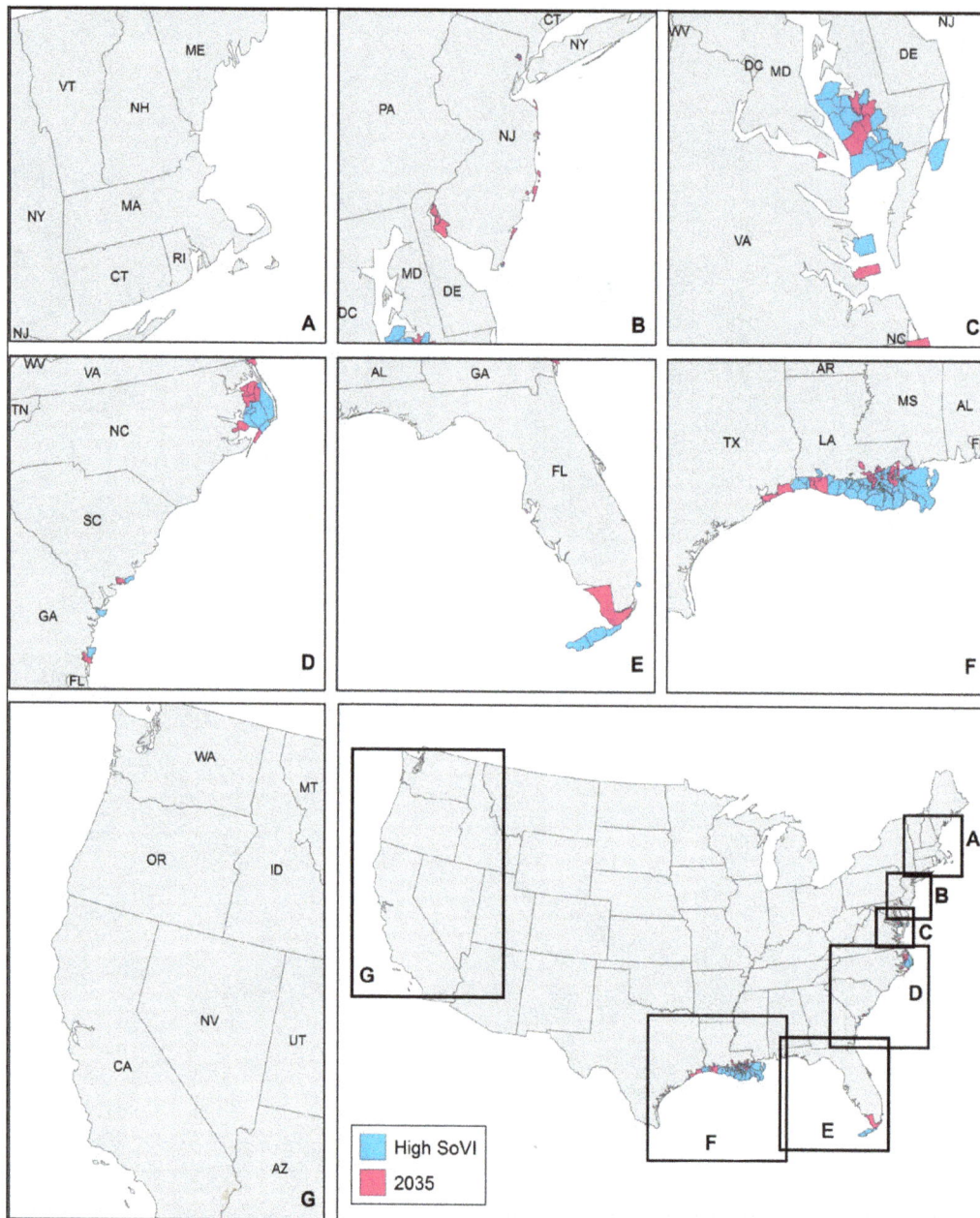

Figure 7: Effectively inundated communities with high socioeconomic vulnerability. Effectively inundated communities in 2035 (pink) with the Intermediate-High scenario. Affected communities with at least one Census tract with a high SoVI score are shown in blue. Note that the regions shown in panels A and G do not have any effectively inundated communities in 2035 with high SoVI. DOI: https://doi.org/10.1525/elementa.234.f7

New Jersey Shore, the Atlantic coast of Florida (including Miami Beach) and the Gulf Coast of Florida off the coast of Cape Coral.

By 2100, the clusters of spared communities mentioned above grow in area (**Figure 9**). Large stretches of the Delaware, Maryland, Virginia, and North Carolina, Florida, and Texas coasts also stand to gain greatly if sea level rise follows the trajectory of Intermediate-Low scenario rather than the Intermediate-High. Large population centers (>100,000 people today) stand to also gain greatly from a slower pace of sea level rise (**Figure 10**). With the Intermediate-Low scenario, only

3 of the 29 large population centers included in the 2100 Intermediate-High cohort would face effective inundation. Communities that would be spared inundation would include four of the five boroughs of New York City, Miami, and San Mateo.

Using the Intermediate-Low scenario as one potential approximation of the magnitude of sea level rise if the goals of the Paris Climate Agreement were met, these results suggest that the emissions choices we make in the coming decades—and ice sheet responses to those choices—could have profound impacts on communities in the coastal US.

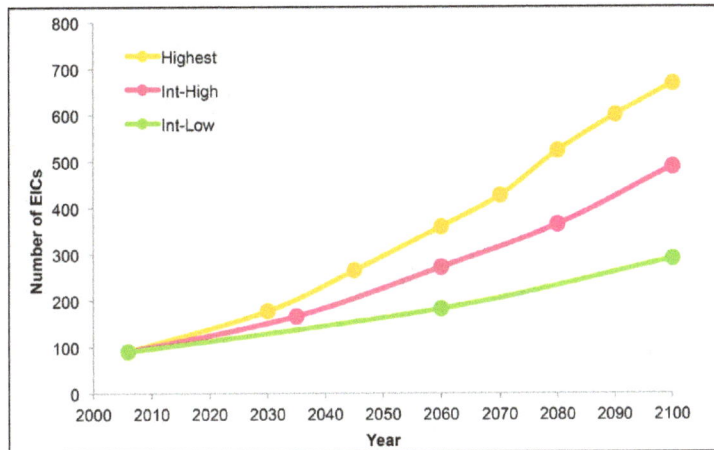

Figure 8: Number of effectively inundated communities for each sea level rise scenario. Number of effectively inundated communities by year for the three scenarios analyzed in this study: Highest (yellow); Intermediate-High (pink); Intermediate-Low (green). DOI: https://doi.org/10.1525/elementa.234.f8

Figure 9: Comparison of effectively inundated communities in 2100 with the Intermediate-Low and Intermediate-High scenarios. Effectively inundated communities with the Intermediate-High scenario are shown in pink. Communities shown in green would be effectively inundated with the Intermediate-High scenario, but spared with the Intermediate-Low. DOI: https://doi.org/10.1525/elementa.234.f9

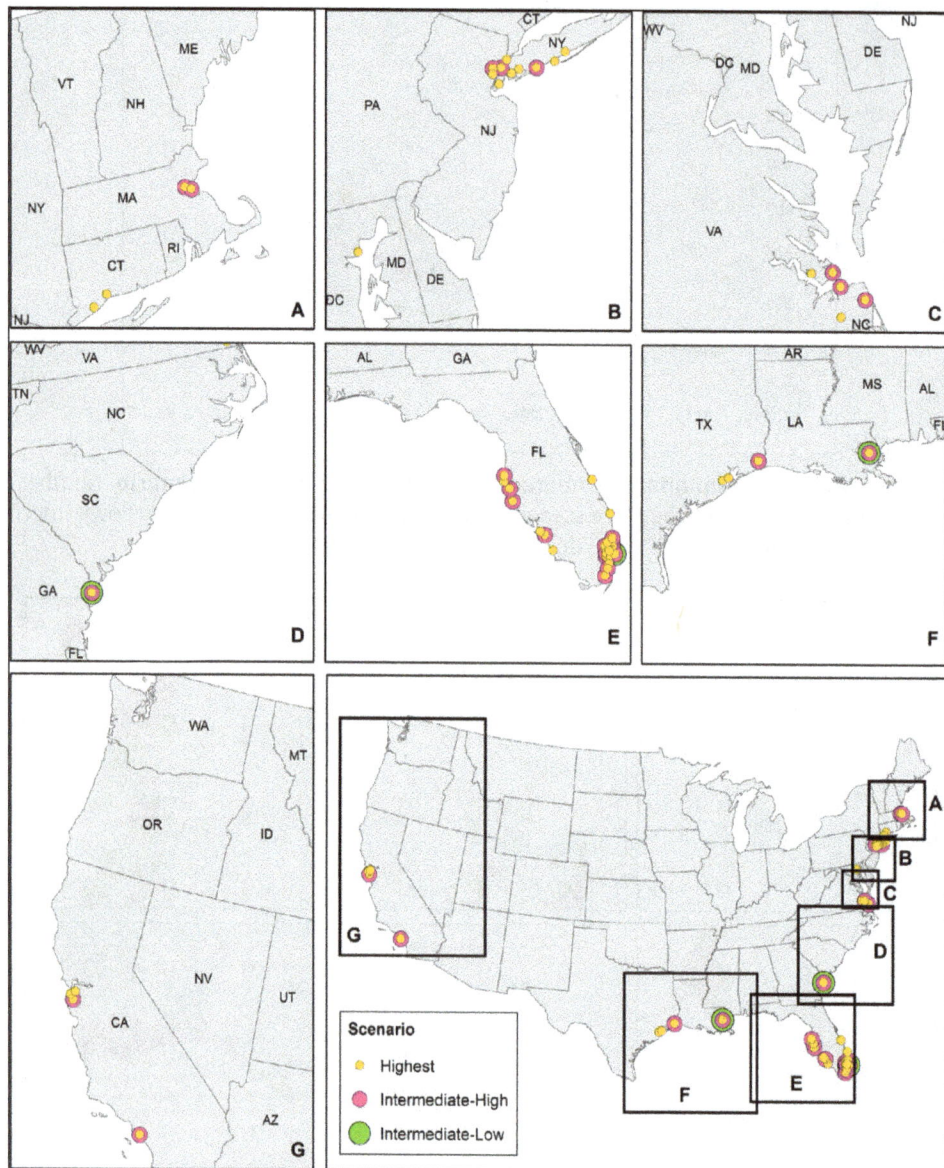

Figure 10: Effectively inundated communities with populations over 100,000. Locations of effectively inundated communities in 2100 with populations over 100,000 with the three scenarios analyzed in this study: Highest (yellow); Intermediate-High (pink); Intermediate-Low (green). DOI: https://doi.org/10.1525/elementa.234.f10

3.10 Online tool

Links to interactive maps showing the extent of inundation at each future time horizon and for each scenario can be found at http://www.ucsusa.org/RisingSeasHitHome (Union of Concerned Scientists 2017). Examples from the tool are shown in Figure S4 and Figure S5.

4. Conclusions

In this study, we have defined effective inundation and mapped its extent for the continental United States for three distinct and localized sea level rise scenarios. Our approach yielded national-level snapshots of the communities most exposed to sea level rise for specific time horizons through the end of this century that can be used for assessing effective inundation at the local level. This community-focused, time horizon-based approach fills a

gap in the existing suite of publicly available tools in that it allows users to visualize future inundation based on specific future time horizons and scenarios.

These results show that, in the absence of measures to manage increased flooding, effective inundation of coastal communities could become widespread within the next 40 years and encompass much of the coast by the end of the century. The growth of effective inundation suggests that communities will face stark choices about their ways of life in the decades to come. From homes and streets being elevated at high cost in Broad Channel, NY, and Norfolk, VA to the value of real estate declining in flood-prone parts of Miami-Dade County, FL, the cost of adapting to rising seas and more frequent flooding is already becoming apparent (Gregory 2013; Urbina 2016; Ruggeri

2017). In places such as Tangier Island, MD, and coastal Louisiana, there are ongoing public discourses about the cost and practicality of saving homes and communities from complete inundation (Gertner 2016; Coastal Protection and Restoration Authority of Louisiana 2017).

Over half of the effectively inundated communities we project for the year 2035 are home to socioeconomically vulnerable populations, which suggests that resources for building climate resilience will need to account for the fact that many communities face not only physical exposure to climate hazards, but also socioeconomic challenges to building resilience.

Using the NCA Intermediate-Low scenario as a proxy for projected sea level rise under a scenario where global warming was capped at 2°C, these results suggest that hundreds of communities in the US could be spared effective inundation were the international community to adhere to the goals of the Paris Agreement.

Whether or not those goals are met, in the coming decades, local, state, and federal governments will need comprehensive plans to provide resources and safe options for communities facing effective inundation, with particular attention to areas with vulnerable populations.

Supplemental Files

The supplemental files for this article can be found as follows:

- **Table S1.** Interview summary regarding present day flooding: uploaded as online supporting information.
- **Table S2.** Tide gauge information: uploaded as online supporting information.
- **Table S3.** Percentage of inundation within communities for all years and scenarios: uploaded as online supporting information.
- **Figure S1.** Schematic diagram of the spatial analyses underlying this study.
- **Figure S2.** Effectively inundated communities in 2060 with the Intermediate-High and Highest scenarios.
- **Figure S3.** Effectively inundated communities in 2060 with the Intermediate-High and Intermediate-Low scenarios.
- **Figure S4.** Inundated areas of Miami Beach, Florida in 2060 as indicated in online tool.
- **Figure S5.** Inundated areas of Oakland, California in 2100 as indicated in online tool.

Acknowledgements

The authors thank a number of people for providing data and methodological support. Matt Pendleton, Doug Marcy, Billy Brooks, and Billy Sweet (NOAA Office for Coastal Management and NOAA CO-OPS) provided DEM data, methodological input, and reviews. Paul Kirshen and Ellen Douglas (University of Massachusetts, Boston) also provided methodological input and reviews. Jeremy Martinich, Lindsay Ludwig, and Stefani Penn (EPA) provided Census tract-level SoVI data, and Susan Cutter (University of South Carolina) provided advice about the development and use of SoVI. Brenda Ekwurzel, Rachel Cleetus, and Nicole Hernandez-Hammer (Union of Concerned Scientists) provided valuable advice and input throughout the course of the research.

Funding information

This work was funded by grants to the Union of Concerned Scientists' Climate and Energy Program from the Barr Foundation, the Energy Foundation, the Common Sense Fund, and members of the Union of Concerned Scientists.

Competing interests

The authors have no competing interests to declare.

Author contributions

- · Contributed to conception and design: KD, ESS
- · Contributed to acquisition of data: KD
- · Contributed to analysis and interpretation of data: KD, ESS, AC, SU
- · Drafted and/or revised the article: KD, ESS, AC, SU
- · Approved submitted version for publication: KD, ESS, AC, SU

References

Adger, WN, Hallie, E and **Winkels, A** 2009 Nested and teleconnected vulnerabilities to environmental change. *Frontiers in Ecology and the Environment* **7**(3): 50–157. DOI: https://doi.org/10.1890/070148

Allen, G 2016 As waters rise, Miami Beach builds higher streets and political willpower. *National Public Radio.*

Applegate, A 2014 For city of Norfolk, park becomes wetlands once again. *PilotOnline.com.* Available at: http://hamptonroads.com/2014/02/city-norfolk-park-becomes-wetlands-once-again [Accessed April 3, 2015].

Church, JA and **White, NJ** 2011 Sea-Level Rise from the Late 19th to the Early 21st Century. *Surveys in Geophysics,* **32**(4–5): 585–602. DOI: https://doi.org/10.1007/s10712-011-9119-1

City of Annapolis 2011 *Flood mitigation strategies for the City of Annapolis, MD: City Dock and Eastport area,* Baltimore, MD: City of Annapolis. Available at: http://www.annapolis.gov/docs/default-source/dnep-documents-pdfs/03–01–2011-sea-level-study.pdf?sfvrsn=6.

City of Charleston 2015 *Sea level rise strategy,* Charleston, SC. Available at: http://www.charleston-sc.gov/

DocumentCenter/View/10089 [Accessed February 22, 2017].

Cleetus, R, Bueno, R and **Dahl, K** 2015 *Surviving and Thriving in the Face of Rising Seas*, Cambridge, MA. Available at: http://www.ucsusa.org/global-warming/prepare-impacts/communities-on-front-lines-of-climate-change-sea-level-rise.

Climate Central 2014 Surging Seas: Sea level rise analysis by Climate Central. Available at: http://sealevel.climatecentral.org/ [Accessed April 4, 2015].

Coastal Protection and Restoration Authority of Louisiana 2017 Louisiana's Comprehensive Master Plan for a Sustainable Coast, 179. Available at: http://coastal.la.gov/wp-content/uploads/2017/04/2017-Coastal-Master-Plan_Web-Book_Final_Compressed-04252017.pdf.

Cutter, SL, Boruff, BJ and **Shirley, WL** 2003 Social vulnerability to environmental hazards. *Social Science Quarterly* **84**(2): 242–261. DOI: https://doi.org/10.1111/1540-6237.8402002

Dahl, KA, Fitzpatrick, MF and **Spanger-Siegfried, E** 2017 Sea level rise drives increased tidal flooding frequency at tide gauges along the U.S. East and Gulf Coasts: Projections for 2030 and 2045. *Plos ONE* **12**(2): e0170949. Available at: http://journals.plos.org/plosone/article/file?id=10.1371/journal.pone.0170949&type=printable. DOI: https://doi.org/10.1371/journal.pone.0170949

Dangendorf, S, et al. 2017 Reassessment of 20th century global mean sea level rise. *Proceedings of the National Academy of Sciences.* DOI: https://doi.org/10.1073/pnas.161007114

Dilling, L, et al. 2015 The dynamics of vulnerability: why adapting to climate variability will not always prepare us for climate change. *WIREs Climate Change.* Available at: http://sciencepolicy.colorado.edu/admin/publication_files/2015.24.pdf [Accessed February 22, 2017]. DOI: https://doi.org/10.1002/wcc.341

Dunning, CM and **Durden, S** 2013 *Social vulnerability analysis: A comparison of tools*, Alexandria, VA: US Army Engineer Institute for Water Resources. Available at: http://www.iwr.usace.army.mil/Portals/70/docs/iwrreports/Social_Vulnerability_Analysis_Tools.pdf.

Emrich, CT and **Cutter, SL** 2011 Social vulnerability to climate-sensitive hazards in the southern United States. *Weather, Climate, and Society* **3**: 193–208. DOI: https://doi.org/10.1175/2011WCAS1092.1

Ezer, T, et al. 2013 Gulf Stream's induced sea level rise and variability along the U.S. mid-Atlantic coast. *Journal of Geophysical Research: Oceans* **118**(2): 685–697. DOI: https://doi.org/10.1002/jgrc.20091

Ezer, T and **Atkinson, LP** 2014 Accelerated flooding along the U.S. East Coast: On the impact of sea-level rise, tides, storms, the Gulf Stream, and the North Atlantic Oscillations. *Earth's Future* **2**(8): 362–382. Available at: http://doi.wiley.com/10.1002/2014EF000252 [Accessed April 3, 2015]. DOI: https://doi.org/10.1002/2014EF000252

FitzGerald, DM, et al. 2008. Coastal Impacts Due to Sea-Level Rise. *Annual Review of Earth and Planetary Sciences* **36**(1): 601–647. DOI: https://doi.org/10.1146/annurev.earth.35.031306.140139

Flick, RE, Murray, JF and **Ewing, LC** 2003 Trends in United States Tidal Datum Statistics and Tide Range. *Journal of Waterway, Port, Coastal, and Ocean Engineering* **129**(4): 155–164. DOI: https://doi.org/10.1061/(ASCE)0733-950X(2003)129:4(155)

Gertner, J 2016 Should the United States save Tangier Island from oblivion? *The New York Times.* Available at: https://www.nytimes.com/2016/07/10/magazine/should-the-united-states-save-tangier-island-from-oblivion.html [Accessed May 17, 2017].

Gesch, DB 2013 Consideration of vertical uncertainty in elevation-based sea-level rise assessments: Mobile Bay, Alabama case study. *Journal of Coastal Research* **63**(sp1): 197–210. DOI: https://doi.org/10.2112/SI63-016.1

Goddard, PB, et al. 2015 An extreme event of sea-level rise along the Northeast coast of North America in 2009–2010. *Nature Communications* **6**: 6346. Available at: http://www.ncbi.nlm.nih.gov/pubmed/25710720. DOI: https://doi.org/10.1038/ncomms7346

Gregory, K 2013 Where streets flood with the tide, a debate over city Aid. *The New York Times.* Available at: http://www.nytimes.com/2013/07/10/nyregion/debate-over-cost-and-practicality-of-protecting-part-of-queens-coast.html?_r=0 [Accessed May 17, 2017].

Haigh, ID, Eliot, M and **Pattiaratchi, C** 2011 Global influences of the 18.61 year nodal cycle and 8.85 year cycle of lunar perigee on high tidal levels. *Journal of Geophysical Research* **116**(C6): 025. Available at: http://doi.wiley.com/10.1029/2010JC006645 [Accessed June 20, 2016]. DOI: https://doi.org/10.1029/2010JC006645

Hamlington, BD, et al. 2015 The effect of the El Niño-Southern Oscillation on U.S. regional and coastal sea level. *Journal of Geophysical Research: Oceans* **120**(6): 3970–3986. Available at: http://doi.wiley.com/10.1002/2014JC010602.

Hauer, ME, Evans, JM and **Mishra, DR** 2016 Millions projected to be at risk from sea-level rise in the continental United States. *Nature Climate Change* **6**(7): 691–695. Available at: http://www.nature.com/doifinder/10.1038/nclimate2961 [Accessed January 26, 2017]. DOI: https://doi.org/10.1038/nclimate2961

Hay, CC, et al. 2015 Probabilistic reanalysis of twentieth-century sea-level rise. *Nature.* DOI: https://doi.org/10.1038/nature14093

Hazards and Vulnerability Research Institute 2013 Social Vulnerability Index for the United States 2006–10. Available at: http://webra.cas.sc.edu/hvri/products/sovi.aspx [Accessed February 22, 2017].

Hinkel, J, et al. 2014 Coastal flood damage and adaptation costs under 21st century sea-level rise.

Proceedings of the National Academy of Sciences **111**(9): 3292–3297. DOI: https://doi.org/10.1073/pnas.1222469111

Holland, L 2016 Study to look at Smith Island flooding solutions. *delmarvanow*. Available at: http://www.delmarvanow.com/story/news/local/maryland/2016/09/06/study-look-smith-island-flooding-solutions/89921176/ [Accessed May 17, 2017].

Horton, R, et al. 2008 Sea level rise projections for current generation CGCMs based on the semi-empirical method. *Geophysical Research Letters* **35**(2): L02715. Available at: http://doi.wiley.com/10.1029/2007GL032486 [Accessed April 3, 2015]. DOI: https://doi.org/10.1029/2007GL032486

Huber, M and **White, K** 2015 *Sea Level Change Curve Calculator (2015.46) User Manual*, United States Army Corps of Engineers. Available at: http://www.corpsclimate.us/docs/Sea_Level_Change_Curve_Calculator_User_Manual_2015_46_FINAL.pdf.

Jankowski, KL, Törnqvist, TE and **Fernandes, AM** 2017 Vulnerability of Louisiana's coastal wetlands to present-day rates of relative sea-level rise. *Nature Communications* **8**: 1–7. DOI: https://doi.org/10.1038/ncomms14792

Jevrejeva, S, Moore, JC and **Grinsted, A** 2010 How will sea level respond to changes in natural and anthropogenic forcings by 2100? *Geophysical Research Letters* **37**(7): 1–5. DOI: https://doi.org/10.1029/2010GL042947

Katz, M 2016 Borough President Katz's Broad Channel street raising task force reviews project progress. *Borough of Queens, City of New York*. Available at: http://www.queensbp.org/borough-president-katzs-broad-channel-street-raising-task-force-reviews-project-progress/ [Accessed February 22, 2017].

Kolker, AS, Allison, MA and **Hameed, S** 2011 An evaluation of subsidence rates and sea-level variability in the northern Gulf of Mexico. *Geophysical Research Letters* **38**(21): 1–6. DOI: https://doi.org/10.1029/2011GL049458

Konikow, LF 2011. Contribution of global groundwater depletion since 1900 to sea-level rise. *Geophysical Research Letters* **38**(17): 1–5. DOI: https://doi.org/10.1029/2011GL048604

Kuhl, L, et al. 2014 Evacuation as a climate adaptation strategy for environmental justice communities. *Climatic Change* **127**(3–4): 493–504. Available at: http://link.springer.com/10.1007/s10584–014–1273–2 [Accessed March 21, 2017]. DOI: https://doi.org/10.1007/s10584-014-1273-2

Kulp, S and **Strauss, BH** 2017 Rapid escalation of coastal flood exposure in US municipalities from sea level rise. *Climatic Change*, 477–489. Available at: http://link.springer.com/10.1007/s10584-017-1963-7. DOI: https://doi.org/10.1007/s10584-017-1963-7

Lane, K, et al. 2013 Health effects of coastal storms and flooding in urban areas: a review and vulnerability assessment. *Journal of Environmental and Public Health*, 913064. Available at: http://www.ncbi.nlm.nih.gov/pubmed/23818911 [Accessed February 22, 2017]. DOI: https://doi.org/10.1155/2013/913064

Lentz, EE, et al. 2016 Evaluation of dynamic coastal response to sea-level rise modifies inundation likelihood. *Nature Climate Change* **6**(March): 1–6. Available at: http://www.nature.com/doifinder/10.1038/nclimate2957. DOI: https://doi.org/10.1038/nclimate2957

Leon, JX, Heuvelink, GBM and **Phinn, SR** 2014 Incorporating DEM uncertainty in coastal inundation mapping. *PLoS ONE* **9**(9): 1–12. DOI: https://doi.org/10.1371/journal.pone.0108727

Maldonado, JK, et al. 2014 The impact of climate change on tribal communities in the US: Displacement, relocation, and human rights. *Climate Change and Indigenous Peoples in the United States: Impacts, Experiences and Actions*, 93–106.

Marcy, DW, et al. 2011 New mapping tool and techniques for visualizing sea level rise and coastal flooding impacts. In: Wallendorf, LW, et al. (eds.), *Proceedings of the 2011 Solutions to Coastal Disasters Conference, June 26–29, 2011*. Reston, Virginia: American Society of Civil Engineers, 474–490.

Marshall, B, Jacobs, B and **Shaw, A** 2014 Losing Ground. *ProPublica*. Available at: http://projects.propublica.org/louisiana/ [Accessed February 22, 2017].

Martinich, J, et al. 2013 Risks of sea level rise to disadvantaged communities in the United States. *Mitigation and Adaptation Strategies for Global Change* **18**(2): 169–185. DOI: https://doi.org/10.1007/s11027-011-9356-0

Milliken, KT, Anderson, JB and **Rodriguez, AB** 2008 A new composite Holocene sea-level curve for the northern Gulf of Mexico. *Geological Society of America Special Papers* **2443**(1): 1–11. DOI: https://doi.org/10.1130/2008.2443(01)

Mitrovica, JX, Gomez, N and **Clark, PU** 2009 The sea-level fingerprint of West Antarctic collapse. *Science* **323**(5915): 753. DOI: https://doi.org/10.1126/science.1166510

Moftakhari, HR, et al. 2015 Increased nuisance flooding along the coasts of the United States due to sea level rise: Past and future. *Geophysical Research Letters* **42**(22): 9846–9852. DOI: https://doi.org/10.1002/2015GL066072

Moftakhari, HR, et al. 2017 Cumulative hazard: The case of nuisance flooding. *Earth's Future* **5**(2): 214–223. Available at: http://doi.wiley.com/10.1002/2016EF000494 [Accessed April 3, 2017].DOI:https://doi.org/10.1002/2016EF000494

Moucha, R, et al. 2008. Dynamic topography and long-term sea-level variations: There is no such thing as a stable continental platform. *Earth and Planetary Science Letters* **271**(1–4): 101–108. DOI: https://doi.org/10.1016/j.epsl.2008.03.056

National Oceanic and Atmospheric Administration 2013a Sea Level Trends – NOAA Tides and Currents. Available at: http://co-ops.nos.noaa.gov/sltrends/sltrends.html [Accessed April 4, 2015].

National Oceanic and Atmospheric Administration 2013b Inundation Analysis – NOAA Tides & Currents. Available at: http://tidesandcurrents. noaa.gov/inundation/ [Accessed April 4, 2015].

NOAA 2016 Inundation Mapping Tidal Surface – Mean Higher High Water – NOAA Data Catalog. Available at: https://data.noaa.gov/dataset/inundation-mapping-tidal-surface-mean-higher-high-water4b2f9 [Accessed February 22, 2017].

NOAA 2017 Digital Coast Sea Level Rise and Coastal Flooding Impacts Viewer: Frequent Questions. Available at: https://coast.noaa.gov/data/digitalcoast/pdf/ slr-faq.pdf [Accessed May 17, 2017].

NOAA Office for Coastal Management 2012 *Mapping coastal inundation primer*, Charleston, South Carolina. Available at: https://coast.noaa.gov/data/ digitalcoast/pdf/coastal-inundation-guidebook.pdf.

Parris, A, et al. 2012 *Global Sea Level Rise Scenarios for the United States National Climate Assessment*, Silver Spring, MD: NOAA Technical Report OAR CPO-1. Available at: http://cpo.noaa.gov/sites/cpo/ Reports/2012/NOAA_SLR_r3.pdf.

Passeri, DL, et al. 2016 Tidal hydrodynamics under future sea level rise and coastal morphology in the Northern Gulf of Mexico. *Earth's Future* 4: 159–176. DOI: https://doi.org/10.1002/2015EF000332

Pfeffer, WT, Harper, JT and **O'Neel, S** 2008 Kinematic constraints on glacier contributions to 21st-century sea-level rise. *Science* **321**(5894): 1340–1343. DOI: https://doi.org/10.1126/science.1159099

Ruggeri, A 2017 Miami's fight against rising seas. *BBC Future*. Available at: http://www.bbc.com/future/ story/20170403-miamis-fight-against-sea-level-rise [Accessed May 17, 2017].

Schaeffer, M, et al. 2012 Long-term sea-level rise implied by 1.5 C warming levels. *Nature Climate Change* **2**: 867–870. DOI: https://doi.org/10.1038/nclimate1584

Schmid, K, Hadley, B and **Waters, K** 2014 Mapping and Portraying Inundation Uncertainty of Bathtub-Type Models. *Journal of Coastal Research* **30**(3): 548–561. Available at: http://www.bioone.org/ doi/abs/10.2112/JCOASTRES-D-13–00118.1. DOI: https://doi.org/10.2112/JCOASTRES-D-13-00118.1

Spanger-Siegfried, E, Fitzpatrick, M and **Dahl, K** 2014 *Encroaching Tides: How Sea Level Rise and Tidal Flooding Threaten U.S. East and Gulf Coast Communities over the Next 30 Years*, Cambridge, MA. Available at: http://ucsusa.org/encroachingtides.

Strauss, BH, et al. 2012 Tidally adjusted estimates of topographic vulnerability to sea level rise and flooding for the contiguous United States. *Environmental Research Letters* **7**(1): 14033. DOI: https://doi. org/10.1088/1748-9326/7/1/014033

Strauss, BH, Kulp, S and **Levermann, A** 2015 Carbon choices determine US cities committed to futures below sea level. *Proceedings of the National Academy of Sciences* **112**(44): 13508–13513. DOI: https:// doi.org/10.1073/pnas.1511186112

Sweet, W, et al. 2014 *Sea Level Rise and Nuisance Flood Frequency Changes around the United States*, Silver Spring, MD: NOAA Technical Report NOS CO-OPS 073.

Sweet, WV and **Park, J** 2014 From the extreme to the mean : Acceleration and tipping points of coastal inundation from sea level rise. *Earth's Future* **2**: 579–600. Available at: http://onlinelibrary.wiley. com/doi/10.1002/2014EF000272/abstract. DOI: https://doi.org/10.1002/2014EF000272

Sweet, WV and **Zervas, C** 2011 Cool-Season Sea Level Anomalies and Storm Surges along the U.S. East Coast: Climatology and Comparison with the 2009/10 El Niño. *Monthly Weather Review* **139**(7): 2290–2299. DOI: https://doi.org/10.1175/MWR-D-10-05043.1

Sweet, W, Zervas, C and **Gill, S** 2009 *Elevated East Coast Sea Level Anomaly: June – July 2009*, Silver Spring, MD. Available at: http://tidesandcurrents. noaa.gov/publications/EastCoastSeaLevelAnomaly_2009.pdf%5Cnhttp://search.proquest.com/ docview/904481884?accountid=10639.

Titus, JG, et al. 2009 *Coastal sensitivity to sea level rise: a focus on Mid-Atlantic Region*, Washington DC: U.S. Climate Change Science Program; Synthesis and Assessment Product 4.1.

TMAC 2015 *Future conditions risk assessment and modeling*, Federal Emergency Management Agency. Available at: https://www.fema.gov/ media-library-data/1454954261186-c348aa9b1768298c9eb66f84366f836e/TMAC_ 2015_Future_Conditions_Risk_Assessment_and_ Modeling_Report.pdf.

TownCharts 2017 Smith Island, Maryland demographics data. Available at: http://www.towncharts.com/ Maryland/Demographics/Smith-Island-CDP-MD-Demographics-data.html [Accessed May 17, 2017].

Union of Concerned Scientists 2015 *Encroaching Tides in the Florida Keys (2015)*, Available at: http://www. ucsusa.org/sites/default/files/attach/2015/10/ encroaching-tides-florida-keys.pdf [Accessed February 22, 2017].

Union of Concerned Scientists 2017 *When Rising Seas Hit Home*. Online at: http://www.ucsusa.org/ RisingSeasHitHome.

Urbina, I 2016 Perils of climate change could swamp coastal real estate. *The New York Times*. Available at: https://www.nytimes.com/2016/11/24/science/ global-warming-coastal-real-estate.html [Accessed May 17, 2017].

USACE 2017 National Levee Database. Available at: http:// nld.usace.army.mil/egis/f?p=471:1: Accessed February 1, 2017].

US Census Bureau 2010 Geographic Terms and Concepts – County Subdivision. Available at: https://www. census.gov/geo/reference/gtc/gtc_cousub.html [Accessed February 22, 2017].

US Fish and **Wildlife Service** 2016 National Wetlands Inventory. Available at: https://www.fws.gov/ wetlands/data/State-Downloads.html [Accessed February 22, 2017].

Vermeer, M and **Rahmstorf, S** 2009 Global sea level linked to global temperature. *Proceedings of the National Academy of Sciences of the United States of America* **106**(51): 21527–21532. DOI: https://doi. org/10.1073/pnas.0907765106

Wadey, MP, Haigh, ID and **Brown, JM** 2014 A century of sea level data and the UK's 2013/14 storm surges: an assessment of extremes and clustering using the Newlyn tide gauge record. *Ocean Science* **10**: 1031–1045. Available at: www.ocean-sci.net/10/1031/2014/ [Accessed June 21, 2016]. DOI: https://doi.org/10.5194/os-10-1031-2014

Walsh, J, et al. 2014 Our Changing Climate. *Climate Change Impacts in the United States: The Third National Climate Assessment*, 19–67. Available at: http://nca2014.globalchange.gov/report/our-changing-climate/introduction.

Weiss, J 2016 Miami Beach's $400 million sea-level rise plan is unprecedented, but not everyone is sold. *Miami New Times*. Available at: http://www.miaminewtimes.com/news/miami-beachs-400-million-sea-level-rise-plan-is-unprecedented-but-not-everyone-is-sold-8398989 [Accessed May 17, 2017].

Weiss, JL, Overpeck, JT and **Strauss, B** 2011 Implications of recent sea level rise science for low-elevation areas in coastal cities of the conterminous U.S.A: A letter. *Climatic Change* **105**(3–4): 635–645. DOI: https://doi.org/10.1007/s10584-011-0024-x

Zervas, C, Gill, S and **Sweet, W** 2013 *Estimating Vertical Land Motion from Long-Term Tide Gauge Records Services Center for Operational Oceanographic Products and Services*, Silver Spring, MD: NOAA Technical Report NOS CO-OPS 065. Available at: https://tidesandcurrents.noaa.gov/publications/Technical_Report_NOS_CO-OPS_065.pdf.

11

Upward nitrate flux and downward particulate organic carbon flux under contrasting situations of stratification and turbulent mixing in an Arctic shelf sea

Ingrid Wiedmann*, Jean-Éric Tremblay[†], Arild Sundfjord[‡] and Marit Reigstad*

Increased sea ice melt alters vertical surface-mixing processes in Arctic seas. More melt water strengthens the stratification, but an absent ice cover also exposes the uppermost part of the water column to wind-induced mixing processes. We conducted a field study in the Barents Sea, an Arctic shelf sea, to examine the effects of stratification and vertical mixing processes on 1) the upward nitrate flux (into surface layers <65 m) and 2) the downward flux of particulate organic carbon (POC) to ≤200 m. In the Arctic-influenced, drift ice-covered northern Barents Sea, we found a low upward nitrate flux into the surface layers (<0.1 mmol nitrate m^{-2} d^{-1}) and a moderate downward POC flux (40–200 m: 150–250 mg POC m^{-2} d^{-1}) during the late phase of a peak bloom. A 1-D residence time calculation indicated that the nitrate concentration in the surface layers constantly declined. In the Atlantic-influenced, ice-free, and weakly stratified southern Barents Sea a high upward nitrate flux was found (into the surface layers ≤25 m: >5 mmol nitrate m^{-2} d^{-1}) during a post bloom situation which was associated with a high downward POC flux (40–120 m: 260–600 mg POC m^{-2} d^{-1}). We suggest that strong wind events during our field study induced vertical mixing processes and triggered upwards nitrate flux, while a combination of down-mixed phytoplankton and fast-sinking mesozooplankton fecal pellets enhanced the downward POC flux. The results of this study underscore the need to further investigate the role of strong, episodic wind events on the upward nitrate and downward POC fluxes in weakly stratified regions of the Arctic that may be ice-free in future.

Keywords: nitrate flux; POC export; sediment trap; sedimentation; warming Arctic; space-for-time substitution

1. Introduction
The declining sea ice cover in Arctic seas (Arrigo and van Dijken, 2015; IPCC, 2013) affects the pelagic marine ecosystem in contrasting ways. Sea ice melt strengthens the water column stratification and hampers the upward nitrate flux into the surface layer (Tremblay et al., 2015), while an absent sea ice cover allows various wind-driven processes (e.g. wind-driven shear, breaking waves) to induce vertical mixing and shelf-break upwelling (Carmack and McLaughlin, 2011; Rainville et al., 2011; Falk-Petersen et al., 2015). Such mixing and upwelling can generate strong upward nitrate fluxes into the surface layers (Hales et al., 2005; Randelhoff et al., 2016). Dependent on the intensity of the nutrient renewal in the surface layer, the plankton abundance and composition may change, which in turn can modify the downward flux

of particulate organic carbon (POC). However, definitive regulating mechanisms of the POC flux are still under discussion (Carmack and Wassmann, 2006; Wassmann and Reigstad, 2011; Forest et al., 2013).

We conducted a field study in the Barents Sea, an Arctic shelf sea, to investigate the upward flux of nitrate and the downward flux of POC in contrasting field situations of ice cover, hydrography, mixing, and plankton abundance and composition. Arctic-derived water masses (temperature T < 0°C, salinity S = 34.4–34.8; Loeng, 1991) influence the northern Barents Sea, which is seasonally ice-covered (annual maximum extension found during March–April; Kvingedal, 2005). In late spring and summer, the sea ice recedes northwards and a phytoplankton bloom commonly occurs in the marginal ice zone, where the waters are well-lit and contain high, winter-accumulated, nutrient concentrations after the ice break-up. This bloom is often associated with a major downward POC flux, because senescent stages, resting stages and aggregates of diatoms, the often prevailing microalgae, have high sinking velocities (Eppley et al., 1967; Bienfang, 1981; Iversen and Ploug, 2013). In addition, an ice edge

* UiT The Arctic University of Norway, Tromsø, NO

[†] Québec-Océan and Takuvik, Biology Department, Université Laval, Québec City, Québec, CA

[‡] Norwegian Polar Institute, Tromsø, NO

Corresponding author: Ingrid Wiedmann (Ingrid.wiedmann@uit.no)

mesozooplankton community has been described to have lower ingestion rates than one in open waters (Wexels Riser et al., 2008). The result is a relatively low POC attenuation in the water column which may contribute to a greater downward POC flux at the ice edge (Wassmann and Reigstad, 2011).

The southern Barents Sea is influenced by Atlantic-derived waters ($T > 3°C$, $S > 35.0$; Loeng, 1991) and is ice-free the whole year. Accordingly, this region is prone to wind mixing, and the water column does not stratify before the sea surface warms in summer (Andreassen and Wassmann, 1998; Reigstad et al., 2002). Due to the absence of sea ice, the enhanced irradiance in surface waters allows the spring phytoplankton bloom to develop earlier than in the seasonally ice-covered northern Barents Sea (Leu et al., 2011). The phytoplankton bloom may thus peak at the ice edge during late June, while the southern Barents Sea already experiences a post-bloom period. The latter is commonly associated with an abundant meso-zooplankton community and enhanced ingestion rates (Wexels Riser et al., 2008), exhibiting a strong POC attenuation in the water column and causing a reduction of the POC export (Wassmann and Reigstad, 2011). However, previous studies also reported major downward POC fluxes in the deep mixed southern Barents Sea during late spring and early summer (Olli et al., 2002; Reigstad et al., 2008). Definitive drivers for this major flux have not been identified yet, but model results suggested that the downward POC flux may be linked to an enhanced upward nutrient flux caused by deep mixing events (Sakshaug and Slagstad, 1992; Sundfjord et al., 2007).

Here, we conducted a field study in the northern, drift ice-covered Barents Sea, the Polar Front, and the ice-free, southern Barents Sea. We assess in particular the link between mixing, upward nitrate flux and downward POC flux at the northernmost and southernmost study sites, because they contrasted in terms of stratification, turbulent mixing, phytoplankton bloom stage and zooplankton abundance. In this way, we could (1) compare the intensity of the upward nitrate flux, (2) study the contribution of this flux to the nitrate stock in the upper water column, and (3) investigate possible mechanisms for the regulation of the downward POC flux under contrasting regimes of hydrography, wind mixing, and spring phytoplankton bloom stage.

2. Materials and methods

Fieldwork was carried out in the Barents Sea with the ice-enforced R/V *Helmer Hanssen* (22–27 June 2011) as part of the Norwegian CONFLUX project. Based on a high resolution northward CTD-F transect along the 30°E longitude (S. Basedow, personal communication), four stations (M1–M4) were chosen for more detailed process studies. The hydrography, vertical mixing, suspended biomass and vertical export were assessed from the marginal ice zone in Arctic-influenced waters (M1), through the Polar Front (M2 and M3), and into Atlantic-influenced waters (M4; **Figure 1**, **Table 1**). These parameters gave important insight into the gradual change in the physical and biological environment from north to south motivating

us to assess the link between the upward nitrate flux and the downward POC flux at the two most contrasting sites (M1 and M4; **Figure 1**).

2.1 Hydrography, sea ice, light conditions, and wind

Physical variables (temperature, salinity) and fluorescence data were obtained at each station from surface to bottom (CTD-F, SeaBird 911*plus*) and processed with the SeaBird standard software package (bin average 0.5 m). Following Brainerd and Gregg (1995), we use the term 'mixed layer' for the weakly stratified surface layer, which was not necessarily actively mixed during the time of data collection. In contrast, we use 'mixing layer' to denote the surface depth interval that was actively mixed with a diffusivity $>10^{-4}$ m^2 s^{-1} during data collection (Wiedmann et al., 2014). Due to our focus on the vertical transport of nitrate and organic matter, we use the term 'mixing layer' instead of the recently suggested term 'turbulent layer' (Franks, 2015). The sea ice conditions were estimated visually based on the scale of the Norwegian Meteorological Institute (11 categories from ice-free to fast ice). The underwater irradiance was measured with a multispectral GMBDH TRIOS light sensor (190–575 nm, 2.15 nm wavelength resolution) at each station between the air-sea interface and 20 m during local noon. The base of the euphotic zone (1% subsurface irradiance) was calculated for the wavelength where chlorophyll *a* (Chl *a*) absorbs the most (430 nm; South and Whittick, 1987) using the equation:

$$I_D = I_0 * \exp\left(-k * z\right) \tag{1}$$

where I_D is the irradiance at depth z, I_0 the subsurface irradiance, and k the diffuse attenuation coefficient. A minor error must be assumed, since the attenuation coefficient did not take into account the shading effects by phytoplankton at the Chl *a* maximum (located below 20 m).

Wind data were noted in the ship's log during each operation, but due to a malfunction of the ship's wind measurement device, data are missing between 23 June (13:00 UTC) and 26 June (08:00 UTC). To interpolate these wind data, we combined our wind measurements with data from three land-based weather stations (Hopen, Bjørnøya, and Edgeøya) of the Norwegian Meteorological Institute (data available at www.yr.no).

2.2 Turbulence, nitrate concentrations and nitrate flux

A loosely tethered microstructure drop sonde (MSS-90 L) with a pair of PNS06 shear probes (Prandke and Stips, 1998) was used to collect sets of 2–3 profiles roughly every four hours during station work, as previously described (Sundfjord et al., 2007; Randelhoff et al., 2016). Only the set of profiles taken closest in time to the CTD and the nitrate profiles are included here. The shear profiles were processed as described in Fer (2006), where data above 8 m are discarded to avoid influence from the ship's keel.

The dissipation ε (W kg^{-1}) was calculated using the equation given by Yamazaki and Osborn (1990):

$$\varepsilon = 7.5 \, \nu \left[(\partial u'/\partial z)^2\right] \tag{2}$$

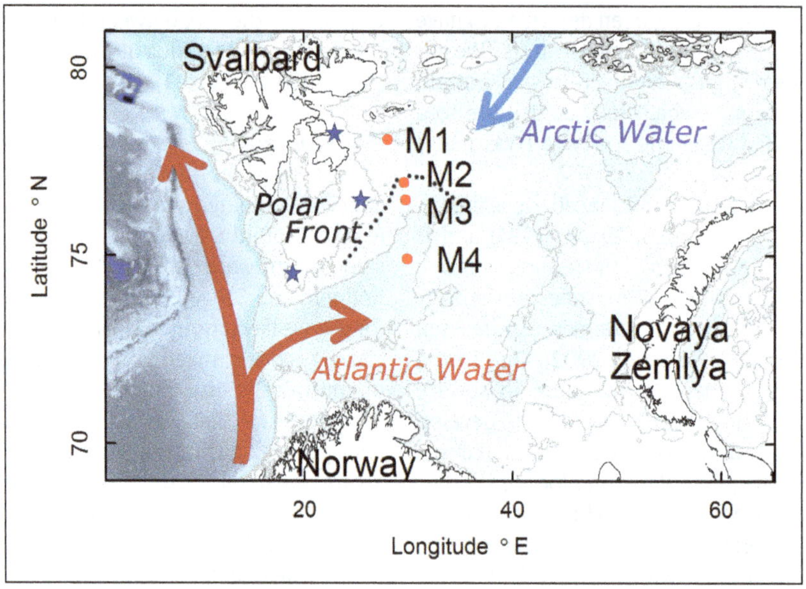

Figure 1: Map showing the Barents Sea with the sampling stations. Hydrography in the Barents Sea is influenced by Atlantic-derived water (red arrows, entering from the southwest) and Arctic-derived water (blue arrow, entering from the northeast). The four sampling stations were in the Arctic-influenced part of the Barents Sea (M1, north of the Polar Front, indicated by the dotted line), in the region of the Polar Front (M2, M3) and in the Atlantic-influenced part (M4). The blue stars designate the land-based weather stations (from north to south: Edgeøya, Hopen and Bjørnøya). DOI: https://doi.org/10.1525/elementa.235.f1

Table 1: Station identity and sampling schedule. DOI: https://doi.org/10.1525/elementa.235.t1

Station	Position	Date (2011)	Depth (m)	Ice cover	Chl a max (m)	Suspended sampling[a] time (UTC)	Deployment trap array[b] time (UTC)	Deployment time (d)
M1	78.0973°N, 28.1258°E	22 June	278	Very open drift ice (30%)	31	16:45	23:30	0.85
M2	76.9493°N, 29.7117°E	24 June	235	Very open drift ice (20%)	44	07:46	16:15	0.94
M3	76.4910°N, 29.8630°E	25 June	282	Open water	32	04:34	21:30	0.82
M4	74.9107°N, 30.0033°E	27 June	371	Open water	45	09:11	16:55	0.98

[a] sampled parameters: chlorophyll a (Chl a), particulate organic carbon (POC) and nitrogen (PON) at 1, 5, 10, 20, 30, 40, 50, 60, 90, 120, 200 m and Chl a maximum.
[b] sampled parameters: POC and PON at 40, 50, 60, 90, 120 m (and 200 m at M1 and M2).

where ν is the temperature-dependent viscosity of seawater and $\partial u'/\partial z$ represents the shear. The horizontal velocity variation u' and the depth z were both resolved to centimeter scale.

In a second step, we calculated the diffusivity K (m^2 s^{-1}) following Osborn (1980) and using:

$$K = \Gamma \varepsilon / N^2 \qquad (3)$$

dividing the product of the dissipation rate Γ (set to a typical value of 0.2; Moum, 1996) and the dissipation ε by the squared Brunt-Väisälä frequency N. Diffusivity data were then averaged over 4-m moving intervals before being used to calculate the nitrate flux.

Continuous depth profiles of nitrate were measured with a Satlantic ISUS V3 ultraviolet spectrophotometer, integrated in the ship-borne CTD system to get simultaneous depth data from the CTD's pressure sensor. The individual nitrate sensor spectra were then processed using the manufacturer's software. The vertical profiles were first objectively adjusted to match near-surface (10 m) nitrate concentration achieved from chemical seawater analysis (procedure following Martin et al., 2010b) and then smoothed using a 10-m moving average before the gradients were obtained for the nitrate flux calculations. Even though individual nitrate measurements have an accuracy of ±2 mmol m^{-3} (Johnson and Coletti, 2002), we expect the vertical gradients to be represented more accurately by using the moving average. The nitracline is here defined as the depth layer in which the nitrate concentration rapidly increases from low surface concentrations to the (subsurface) maximum concentration (Omand and Mahadevan, 2015).

Computation of the nitrate flux F_N was based on the nitrate (N) concentration with depth z and the diffusivity (K), using the equation:

$$F_N = -K*(dN/dz) \qquad (4)$$

where here the upward nitrate flux always represents the flux from a depth layer below (e.g. 200 m – depth x) into a depth layer above (depth x – surface).

2.3 Nitrate uptake rates

Nitrate uptake rates are strongly dependent on the available photosynthetically active radiation (PAR). To assess this relationship, water from the surface and the subsurface Chl a maximum (SCM) was collected at each station, split into ten 500 mL tissue culture flasks and each spiked with a trace amount of ^{15}N-potassium nitrate (0.1 μM). Each set of ten flasks was placed in a separate ten-position, linear light-gradient incubator designed to minimize spectral shift (Babin et al., 1994). The incubators were illuminated by a single full-spectrum 400 W Optimarc metal-halide lamp mimicking solar irradiance. Optically neutral filters (Lee Filters) were placed in front of the incubator with the surface samples to yield measured irradiances ranging from 5 to 630 μmol quanta m^{-2} s^{-1}. For the incubator with SCM samples, one layer of a blue filter (118 Light Blue Lee Filters Ltd.) was combined with optically-neutral filters (Lee Filters) to provide irradiances ranging from 3 to 365 μmol quanta m^{-2} s^{-1}. Temperature was maintained at in situ levels with a chilling circulator. In order to minimize isotopic dilution and photo-acclimation to experimental conditions, the incubations were kept as short as possible (5–6 h) to ensure detection of the ^{15}N label in particulate organic nitrogen (PON). Incubations were terminated by filtration onto 24-mm pre-combusted Whatman GF/F filters. All filters were desiccated at 60°C and stored dry for analysis ashore. An elemental analyzer (ECS 4010, Costech Analytical Technologies Inc.) coupled to a mass spectrometer (Delta V Advantage, Thermo Finnigan) was used to determine PON and its isotopic enrichment using a modified Dumas method (for details see Blais et al., 2012).

Specific nitrate uptake (V) was calculated using equation 3 of Collos (1987), and uptake-irradiance parameters (and standard errors on these parameters) were calculated on specific uptake data using the double exponential model of Platt et al. (1980):

$$V = V_d + V_s \left[1 - \exp\left(-\alpha E/V_s\right) \right] \left[\exp\left(-\beta E/V_s\right) \right] \qquad (5)$$

and

$$V_{max} = V_s \left[\alpha/(\alpha+\beta) \right] \left[\beta/(\alpha+\beta) \right]^{\beta/\alpha} \qquad (6)$$

where V_d is the dark uptake (h^{-1}), V_s is the theoretical maximum uptake in the absence of photoinhibition (h^{-1}), V_{max} is the maximum observed uptake (h^{-1}), E is the incubation irradiance (PAR, μmol quanta m^{-2} s^{-1}), and α and β [h^{-1} (μmol quanta m^{-2} s^{-1})$^{-1}$] are the photosynthetic efficiency at low irradiance (initial slope of the relationship) and the photoinhibition parameter, respectively.

The model formulated by Platt et al. (1980) was initially developed to describe the relationship between primary production and light intensity, but Tremblay et al. (2006) and Martin et al. (2012) have shown that this approach can also be used successfully to assess nitrate uptake. Compared to other set-ups, this approach (1) allows short incubation times (minimizing bottle effects and artefacts) and (2) provides dynamic parameters, which can be used to run a simulation spanning a few days, enabling us to compare the nitrate 'demand' at the two most contrasting stations. In order to estimate nitrate uptake in a given layer, we combined the continuous record of PAR on deck with the vertical attenuation coefficient of underwater irradiances (k), measured at local noon, to estimate E and compute equation 5 at each 1-m depth bin throughout the day. While absolute nitrate uptake would normally be obtained by multiplying equation 5 by PON, the latter was not available at a 1-m resolution. To circumvent this limitation, the specific uptake parameters were first converted into Chl a-specific, absolute values through multiplication by PON and division by the Chl a concentration of the incubated samples (noting that concentrations of Chl a and PON were well correlated during our field study; $R^2 = 0.73 \pm$ std dev 0.14). Uptake-irradiance parameters thus established with the surface sample were assigned to all depths in the upper mixed layer, whereas parameters established for the SCM were used at the SCM and below it. Between the base of the mixed layer and the SCM, parameters were interpolated according to the vertical gradient of nitrate concentration for V_d and V_{max}, and according to depth for α and β. This procedure is justified by the fact that the nitrate concentration and depth were robust predictors of V_{max} and α, respectively, for the set of eight curves obtained for stations M1, M2, M3, and M4 at the surface and the SCM (Table S1), though we note that this procedure could not take into account possible changes in taxonomy and shade acclimation below the SCM.

For M1 and M4, the stations that we investigated in detail, absolute nitrate uptake rates (μmol N L^{-1} h^{-1}) were then estimated for each depth bin and time of day by multiplying Chl a-normalized absolute N uptake by Chl a. The latter was estimated from post-calibrated in vivo fluorescence data from the CTD. By iterating this numerical approach, vertically integrated nitrate uptake in a given layer at M1 and M4 was averaged over five days to prevent giving too much importance to short-term conditions at the time of sampling.

2.4 Suspended and sedimenting biomass (Chl a, POC, PON, C/N ratio)

Suspended biomass was collected with Niskin bottles attached to the CTD rosette at 12 sampling depths between subsurface and 200 m (**Table 1**) to construct depth profiles of Chl a, POC, PON and the atomic C/N ratio. Collected water was gently transferred from Niskin bottles into carboys and stored cool and dark until filtration within few hours. Triplicates (50–200 mL) of each depth were vacuum-filtered onto Whatman GF/F filters (pore size 0.7 μm) and Whatman Nucleopore membrane filters (pore size 10 μm) to achieve a size-fractionation of the Chl *a* containing material (total and >10 μm). Chl *a* was extracted in 5 mL methanol (12 h, room temperature, darkness) and the Chl *a* concentration was measured using a Turner Design 10-AU fluorometer (calibrated with Chl *a*, Sigma C6144), applying the acidification method (Holm-Hansen and Riemann, 1978). For POC and PON, triplicates (200 mL) of each sampling depth were filtered on pre-combusted Whatman GF/F filters. Larger organisms such as copepods or chaetognats were removed before the filters were frozen (−20°C) until analyses (<6 months). Analyses were carried out using a Leeman Lab CHN Elemental Analyzer (for details see Reigstad et al., 2008). A C/N ratio of 6.6 (Redfield ratio; Redfield, 1934, 1958) has traditionally been an indicator for very recently produced ("fresh") phytoplankton material, while higher ratios have been assumed to .indicate more degraded biological material or material of terrestrial origin. Research during the last years, however, has shown that the C/N ratio of suspended material can vary with the physiological state of phytoplankton (e.g. nitrate depletion, as in Mei et al., 2005, and light limitation, as in Geider and La Roche, 2002) or the species composition (Fernández-Méndez et al., 2014). In addition, Frigstad et al. (2014) reported that the C/N ratio was considerably different in the Atlantic-influenced part of the Barents Sea (C/N = 6.7) and the Arctic-influenced part (C/N = 7.9). Taking this uncertainty into account, we assessed the "freshness" of the suspended biomass by a combination of microscopic cell counts and the C/N ratio.

A surface-tethered sediment trap array was deployed for 20–24 h at each of the four sampling stations (**Table 1**). Semi-Lagrangian drifting was ensured by anchoring the trap array on an ice-floe at M1 and M2. At M3 and M4 the trap array was freely drifting in open waters, but with the buoyancy located below the surface to minimize wind drift and potential pumping caused by wave action. Paired trap cylinders (KC Denmark, inner diameter 72 mm, length 450 m) were mounted at each sampling depth, determining the downward flux at 40, 50, 60, 90, and 120 m at all stations, as well as at 200 m at M1 and M2. The content of the cylinders was transferred into carboys after trap array recovery and stored cool and in darkness until filtered in triplicates (200 mL; swimmers were removed to the extent possible) and analyzed as described previously for suspended POC and PON. To be able to compare the attenuation of the POC flux at all four stations, we calculated the percentage relative to 120 m following the equation:

Attenuation of sinking POC (120)

$$= \frac{strongest\ POC\ flux\ at\ station - POC\ flux\ at\ 120\ m}{POC\ flux\ at\ 120\ m} \qquad (7)$$

2.5 1-D residence time calculations
The upward nitrate flux and the nitrate uptake by autotrophs affect the nitrate stock in the upper water column. We conducted simple 1-D residence time calculations for M1 and M4 (**Table 2**) to investigate how

Table 2: Integrated nitrate stock in biologically important layers and the upward nitrate flux into them[a]. DOI: https://doi.org/10.1525/elementa.235.t2

Station	Biologically important layer	Depth interval (m)	Integrated nitrate stock (mmol m^{-2})	Upward nitrate flux into base of layer		Integrated nitrate uptake[b] (mmol m^{-2} d^{-1})	Time to nitrate exhaustion (d)	
				(mmol m^{-2} d^{-1})	(% d^{-1})		Without upward nitrate flux	With upward nitrate flux[c]
M1	Low nitrate surface layer (<0.6 mmol m^{-3})	0–11	6.1	0.01	0.2	0.4	15	16
	Mixing layer	0–13	7.3	0.00	0.0	0.5	16	16, 45[d]
	Mixed layer	0–23	15.7	0.06	0.4	1.0	16	17
	Above SCM[e] layer	0–40	99.8	0.07	0.1	14.6	7	7
	Euphotic zone	0–65	277.9	0.01	0.0	17.4	16	16
M4	Low nitrate surface layer (<0.6 mmol m^{-3})	0–17	8.8	30.00	341.0	0.9	10	-[f]
	Mixing layer	0–25	14.4	5.40	37.5	1.5	10	-[f], 16[g], 21[h], 25[i]
	Mixed layer	0–38	34.2	0.11	0.3	3.5	10	10
	Above SCM layer	0–45	52.1	0.34	0.7	5.0	10	11
	Euphotic zone	0–45	52.1	0.34	0.7	5.0	10	11

[a] From shallowest to deepest layer; see section 2.4 for explanation of the calculations.
[b] 5-day mean.
[c] Nitrate upward flux as given in this table (column for upward nitrate flux into base of layer) unless specified otherwise.
[d] Nitrate upward flux 0.004 mmol m^{-2} d^{-1} for 5 days, then upward nitrate flux of 0.350 mmol m^{-2} d^{-1}.
[e] Subsurface chlorophyll *a* maximum.
[f] The layer would not be replenished if assuming the upward nitrate flux in the table.
[g] Nitrate upward flux of 5.395 mmol m^{-2} d^{-1} for 1 day, then upward nitrate flux of 0.300 mmol m^{-2} d^{-1}.
[h] Nitrate upward flux of 5.395 mmol m^{-2} d^{-1} for 2 days, then upward nitrate flux of 0.300 mmol m^{-2} d^{-1}.
[i] Nitrate upward flux of 5.395 mmol m^{-2} d^{-1} for 3 days, then upward nitrate flux of 0.300 mmol m^{-2} d^{-1}.

the integrated nitrate concentration in a certain layer of biological relevance is influenced by the nitrate uptake and the upward nitrate flux into this layer from below. The following five depth layers were investigated in detail:

1) the "low nitrate surface layer" with nitrate concentrations ≤ 0.6 mmol nitrate m^{-3} (A limiting nitrate surface layer with <0.5 mmol nitrate m^{-3} could not be investigated, because nitrate concentrations <0.5 mmol m^{-3} were found only in the uppermost 5–6 m at M1. Turbulence measurements from this depth interval were omitted, however (Section 2.1), and thus no upward nitrate flux into this layer could be calculated. As an alternative, we studied the "low nitrate surface layer" defined by ≤ 0.6 mmol nitrate m^{-3}.),
2) the depth interval above the SCM,
3) the euphotic zone (irradiance $>1\%$ of the subsurface irradiance),
4) the mixed layer, and
5) the mixing layer.

By coincidence, the depth range of some layers overlapped (e.g. euphotic zone and SCM at M4), but we chose to keep both to provide a holistic picture in **Table 2**. The contribution of the upward nitrate flux to the stock (% input from below; **Table 2**) was calculated as the ratio of the upward nitrate flux into a layer to the integrated nitrate concentration in this layer. The time to nitrate exhaustion in each layer (without taking upward nitrate flux into account) equals the ratio of the integrated nitrate stock in the layer to the integrated nitrate uptake in it. To calculate the time to nitrate exhaustion with the upward nitrate flux, we started with the integrated nitrate stock in the depth layer and assumed, for each consecutive day, a constant nitrate uptake and a certain upward nitrate flux.

3. Results

3.1 Hydrography and wind

Station M1 in the northern Barents Sea was covered with very open drift ice (30%, **Table 1**). A halocline (7–23 m) structured the water column in a well-mixed, meltwater-affected, surface layer (upper 7 m: temperature $T = -1.2°C$, salinity $S = 32.9$; **Figure 2a**) and a zone below, which con-

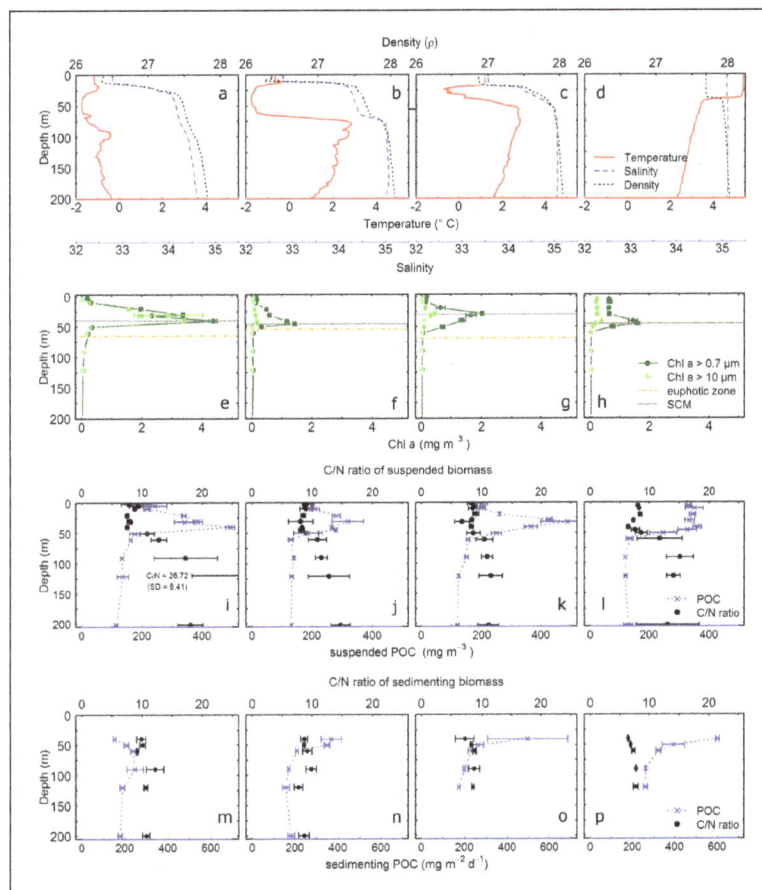

Figure 2: Hydrography and suspended and sedimenting biomass at the sampling stations. Hydrography with temperature (red line), salinity (blue stippled) and density (black dotted) in first row **(a, b, c, d)**. Subsurface chlorophyll *a* maximum (SCM, grey line), euphotic zone (orange stippled line) and suspended chlorophyll *a* (Chl *a*, dark green: >0.7 µm; light green: >10 µm) in second row **(e, f, g, h)**. Suspended particulate organic carbon (POC, blue crosses and blue stippled line) and atomic C/N ratio (black dots) in the third row **(i, j, k, l)** and the sedimenting POC and its atomic C/N ratio in the fourth row **(m, n, o, p)**. Left column: station M1, second column from left: M2, second column from right: M3, right column: station M4. DOI: https://doi.org/10.1525/elementa.235.f2

sisted of Arctic-originating water gradually mixed with Atlantic water at depth (25–200 m: $T < 0°C$, $S = 34.0$–34.7).

Station M2 was located at the Polar Front in very open drift ice (20%; **Table 1**). In this area, colder and fresher Arctic-derived water masses tend to cover warmer and more saline Atlantic-derived waters (Loeng, 1991) which was also observed during our study (**Figure 2b**). A well-mixed melt-water layer (0–15 m: $T < 0.0°C$, $S = 32.6$) was separated by a strong halocline (15–20 m) from the lower part of the water column, which was influenced, increasingly with depth, by Atlantic water (200 m: $T = 0.9°C$, $S = 35.0$).

Station M3 was in ice-free waters. Its well-mixed surface layer (0–16 m) showed an influence of recent ice melt ($S < 33.6$; **Figure 2c**) and atmospheric warming (>1°C). Arctic-derived water masses prevailed at 20–30 m ($T < 0°C$, $S < 34.5$), but warmer and more saline water was found below 50 m ($T > 1.6°C$, $S > 35.0$). This layering indicated that M3 was also situated in the Polar Front.

Station M4 was located in the ice-free, Atlantic-influenced, southern Barents Sea. The stratification was relatively weak and dominated by a temperature gradient (**Figure 2d**). Water masses were characterized by $T > 5.0°C$ (salinity of 35.09) above the thermocline, but $T = 3.5°C$ below it (40 m), and gradually decreasing with depth (200 m: $T = 2.3°C$). The salinity was fairly constant below the thermocline ($S = 35.10$–35.13).

A fresh to strong breeze was recorded at M1 (9.5–13.3 m s^{-1}) and M3 (9.3–13.5 m s^{-1}), while a moderate to fresh breeze prevailed at M4 (5.5–10.9 m s^{-1}; **Figure 3**). Wind data from M2 are lacking due to the malfunction of our vessel's wind current meter, but land-based permanent weather stations in the area (Bjørnøya, Hopen and Edgeøya; **Figure 1**) recorded a wind speed

of 3–10 m s^{-1} during this period. Following Coelingh et al. (1996), who concluded that the wind speed in the southern North Sea tended to be higher at sea-based stations compared to coastal ones, we presume that the wind speed at M2 was in the range of wind speed encountered at M1 and M3.

3.2 Euphotic zone and suspended biological parameters (size fractionated Chl a, POC, C/N ratio)

At station M1 the base of the euphotic zone (1% irradiance of surface irradiance at 430 nm) was at 65 m (**Figure 2e**). The SCM at 40 m had the highest Chl a concentration of the present study (4.4 mg Chl a m^{-3}). The Chl a was dominated by small cells in the upper 5 m (approximately 90% < 10 µm; **Figure 2e**), but larger cells prevailed between 10 and 120 m (66–90% > 10 µm; **Figure 2e**). The most abundant taxa were the phytoplankton genera *Chaetoceros, Thalassiosira* and *Phaeocystis* (Wiedmann et al., 2014: 265 × 10^3 cells L^{-1}, 156 × 10^3 cells L^{-1}, and 107 × 10^3 cells L^{-1}, respectively), and despite the presence of sea ice, no ice algae were found. The Chl a peak was well correlated to the depth distribution of suspended POC and PON ($R^2 = 0.91$ and 0.86), and the C/N ratio of 7.5–9.5 suggested little to moderately degraded suspended biomass down to 50 m (**Figure 2i**), which was confirmed by microscopic cell counts (healthy cells and few resting spores).

At M2 the euphotic zone reached to a depth of 54 m (**Figure 2f**) and the SCM was found at 44 m (1.5 mg Chl a m^{-3}). The microalgae community was dominated by small phytoplankton (50–80% < 10 µm; **Figure 2f**) down to 200 m, and the species *Phaeocystis pouchetii* (274 × 10^3 cells L^{-1}; Wiedmann et al., 2014) prevailed, though larger pelagic taxa were also found

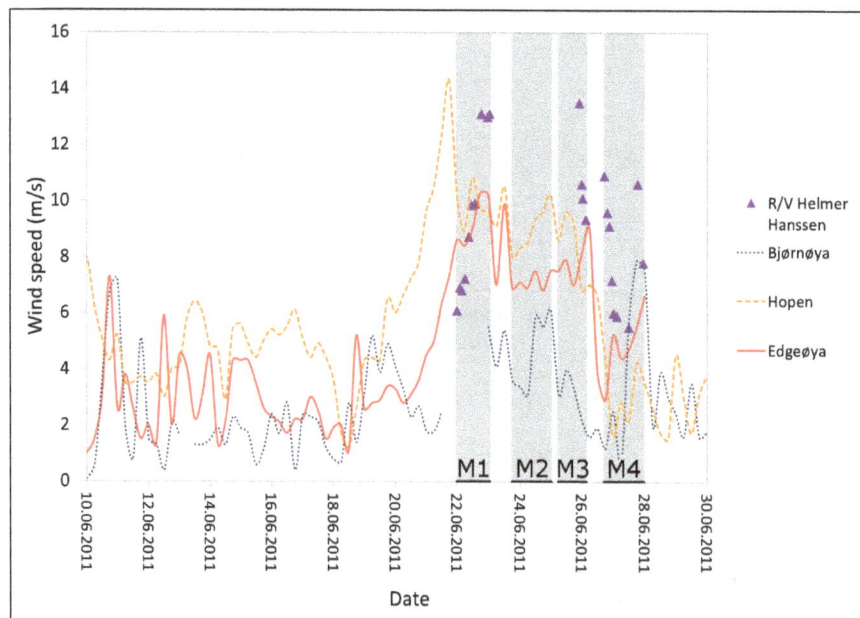

Figure 3: Wind speed measurements recorded on land-based stations and on the research vessel. Wind data from the land-based measurements stations on Bjørnøya (blue dotted line), Hopen (orange stippled line) and Edgeøya (red line) between 10 June 2011 and 30 June 2011 and the wind speed recordings on the R/V *Helmer Hanssen* (violet triangles). Grey shaded areas indicate the time spent at the four sampling stations (M1–M4). DOI: https://doi.org/10.1525/elementa.235.f3

(*Chaetoceros* sp., 3700 cells L⁻¹, and *Thalassiosira* sp., 30 × 10³ cells L⁻¹). The Chl *a* depth profile correlated moderately with the POC and PON depth distribution ($R^2 = 0.56$ and 0.54, respectively) and the C/N ratio of 8.1–9.2 (top 50 m) indicated that the suspended biomass was in a similar state of "freshness" as at M1 (**Figure 2j**).

The deepest euphotic zone (approximately 70 m) and the shallowest SCM (30 m: 2.0 mg m⁻³; **Figure 2g**) was observed at station M3. No cell counts are available for this station, but M3 was also dominated by small cells (70–90%) and the Chl *a* depth profile correlated well with the POC and PON depth distribution ($R^2 = 0.93$ and 0.82, respectively). Apart from the 32 m sample directly under the SCM (C/N = 6.8), the C/N range at M3 was similar to the range observed at M2 and M1 (**Figure 2k**).

At station M4, the base of the euphotic zone coincided with the SCM (both at 45 m; Chl *a* concentration: 1.6 mg m⁻³; **Figure 2h**). The phytoplankton taxon *Phaeocystis* clearly dominated (1.810 × 10⁶ cells L⁻¹; Wiedmann et al., 2014), and only few large cells (*Thalassiosira* sp. 4050 cells L⁻¹) were found. In contrast to the Chl *a* depth distribution, the suspended POC was rather evenly distributed in the upper 40 m (**Figure 2h** and **l**) and thus the two parameters were only moderately correlated ($R^2 = 0.64$). A better correlation was found between the Chl *a* and the PON depth profile ($R^2 = 0.73$). The C/N ratio of the suspended biomass in the top 50 m (consisting almost exclusively of single cells of *Phaeocystis pouchetii*) was somewhat lower than at the previous stations (C/N = 6.4–8.6; **Figure 2l**).

Based on the integrated nitrate concentrations (**Figure 4a, b**; M2 and M3 not shown), the Chl *a* concentration, the relative abundance of small cells, and the composition of phytoplankton and zooplankton (C. Svensen, unpublished data), the bloom stage was determined at our four sampling stations. We classified them as a late peak bloom stage (M1), a late bloom stage (M2, M3) and post-bloom stage (M4), respectively. The term "late peak bloom" (M1) may not be very common, but we use it here to describe that M1 was in a late stage of a peak bloom, because we found in the surface waters 1) low nitrate concentrations, 2) high Chl *a* concentration, dominated by large phytoplankton at ≥10 m), yet 3) an abundant zooplankton community, mainly consisting of copepod eggs and nauplii (C. Svensen, unpublished data).

3.3 Nitrate concentration and nitrate flux at M1 and M4

The suspended biological variables and hydrography, as well as the data on vertical export, implied a gradual change from north to south, which we used as a space-for-time substitution to investigate the impact of the upward nitrate flux at the two most contrasting

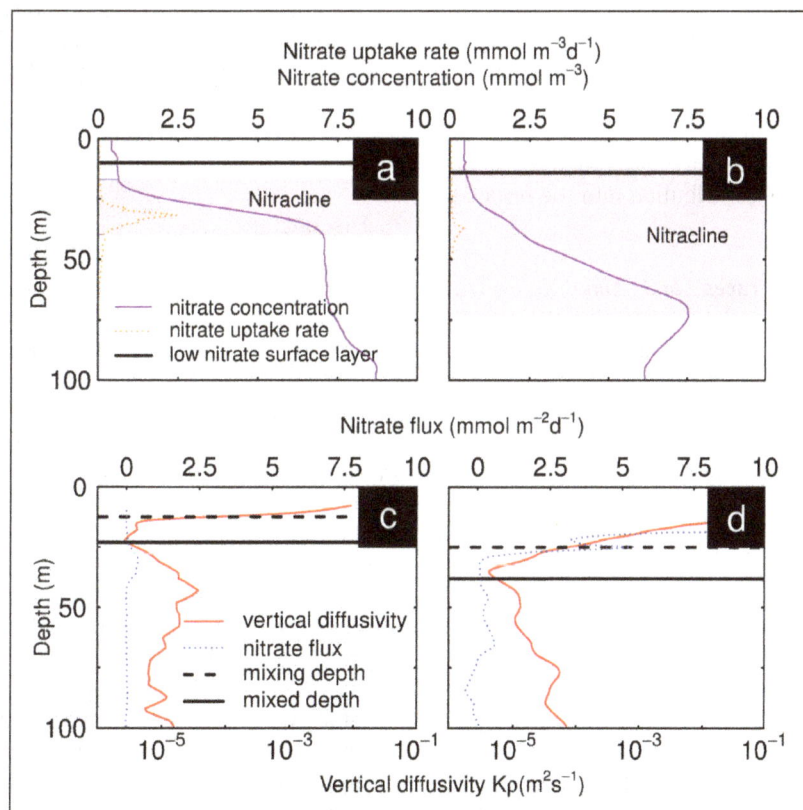

Figure 4: Nitrate concentration, nitrate uptake rates and mixing depth at the two most contrasting stations. Nitrate concentrations (purple line), nitrate uptake rates (orange dotted line) and nitracline (purple shaded field) at station M1 **(a)** and M4 **(b)** in the upper row. Vertical diffusivity (red line) and the upward nitrate flux (blue dotted line) in the lower row **(c, d)**. The black lines indicate the layers used in the 1-D residence time calculation (a, b: solid line, low nitrate surface layer with a nitrate concentration ≤0.6 mmol nitrate m⁻³; c, d: dashed line, mixing depth; solid line, mixed depth). DOI: https://doi.org/10.1525/elementa.235.f4

situations M1 and M4. At M1, the nitrate concentration was low in the upper 11 m (<0.6 mmol m^{-3}, **Figure 4a**) and, with surface-enhanced mixing protruding only to 13 m (mixing depth; **Figure 4c**), a weak upward nitrate flux of 0.01 mmol nitrate m^{-2} d^{-1} was estimated into the low nitrate surface layer (0–11 m; **Figure 4a, Table 2**). Because of the strong stratification of the water column (**Figure 2a**), the upward nitrate flux into the upper part of the nitracline (20–25 m) was also low (**Figure 4a, c**). However, between 25 m and nearly 40 m, the stratification was weaker (**Figure 2a**), and the upward nitrate flux was 0.1–0.4 mmol m^{-2} d^{-1} (**Figure 4c**). Below 40 m, a combination of declining vertical diffusivity and a vertically rather stable nitrate concentration resulted in a weak upward nitrate flux (<0.1 mmol m^{-2} d^{-1}; **Figure 4c**). Overall, the nitrate flux contributed little to the nitrate stock in our five depth layers, which we investigated in detail (**Table 2**).

At M4, the nitrate concentration increased from the surface (1 m: <0.5 mmol nitrate m^{-3}) to a subsurface maximum at approximately 73 m (7.54 mmol nitrate m^{-3}; **Figure 4b**). The minor decline in nitrate concentration below the subsurface maximum values (**Figure 4b**) likely reflects differences in advection history at the different subsurface depths. The enhanced diffusivity (>10^{-4} m^2 d^{-1}) in the uppermost 25 m resulted in an upward nitrate flux of 30.0 mmol nitrate m^{-2} d^{-1} into the low nitrate surface layer and 5.4 mmol nitrate m^{-2} d^{-1} into the mixing layer (**Figure 4d, Table 2**). This flux replenished the nitrate stock in the low nitrate surface layer (0–17 m) and the mixing layer (0–25 m) by approximately 341% d^{-1} and 38% d^{-1} (**Table 2**). The contribution of the upward nitrate flux to the nitrate stock in the other layers at M4 was <1%, which still exceeded the contribution into the respective layers at M1.

3.4 Nitrate uptake rates and time to nitrate exhaustion

The nitrate uptake rate peaked at 32 m at M1 (2.5 mmol nitrate m^{-3} d^{-1}; **Figure 4a**), and at 37 m at M4 (0.4 mmol nitrate m^{-3} d^{-1}; **Figure 4b**). By using 1-D residence time calculations, we aimed to assess changes in the integrated nitrate concentration by nitrate uptake and the upward nitrate flux in our five layers. At M1, the nitrate concentration declined in the five depth layers, whether or not the upward nitrate flux into this layer was taken into account (**Table 2**). The nitrate flux was too weak to compensate for the nitrate uptake, and including it into our 1-D calculations prolonged the time to nitrate exhaustion by only one day at maximum. The 1-D calculations for the above SCM layer, the euphotic zone and the mixed layer at M4 gave the same result (**Table 2**). Considerably different results, however, were obtained for the low nitrate surface layer and the mixing layer at M4. Because of the intense upward nitrate flux, which exceeded the nitrate uptake rate, the nitrate stock could be replenished at M4. We conducted additional calculations for the mixing layer at M4 (**Table 2**), because Sakshaug and Slagstad (1992) suggested that strong wind events may occur at 10-day intervals in the southern Barents Sea, causing episodic periods

of enhanced mixing. We recorded wind speed >12 m s^{-1} at M3, and thus presumed that a strong wind event had also occurred at M4 prior to our arrival there (**Figure 3**). This event would have likely induced vertical mixing processes and resulted in high diffusivity and strong upward nitrate flux, which we observed at M4 (**Figure 4d**). We aimed to transfer this situation into our 1-D calculation by using a high upward nitrate flux into the mixing layer for 1–3 days, followed by a weaker upward nitrate flux of 0.30 mmol nitrate m^{-2} d^{-1}. The latter equals an average flux at 50–70 m, a depth interval not influenced by surface mixing processes. Our 1-D residence time calculation suggested that, for a situation of strong mixing (1, 2, or 3 days) followed by a period of relaxation, the integrated nitrate concentration in the mixing layer was replenished considerably and the period to nitrate depletion in the mixing layer was prolonged to 16, 21 or 25 days (**Table 2**).

3.5 Characterization of the vertical flux (POC, C/N ratio)

The downward POC flux (at 120 m) was higher at M4 (261 mg POC m^{-2}) than at the other stations (156–187 mg POC m^{-2} d^{-1}; **Figure 2m–p**), and the attenuation of the flux was weakest at M1 (M1: 22%; M2–M4: 58–65%). The C/N ratio of the sinking biomass was highest at the northernmost station M1, intermediate at M2 and M3 and lowest at M4 (**Figure 2m–p**). The composition of the sinking material was assessed in a previous study (Wiedmann et al., 2014) and showed that diatom colonies contributed most to the sinking biomass at M1. The sinking biomass at M2 was a mixture of diatoms, fecal pellets and unidentified matter, while it was dominated by fecal pellets at M4. No data are available from M3.

4. Discussion

4.1 Impact of water column stratification and vertical turbulent mixing on the upward nitrate flux

In marine ecosystems, the nitrate flux is commonly oriented upwards (Mann and Lazier, 2006), and the strength of this flux can vary considerably. Strongly stratified waters hamper it (Moum, 1996; Osborn, 1980), while wind-induced vertical mixing processes such as water shear, waves, Ekman pumping, frontal and eddy-induced upwelling (Omand and Mahadevan, 2015) can promote it, because these processes break down the stratification.

At station M1 in the northern Barents Sea, a combination of a moderate halocline and the presence of drift ice apparently hindered deep turbulent mixing. This hindering resulted in a low upward nitrate flux (<0.04 mmol nitrate m^{-2} d^{-1}) into the five layers, which we investigated in detail in our 1-D residence time calculations (**Table 2**). The intensity of the upward nitrate flux at M1 was similar to previous reports from 1) the stratified, partly ice-covered, northern Barents Seas (upward nitrate flux into the upper mixed layer during a summer ice edge bloom: 0.14 mmol nitrate m^{-2} d^{-1}; Sundfjord et al., 2007), 2) drift ice-covered waters northwest of Svalbard (May and August: nitrate flux of 0.2–0.3 mmol nitrate m^{-2} d^{-1}; Randelhoff et al., 2016), 3) the ice-free, northeast Atlantic subpolar gyre (upward nitrate flux into the upper mixed layer during summer: 0.02–0.60 mmol nitrate $m^{-2} d^{-1}$; Painter et al., 2014), and 4) the Porcupine Abyssal Plain, NE Atlantic (upward nitrate flux into the euphotic zone during a weakly stratified summer situation: 0.09 mmol nitrate $m^{-2} d^{-1}$; Martin et al., 2010a).

In contrast, the ice-free, weakly stratified waters at M4 in the southern Barents Sea were more prone to wind-induced mixing, and the upward nitrate flux into the base of the mixing layer exceeded the one at M1 by up to two orders of magnitude (>5 mmol $m^{-2} d^{-1}$ into the base of the mixing layer; **Table 2**, **Figure 4c, d**). Comparatively high fluxes have been reported previously, into the base of the upper mixed layer in the southern Barents Sea during July (Sundfjord et al., 2007) and into the base of the SCM in the tidally mixed Celtic Sea during summer (Sharples et al., 2007). However, these high upward nitrate fluxes diverge from the common understanding of nitrate replenishment in the Arctic Ocean and sub-Arctic seas. Convective winter mixing is usually assumed to be the major driver of the upward nutrient flux and replenishment of the nitrate concentrations in the surface layers (Louanchi and Najjar, 2001), because thermal and meltwater stratification tends to hamper deep vertical mixing during summer (Martin et al., 2010a; Bourgault et al. 2011; Painter et al., 2014; Randelhoff et al. 2016).

Nonetheless, we consider our result of high upward nitrate flux in the southern Barents Sea during early summer as reliable, because a combination of factors seemed to facilitate the flux in this area then. The stratification in the southern, ice-free Barents Sea is not influenced by sea ice melt water, and this shelf sea, due to its location far north, is less exposed to atmospheric warming than shelf seas further south; thus, a weakly stratified water column can be found during late spring and early summer. Previous studies suggest that episodic wind events have a major effect on the southern Barents Sea (Sakshaug and Slagstad, 1992; Le Fouest et al., 2011), and a wind-mixed layer of 50–100 m depth during May has been proposed (Slagstad and McClimans, 2005). In combination with the nitracline being located at 20–70 m (Reigstad et al., 2002; Hodal et al., 2008), an intense upward nitrate flux, as observed at our station M4, is a likely event. As nitrate entrainment by a storm during a post-bloom situation has also been reported from the Bering Sea (Sambrotto et al., 1986), we hypothesize that a strong upward nitrate flux during a post-bloom situation may be a more wide-spread phenomenon in Arctic shelf seas than previously realized. More research on this phenomenon is needed.

4.2 Factors impacting the nitrate stock in the upper water column

The spring phytoplankton bloom at high latitudes is commonly assumed to cause a decline in the nitrate surface concentration and a deepening of the nitracline. Our observations during a late peak bloom, a late bloom and a post-bloom situation in the Barents Sea generally corresponded with this pattern. A minor modification was observed in the form of a subsurface nitrate peak at 70 m at station M4, but we assume this peak was linked to advection of nutrient-rich Atlantic water (Torres-Valdés et al.,

2013). Similar impacts on the nitrate depth profile have been reported from the Japan Sea and open oceanic sites (Kaplunenko et al., 2013; Omand and Mahadevan, 2015); such subsurface peaks do not contradict the general pattern of a nitrate decline during the spring phytoplankton bloom.

In our 1-D residence time calculation we investigated how the upward nitrate flux and the nitrate uptake modified the nitrate stock in five different layers of biological relevance at M1 and M4. These calculations have some limitations. First, the upward nitrate fluxes are based on few turbulence measurements and nitrate depth profiles; they represent only a snap-shot of the situation in the water column. Accordingly, a higher upward nitrate flux can occur in shallower waters than at greater depth (e.g. M4, 25 m: 5.40 mmol nitrate $m^{-2} d^{-1}$ vs. M4, 45 m: 0.34 mmol nitrate $m^{-2} d^{-1}$), but this situation can only be assumed to be sustainable for hours to few days. Second, the nitrate uptake rates represent also only snap-shot measurements because they were based on short incubations to minimize bottle effects and artefacts, but we extrapolated the uptake rates with a model, thus giving a reasonable estimate for a period of several days. Third, a potential impact of shallow water nitrification has not been included in our 1-D calculations, because relevant data for the Barents Sea were not available. In the northern Bering Sea and southern Chukchi Sea, however, Shiozaki et al. (2016) reported only a minor impact of nitrate production via nitrification, which contributed <5% to nitrate assimilation.

Given these limitations, our 1-D calculations nevertheless illustrate a potentially important set of scenarios for the Barents Sea. At M1, mixing processes were hampered by drift ice and a moderate stratification. This hampering resulted in a minor upward nitrate flux, which contributed very little to the nitrate standing stock in the layers, which we investigated in detail (**Table 2**). Similarly low rates of daily nitrate injections have been observed in the subpolar Atlantic Ocean gyre (Painter et al., 2014). In an additional calculation example (see last column of **Table 2**), we aimed to reproduce the situation of a northwards moving marginal ice zone, because ice maps from the Norwegian Meteorological Institute (www.polarview.met.no) indicate this development subsequent to our work at station M1. However, using a stronger upward nitrate flux (0.350 mmol $m^{-2} d^{-1}$) only prolonged the time to nitrate depletion at M1 (**Table 2**), because the upward nitrate flux could not counterbalance the nitrate uptake rates. Thus, the situation at M1 agreed well with the common understanding of a constant decline of the nitrate surface concentration during the spring phytoplankton bloom.

At M4, the situation was different. The upward nitrate flux into the mixing layer and the low nitrogen surface layer was high enough to replenish the nitrate concentration in these layers. Our 1-D calculation suggested that this replenishment occurred even when we mirrored the pattern of 1–3 days of deep mixing (strong upward nitrate flux of >5 mmol $m^{-2} d^{-1}$) followed by weak mixing (resulting in a weak upward nitrate flux). Our simple 1-D calculations thus point to an interesting concept of two possible scenarios: a constant decline of the surface nitrate concentration obviously takes place in many high latitude regions during the spring phytoplankton bloom, but there may also be areas where an episodic replenishment of surface nitrate concentration occurs during late spring and early summer.

4.3 Impact of water column stratification and turbulent mixing on the downward POC flux

The intensity of the downward POC flux reflects the hydrographical situation and the ecological interactions of the plankton in the upper water column. High POC sedimentation events tend to occur when a temporal, weak coupling of primary production and grazer activity allows sinking of the produced biomass.

This situation has been suggested to occur during the ice edge phytoplankton bloom in the Barents Sea (Sakshaug et al., 1991, 2009; Wassmann and Reigstad, 2011), and our study confirms that. The downward POC flux we encountered at station M1 was comparable to previous measured exports in this region during spring (Andreassen and Wassmann, 1998; Coppola et al., 2002; Olli et al., 2002; Reigstad et al., 2008). The high Chl a: POC ratio and the cell counts suggest that suspended autotrophs were the prevailing form of POC in the water column, while aggregates of large diatoms (>10 μm) have been identified as the prevailing vehicle of vertically exported biomass to ≤60 m at M1 (Wiedmann et al., 2014). These aggregates can sink a few hundred meters per day, depending on species and physiological stage (Bienfang et al., 1982; Iversen and Ploug, 2013), and as mesozooplankton abundances were low at M1 (Wiedmann et al., 2014; C. Svensen, unpublished data), a weak attenuation of the sinking biomass occurred at this station in the northern Barents Sea.

Despite the common conception of a post bloom situation being associated with a minor POC sedimentation (Sakshaug et al, 2009), we found a stronger downward POC flux during the post bloom situation at M4 than during the late peak bloom at M1 (**Figure 2m, p**). Similarly high downward POC fluxes have been observed previously in the southern Barents Sea during late spring and early summer (Olli et al., 2002: in July, downward POC flux up to 400 mg POC m^{-2} d^{-1}; Reigstad et al., 2008: in late May, 400–750 mg POC m^{-2} d^{-1} at 40–200 m), but the underlying mechanisms have not been fully understood.

During our study, single cells of *Phaeocystis pouchetii* (5 μm diameter) were highly abundant at M4 (Wiedmann et al., 2014). They have a low sinking velocity, but may contribute to the downward POC flux when down-mixing occurs (Reigstad and Wassmann, 2007). Because we found deep vertical mixing at M4 (**Figure 4d**), and a low C/N ratio of the sedimenting material (C/N = 6.4–7.7; **Figure 2l**) suggests a fast downward transport of recently produced biomass, we presume that down-mixing of *Phaeocystis* cells occurred at M4. Further, the mesozooplankton abundance increased during our field study from north to south (C. Svensen, unpublished data), so that an intense top-down control can be assumed at M4. Such top-down control may at a first seem to contradict the moderate biomass attenuation and strong downward POC flux observed at

M4. However, pulsed nitrate supply can stimulate primary production (southeastern Bering Sea; Sambrotto et al., 1986), which in turn may enhance feeding rates of copepods and result in the production of larger fecal pellets (Turner and Ferrante, 1979, and references therein; Wexels Riser et al., 2007). Fecal pellets have been observed frequently in the gel traps at M4 (Wiedmann et al., 2014) and, as they have a considerable sinking velocity (5–220 m d^{-1}; Turner, 2002), they may contribute to a fast transport from surface layers to depth. Thus, the strong downward POC flux we found at M4 may be a result of intense grazing and repackaging of slowly sinking biomass into rapidly sinking pellets. Emerging from this study are two scenarios for strong downward POC flux: one associated with the flux of large diatoms during a late peak bloom situation, and another associated with a post-bloom situation in deep mixed waters, where down-mixing of small, slowly-sinking phytoplankton cells and fast-sinking fecal pellets help to explain minor biomass attenuation in the surface layers.

5. Conclusion

During a field study with four stations, we observed a gradual change in the hydrography and phytoplankton bloom stage in the Barents Sea, an Arctic shelf sea. In the moderately stratified water column in the northern Barents Sea a late peak bloom prevailed, while a late bloom was found in the stratified waters of the Polar Front and a post-bloom situation in the deep-mixed waters of the southern Barents Sea. We used this space-for-time substitution to investigate the northernmost (late peak bloom) and southernmost (post bloom) station in detail as they represented the most contrasting situations. A weak upward nitrate flux characterized the stratified waters in the marginal ice zone in the north, where the flux could not counterbalance the nitrate uptake rate and our 1-D residence time calculation implied a constant decline of the nitrate concentration in the surface layers (<65 m). In contrast, a substantial upward nitrate flux was found into the surface layers (<23 m) in the Atlantic-influenced, ice-free waters of the southern Barents Sea, where the 1-D calculations suggested that the nitrate concentrations in the layers were replenished. Though a high downward POC flux is commonly associated with a peak phytoplankton bloom stage, we found a higher downward POC flux during the post-bloom situation of the southern Barents Sea compared to the late peak bloom stage in the northern Barents Sea. We suggest that the intense upward nitrate flux during the post-bloom situation stimulated the pelagic system and that a combination of downward mixed phytoplankton cells and fast-sinking fecal pellets enhanced the POC export. From the perspective of amplified climate warming in the Arctic, we see an urgent need for further investigation of the effect of deep-mixing events on the downward POC flux in ice-free, weakly stratified Arctic shelf regions with a shallow nitracline. The results of such studies will help to improve our understanding of the food supply for benthic ecosystems and bottom-associated fish stocks as well as the potential for carbon sequestration in future, ice-free Arctic regions.

Supplemental File

The supplemental file for this article can be found as follows:

· **Table S1.** Parameters used in the calculation of nitrate uptake.

Acknowledgements

We thank the captain and the crew of the R/V *Helmer Hanssen* for practical support during the fieldwork. Sigrid Øygarden, Christian Wexels Riser and Camilla Svensen helped with field and laboratory work, and two anonymous reviewers and the editor-in-chief improved the manuscript substantially. All help was highly appreciated.

Funding information

The present work is a part of the CONFLUX project, funded by Tromsø Forskningsstiftelse, but also financially supported by the CarbonBridge project (Research Council of Norway, no. 226415). Wiedmann's contribution was partially funded by the Research Center for ARCtic Petroleum Exploration (ARCEx) (Research Council of Norway, no. 228107 and 8 industry partners) and Sundfjords's participation was partially funded by the Center of Ice, Climate and Ecosystem (ICE) at the Norwegian Polar Institute.

Competing interests

The authors have no competing interests to declare.

Author contribution

· Contributed to conception and design: IW, JET, AS, MR
· Contributed to acquisition of data: IW, JET, AS, MR
· Contributed to analysis and interpretation of data: IW, JET, AS, MR
· Drafted and/or revised the article: IW, JET, AS, MR
· Approved the submitted version for publication: IW, JET, AS, MR

References

Andreassen, IJ and **Wassmann, P** 1998 Vertical flux of phytoplankton and particulate biogenic matter in the marginal ice zone of the Barents Sea in May 1993. *Mar Ecol Prog Ser* **170**: 1–14. DOI: https://doi.org/10.3354/meps170001

Arrigo, KR and **van Dijken, GL** 2015 Continued increases in Arctic Ocean primary production. *Prog Oceanogr* **136**: 60–70. DOI: https://doi.org/10.1016/j.pocean.2015.05.002

Babin, M, Morel, A and **Gagnon, R** 1994 An incubator designed for extensive and sensitive measurements of phytoplankton photosynthetic parameters. *Limnol Oceanogr* **39**(3): 694–702. DOI: https://doi.org/10.4319/lo.1994.39.3.0694

Bienfang, PK 1981 Sinking rates of heterogeneous, temperate phytoplankton populations. *J Plankton Res* **3**(2): 235–253. DOI: https://doi.org/10.1093/plankt/3.2.235

Bienfang, PK, Harrison, PJ and **Quarmby, LM** 1982 Sinking rate response to depletion of nitrate, phosphate and silicate in four marine diatoms. *Mar Biol* **67**(3): 295–302. DOI: https://doi.org/10.1007/BF00397670

Blais, M, Tremblay, J-É, Jungblut, AD, Gagnon, J, Martin, J, et al. 2012 Nitrogen fixation and identification of potential diazotrophs in the Canadian Arctic. *Global Biogeochem Cy* **26**(3): GB3022. DOI: https://doi.org/10.1029/2011GB004096

Bourgault, D, Hamel, C, Cyr, F, Tremblay, JÉ, Galbraith, PS, et al. 2011 Turbulent nitrate fluxes in the Amundsen Gulf during ice-covered conditions. *Geophys Res Lett* **38**(15): L15602. DOI: https://doi.org/10.1029/2011GL047936

Brainerd, KE and **Gregg, MC** 1995 Surface mixed and mixing layer depths. *Deep-Sea Res I* **42**(9): 1521–1543. DOI: https://doi.org/10.1016/0967-0637(95)00068-H

Carmack, E and **McLaughlin, F** 2011 Towards recognition of physical and geochemical change in Subarctic and Arctic Seas. *Prog Oceanogr* **90**(1–4): 90–104. DOI: https://doi.org/10.1016/j.pocean.2011.02.007

Carmack, E and **Wassmann, P** 2006 Food webs and physical–biological coupling on pan-Arctic shelves: Unifying concepts and comprehensive perspectives. *Prog Oceanogr* **71**(2–4): 446–477. DOI: https://doi.org/10.1016/j.pocean.2006.10.004

Coelingh, JP, van Wijk, AJM and **Holtslag, AAM** 1996 Analysis of wind speed observations over the North Sea. *J Wind Eng Ind Aerodyn* **61**(1): 51–69. DOI: https://doi.org/10.1016/0167-6105(96)00043-8

Collos, Y 1987 Calculations of ^{15}N uptake rates by phytoplankton assimilating one or several nitrogen sources. *Int J Radiat Appl Instrum Part A* **38**(4): 275–282. DOI: https://doi.org/10.1016/0883-2889(87)90038-4

Coppola, L, Roy-Barman, M, Wassmann, P, Mulsow, S and **Jeandel, C** 2002 Calibration of sediment traps and particulate organic carbon export using ^{234}Th in the Barents Sea. *Mar Chem* **80**(1): 11–26. DOI: https://doi.org/10.1016/S0304-4203(02)00071-3

Eppley, RW, Holmes, RW and **Strickland, JDH** 1967 Sinking rates of marine phytoplankton measured with a fluorometer. *J Exp Mar Biol Ecol* **1**(2): 191–208. DOI: https://doi.org/10.1016/0022-0981(67)90014-7

Falk-Petersen, S, Pavlov, V, Berge, J, Cottier, F, Kovacs, K, et al. 2015 At the rainbow's end: high productivity fueled by winter upwelling along an Arctic shelf. *Polar Biol* **38**(1): 5–11. DOI: https://doi.org/10.1007/s00300-014-1482-1

Fer, I 2006 Scaling turbulent dissipation in an Arctic fjord. *Deep-Sea Res I* **53**(1–2): 77–95. DOI: https://doi.org/10.1016/j.dsr2.2006.01.003

Fernández-Méndez, M, Wenzhöfer, F, Peeken, I, Sørensen, HL, Glud, RN, Boetius, A 2014 Composition, Buoyancy Regulation and Fate of Ice Algal Aggregates in the Central Arctic Ocean. *PLoS ONE* **9**(9): e107452. DOI: https://doi.org/10.1371/journal.pone.0107452

Forest, A, Babin, M, Stemmann, L, Picheral, M, Sampei, M, et al. 2013 Ecosystem function and particle flux dynamics across the Mackenzie Shelf (Beaufort Sea, Arctic Ocean): an integrative analysis of spatial variability and biophysical forcings. *Biogeosciences* **10**(5): 2833–2866. DOI: https://doi.org/10.5194/bg-10-2833-2013

Franks, PJS 2015 Has Sverdrup's critical depth hypothesis been tested? Mixed layers vs. turbulent layers. *ICES J Mar Sci* **72**(6): 1897–1907. DOI: https://doi.org/10.1093/icesjms/fsu175

Frigstad, H, Andersen, T, Bellerby, RGJ, Silyakova, A and **Hessen, DO** 2014 Variation in the seston C:N ratio of the Arctic Ocean and pan-Arctic shelves. *J Mar Sys* **129**: 214–223. DOI: https://doi.org/10.1016/j.jmarsys.2013.06.004

Geider, R and **La Roche, J** 2002 Redfield revisited: variability of C:N:P in marine microalgae and its biochemical basis. *Eur J Phycol* **37**(1): 1–17. DOI: https://doi.org/10.1017/S0967026201003456

Hales, B, Moum, JN, Covert, P and **Perlin, A** 2005 Irreversible nitrate fluxes due to turbulent mixing in a coastal upwelling system. *JGR: Oceans* **110**(C10): S11. DOI: https://doi.org/10.1029/2004JC002685

Hodal, H and **Kristiansen, S** 2008 The importance of small-celled phytoplankton in spring blooms at the marginal ice zone in the northern Barents Sea. *Deep-Sea Res II* **55**(20–21): 2176–2185. DOI: https://doi.org/10.1016/j.dsr2.2008.05.012

Holm-Hansen, O and **Riemann, B** 1978 Chlorophyll *a* Determination: Improvements in Methodology. *Oikos* **30**(3): 438–447. DOI: https://doi.org/10.2307/3543338

IPCC (ed.) 2013 *Climate Change 2013: The Physical Science Basis. Contribution of Working Group I to the Fifth Assessment Report of the Intergovernmental Panel on Climate Change.* Cambridge, United Kingdom and New York, NY, USA: Cambridge University Press.

Iversen, MH and **Ploug, H** 2013 Temperature effects on carbon-specific respiration rate and sinking velocity of diatom aggregates – potential implications for deep ocean export processes. *Biogeosciences* **10**(6): 4073–4085. DOI: https://doi.org/10.5194/bg-10-4073-2013

Johnson, KS and **Coletti, LJ** 2002 In situ ultraviolet spectrophotometry for high resolution and long-term monitoring of nitrate, bromide and bisulfide in the ocean. *Deep-Sea Res I* **49**(7): 1291–1305. DOI: https://doi.org/10.1016/S0967-0637(02)00020-1

Kaplunenko, DD, Lobanov, VB, Tishchenko, PY and **Shvetsova, MG** 2013 Nitrate in situ measurements in the northern Japan Sea. *Deep-Sea Res Pt II* **86–87**: 10–18. DOI: https://doi.org/10.1016/j.dsr2.2012.08.005

Kvingedal, B 2005 Sea-Ice Extent and Variability in the Nordic Seas, 1967–2002. In: Drange, H, Dokken, T, Furevik, T, Gerdes, R and Berger, W (eds.), *The Nordic Seas: An Integrated Perspective*, 39–49. Washington, D. C.: American Geophysical Union.

Le Fouest, V, Postlethwaite, C, Morales Maqueda, MA, Bélanger, S and **Babin, M** 2011 On the role of tides and strong wind events in promoting summer primary production in the Barents Sea. *Cont Shelf Res* **31**(17): 1869–1879. DOI: https://doi.org/10.1016/j.csr.2011.08.013

Leu, E, Søreide, JE, Hessen, DO, Falk-Petersen, S and **Berge, J** 2011 Consequences of changing sea-ice cover for primary and secondary producers in the European Arctic shelf seas: Timing, quantity, and quality. *Progr Oceanogr* **90**(1–4): 18–32. DOI: https://doi.org/10.1016/j.pocean.2011.02.004

Loeng, H 1991 Features of the physical oceanographic conditions of the Barents Sea. *Polar Res* **10**(1): 5–18. DOI: https://doi.org/10.1111/j.1751-8369.1991.tb00630.x

Louanchi, F and **Najjar, RG** 2001 Annual cycles of nutrients and oxygen in the upper layers of the North Atlantic Ocean. *Deep-Sea Res II* **48**(10): 2155–2171. DOI: https://doi.org/10.1016/S0967-0645(00)00185-5

Mann, K and **Lazier, J** 2006 *Dynamics of Marine Ecosystems: Biological-Physical Interactions in the Oceans*, Third Edition. Blackwell Publishing.

Martin, AP, Lucas, MI, Painter, SC, Pidcock, R, Prandke, H, et al. 2010a The supply of nutrients due to vertical turbulent mixing: A study at the Porcupine Abyssal Plain study site in the northeast Atlantic. *Deep-Sea Res II* **57**(15): 1293–1302. DOI: https://doi.org/10.1016/j.dsr2.2010.01.006

Martin, J, Tremblay, J-É, Gagnon, J, Tremblay, G, Lapoussiére, A, et al. 2010b Prevalence, structure and properties of subsurface chlorophyll maxima in Canadian Arctic waters. *Mar Ecol Prog Ser* **412**: 69–84. DOI: https://doi.org/10.3354/meps08666

Martin, J, Tremblay, J-È and **Price, NM** 2012 Nutritive and photosynthetic ecology of subsurface chlorophyll maxima in Canadian Arctic waters. *Biogeosciences* **9**: 5353–5371. DOI: https://doi.org/10.5194/bg-9-5353-2012

Mei, ZP, Legendre, L, Tremblay, JÉ, Miller, LA, Gratton, Y, et al. 2005 Carbon to nitrogen (C:N) stoichiometry of the spring–summer phytoplankton bloom in the North Water Polynya (NOW). *Deep-Sea Res I* **52**(12): 2301–2314. DOI: https://doi.org/10.1016/j.dsr.2005.07.001

Moum, JN 1996 Efficiency of mixing in the main thermocline. *J Geophys Res* **101**(C5): 12057–12069. DOI: https://doi.org/10.1029/96JC00508

Olli, K, Rieser, CW, Wassmann, P, Ratkova, T, Arashkevich, E, et al. 2002 Seasonal variation in vertical flux of biogenic matter in the marginal ice zone and the central Barents Sea. *J Mar Syst* **38**: 189–204. DOI: https://doi.org/10.1016/S0924-7963(02)00177-X

Omand, MM and **Mahadevan, A** 2015 The shape of the oceanic nitracline. *Biogeosciences* **12**(11): 3273–3287. DOI: https://doi.org/10.5194/bg-12-3273-2015

Osborn, TR 1980 Estimates of the Local Rate of Vertical Diffusion from Dissipation Measurements. *J Phys Oceanogr* **10**(1): 83–89. DOI: https://doi.org/10.1175/1520-0485(1980)010<0083:EOTLRO>2.0.CO;2

Painter, SC, Henson, SA, Forryan, A, Steigenberger, S, Klar, J, et al. 2014 An assessment of the vertical diffusive flux of iron and other nutrients to the surface waters of the subpolar North Atlantic Ocean. *Biogeosciences* **11**(8): 2113–2130. DOI: https://doi.org/10.5194/bg-11-2113-2014

Platt, T, Gallegos, CL and Harrison, WG 1980 Photoinhibition of photosynthesis in natural assemblages of marine phytoplankton. *J Mar Res* **38**: 687–701.

Prandke, H and Stips, A 1998 Test measurements with an operational microstructure-turbulence profiler: Detection limit of dissipation rates. *Aquat Sci* **60**(3): 191–209. DOI: https://doi.org/10.1007/s000270050036

Rainville, L, Lee, CM and Woodgate, RA 2011 Impact of wind-driven mixing in the Arctic Ocean. *Oceanography* **24**(3): 136. DOI: https://doi.org/10.5670/oceanog.2011.65

Randelhoff, A, Fer, I, Sundfjord, A, Tremblay, J-É and Reigstad, M 2016 Vertical fluxes of nitrate in the seasonal nitracline of the Atlantic sector of the Arctic Ocean. *JGR: Oceans* **121**(7): 5282–5295. DOI: https://doi.org/10.1002/2016JC011779

Redfield, AC 1934 On the proportion of organic derivates in sea water and their relation to the composition of plankton. *James Johnstone Memorial Volume,* 177–192. Liverpool: Liverpool University Press.

Redfield, AC 1958 The biological control of chemical factors in the environment. *Am Sci* **46**(3): 205–221.

Reigstad, M, Riser, CW, Wassmann, P and Ratkova, T 2008 Vertical export of particulate organic carbon: Attenuation, composition and loss rates in the northern Barents Sea. *Deep-Sea Res II* **55**: 2308–2319. DOI: https://doi.org/10.1016/j.dsr2.2008.05.007

Reigstad, M and Wassmann, P 2007 Does *Phaeocystis* spp. contribute significantly to vertical export of organic carbon? *Biogeochemistry* **83**(1–3): 217–234. DOI: https://doi.org/10.1007/s10533-007-9093-3

Reigstad, M, Wassmann, P, Wexels Riser, C, Øygarden, S and Rey, F 2002 Variations in hydrography, nutrients and chlorophyll a in the marginal ice-zone and the central Barents Sea. *J Mar Syst* **38**(1–2): 9–29. DOI: https://doi.org/10.1016/S0924-7963(02)00167-7

Sakshaug, E, Johnsen, G, Kristiansen, S, von Quillfeldt, C, Rey, F, et al. 2009 Phytoplankton and primary production. In: Sakshaug, E, Johnsen, G and Kovacs, K (eds.), *Ecosystem Barents Sea,* 167–208. Trondheim, Norway: Tapir Academic Press.

Sakshaug, E, Kristiansen, S and Syvertsen, E 1991 Planktonalger. In: Sakshaug, E, Bjørge, A, Gulliksen, F, Loeng, H and Melhum, F (eds.), *Økosystem Barentshav.* Oslo: Universitetsforlaget.

Sakshaug, E and Slagstad, D 1992 Sea ice and wind: Effects on primary productivity in the Barents Sea. *Atmos Ocean* **30**(4): 579–591. DOI: https://doi.org/10.1080/07055900.1992.9649456

Sambrotto, RN, Niebauer, HJ, Goering, JJ and Iverson, RL 1986 Relationships among vertical mixing, nitrate uptake, and phytoplankton growth during the spring bloom in the southeast Bering Sea middle shelf. *Cont Shelf Res* **5**(1–2): 161–198. DOI: https://doi.org/10.1016/0278-4343(86)90014-2

Sharples, J, Tweddle, JF, Mattias Green, J, Palmer, MR, Kim, Y-N, et al. 2007 Spring-neap modulation of internal tide mixing and vertical nitrate fluxes at a shelf edge in summer. *Limnol Oceanogr* **52**(5): 1735–1747. DOI: https://doi.org/10.4319/lo.2007.52.5.1735

Shiozaki, T, Ijichi, M, Isobe, K, Hashihama, F, Nakamura, K-i, et al. 2016 Nitrification and its influence on biogeochemical cycles from the equatorial Pacific to the Arctic Ocean. *ISME J* **10**(9): 2184–2197. DOI: https://doi.org/10.1038/ismej.2016.18

Slagstad, D and McClimans, TA 2005 Modeling the ecosystem dynamics of the Barents sea including the marginal ice zone: I. Physical and chemical oceanography. *J Mar Sys* **58**(1–2): 1–18. DOI: https://doi.org/10.1016/j.jmarsys.2005.05.005

South, GR and Whittick, A 1987 *An Introduction to Phycology.* Wiley-Blackwell.

Sundfjord, A, Fer, I, Kasajima, Y and Svendsen, H 2007 Observations of turbulent mixing and hydrography in the marginal ice zone of the Barents Sea. *J Geophys Res-Oceans* **112**(C5): 008. DOI: https://doi.org/10.1029/2006JC003524

Torres-Valdés, S, Tsubouchi, T, Bacon, S, Naveira-Garabato, AC, Sanders, R, et al. 2013 Export of nutrients from the Arctic Ocean. *J Geophys Res-Oceans* **118**(4): 1625–1644. DOI: https://doi.org/10.1002/jgrc.20063

Tremblay, J-É, Anderson, LG, Matrai, P, Coupel, P, Bélanger, S, et al. 2015 Global and regional drivers of nutrient supply, primary production and CO_2 drawdown in the changing Arctic Ocean. *Prog Oceanogr* **139**: 171–196. DOI: https://doi.org/10.1016/j.pocean.2015.08.009

Tremblay, J-É, Michel, C, Hobson, KA, Gosselin, M and Price, NM 2006 Bloom dynamics in early opening waters of the Arctic Ocean, *Limnol Oceanogr* **51**(2): 900–912. DOI: https://doi.org/10.4319/lo.2006.51.2.0900

Turner, JT 2002 Zooplankton fecal pellets, marine snow and sinking phytoplankton blooms. *Aquat Microb Ecol* **27**: 57–102. DOI: https://doi.org/10.3354/ame027057

Turner, JT and Ferrante, JG 1979 Zooplankton fecal pellets in aquatic ecosystems. *Bioscience* **29**(11): 670–677. DOI: https://doi.org/10.2307/1307591

Wassmann, P and Reigstad, M 2011 Future Arctic Ocean seasonal ice zones and implications for pelagic-benthic coupling. *Oceanography* **24**(3): 220–231. DOI: https://doi.org/10.6570/oceanog.2011.74

Wexels Riser, C, Reigstad, M, Wassmann, P, Arashkevich, E and Falk-Petersen, S 2007 Export or retention? Copepod abundance, faecal pellet production and vertical flux in the marginal ice zone through snap shots from the northern Barents Sea. *Polar Biol* **30**(6): 719–730. DOI: https://doi.org/10.1007/s00300-006-0229-z

Wexels Riser, C, Wassmann, P, Reigstad, M and **Seuthe, L** 2008 Vertical flux regulation by zooplankton in the northern Barents Sea during Arctic spring. *Deep-Sea Res II* **55**(20–21): 2320–2329. DOI: https://doi.org/10.1016/j.dsr2.2008.05.006

Wiedmann, I, Reigstad, M, Sundfjord, A and **Basedow, S** 2014 Potential drivers of sinking particle's size spectra and vertical flux of particulate organic carbon (POC): Turbulence, phytoplankton, and zooplankton. *J Geophys Res-Oceans* **119**(10): 6900–6917. DOI: https://doi.org/10.1002/2013JC009754

Yamazaki, H and **Osborn, T** 1990 Dissipation estimates for stratified turbulence. *J Geophys Res* **95**(C6): 9739–9744. DOI: https://doi.org/10.1029/JC095iC06p09739

12

Regional trend analysis of surface ozone observations from monitoring networks in eastern North America, Europe and East Asia

Kai-Lan Chang*, Irina Petropavlovskikh*,†, Owen R. Cooper*,†, Martin G. Schultz‡ and Tao Wang§

Surface ozone is a greenhouse gas and pollutant detrimental to human health and crop and ecosystem productivity. The Tropospheric Ozone Assessment Report (TOAR) is designed to provide the research community with an up-to-date observation-based overview of tropospheric ozone's global distribution and trends. The TOAR Surface Ozone Database contains ozone metrics at thousands of monitoring sites around the world, densely clustered across mid-latitude North America, western Europe and East Asia. Calculating regional ozone trends across these locations is challenging due to the uneven spacing of the monitoring sites across urban and rural areas. To meet this challenge we conducted a spatial and temporal trend analysis of several TOAR ozone metrics across these three regions for summertime (April–September) 2000–2014, using the generalized additive mixed model (GAMM). Our analysis indicates that East Asia has the greatest human and plant exposure to ozone pollution among investigating regions, with increasing ozone levels through 2014. The results also show that ozone mixing ratios continue to decline significantly over eastern North America and Europe, however, there is less evidence for decreases of daytime average ozone at urban sites. The present-day spatial coverage of ozone monitors in East Asia (South Korea and Japan) and eastern North America is adequate for estimating regional trends by simply taking the average of the individual trends at each site. However the European network is more sparsely populated across its northern and eastern regions and therefore a simple average of the individual trends at each site does not yield an accurate regional trend. This analysis demonstrates that the GAMM technique can be used to assess the regional representativeness of existing monitoring networks, indicating those networks for which a regional trend can be obtained by simply averaging the trends of all individual sites and those networks that require a more sophisticated statistical approach.

Keywords: tropospheric ozone; regional variation; irregular monitoring network; space-time model

1. Introduction

Tropospheric ozone is a greenhouse gas and pollutant detrimental to human health and crop and ecosystem productivity (REVIHAAP, 2013; U.S. Environmental Protection Agency, 2013; LRTAP Convention, 2015; Monks et al., 2015). Since 1990 a large portion of the anthropogenic reactive gas emissions that produce ozone have shifted from North America and Europe to Asia (Granier et al., 2011; Cooper et al., 2014; Zhang et al., 2016). This rapid shift, coupled with limited ozone monitoring in developing nations, presents a challenge to the scientists trying to summarize and understand recent changes in ozone at the global scale. To address this challenge the International Global Atmospheric Chemistry Project (IGAC) developed the *Tropospheric Ozone Assessment Report (TOAR): Global metrics for climate change, human health and crop/ecosystem research* (www.igacproject.org/TOAR). Initiated in 2014, TOAR's mission is to provide the research community with an up-to-date scientific assessment of tropospheric ozone's global distribution and trends from the surface to the tropopause. TOAR has produced the world's largest database of surface ozone metrics from hourly observations at over 9000 sites around the globe. These ozone metrics are freely accessible for research on the global-scale impact of ozone on climate, human health and crop/ecosystem productivity (Schultz et al., 2017).

Figure 1 illustrates the coverage of the TOAR surface ozone database across North America, Europe and East Asia (additional sites are available for other regions of the

* NOAA Earth System Research Laboratory, Boulder, US

† Cooperative Institute for Research in Environmental Sciences, University of Colorado, Boulder, US

‡ Institute for Energy and Climate Research (IEK-8), Forschungszentrum Jülich, Jülich, DE

§ Department of Civil and Environmental Engineering, The Hong Kong Polytechnic University, Hong Kong, CN

Corresponding author: Kai-Lan Chang (kai-lan.chang@noaa.gov)

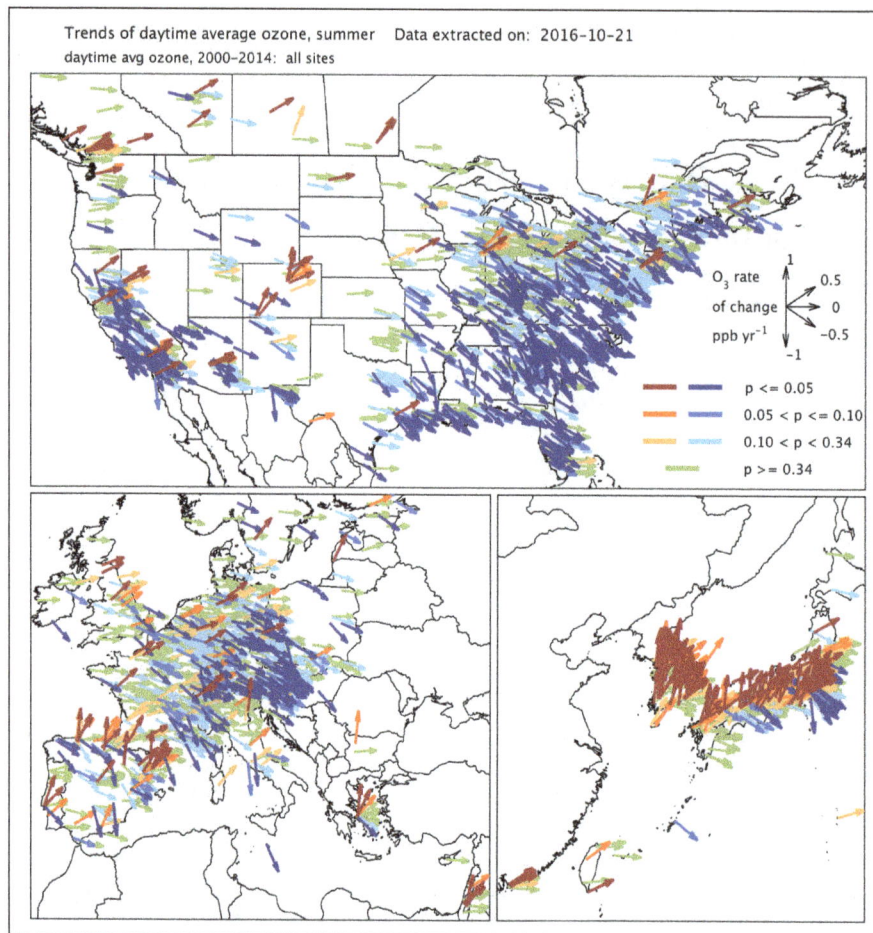

Figure 1: Trends (2000–2014) of summertime (April–September) daytime average ozone at available ozone monitoring stations. Vector colors indicate the p-values on the linear trend for each site: blues indicate negative trends, oranges indicate positive trends and green indicates weak or no trend; lower p-values have greater color saturation. This and other TOAR trend figures can be downloaded from: https://doi.pangaea.de/10.1594/PANGAEA.876108. DOI: https://doi.org/10.1525/elementa.243.f1

world) and shows the annual ozone trends at each station of the April–September average daytime ozone value for the period 2000–2014. In general these observed trends reflect recent changes in ozone precursor emissions. Several studies have documented the decrease of surface ozone across the eastern United States in response to decreases of domestic ozone precursor emissions (Kim et al., 2006; Lefohn et al., 2010; Cooper et al., 2012; Simon et al., 2015; Lin et al., 2017) and also across much of western Europe (Derwent et al., 2010; Simpson et al., 2014; European Environment Agency, 2016). In contrast, China has experienced decades of emissions increases (Zhao et al., 2013) and several studies have documented increasing ozone at the few sites available for assessing long term trends (Ma et al., 2016; Sun et al., 2016; Xu et al., 2016; Li et al., 2017; Wang et al., 2017). However, some regions of East Asia have experienced emissions decreases in recent years such as Beijing, the Pearl River Delta, Taiwan and Japan, and more works is required to understand the response of surface ozone (Duncan et al., 2016; Krotkov et al., 2016; Liu et al., 2016; Miyazaki et al., 2017; Van der A et al., 2017).

While **Figure 1** provides a great deal of detail regarding the regional distribution of trends, the wide range of

trend values across urban and rural areas, especially for Europe, makes it difficult to describe the overall regional trend. If one were to simply average all trend values across a region, how should they be weighted in terms of their spatial representation, and what is our confidence that a regional mean trend would be statistically significant?

This paper aims to answer these questions by applying a generalized additive mixed model (GAMM) to determine the systematic regional variations of several ozone metrics across eastern North America, Europe and East Asia. Quantifying a regional ozone trend is complicated by temporal and spatial variabilities. Also, this estimation is vigorously challenged by data inhomogeneity in time and by the irregularity of the spatial distribution of stations, as well as by interruptions in observational records. Furthermore, measurement practices may change over the years at a given site, impacting the observed quantity of ozone. The typical problems in the analysis of multi-site datasets are discussed by Chandler (2005); Wagner and Fortin (2005); Paciorek et al. (2009).

Our main focus is to derive regional trends using an advanced and more accurate GAMM approach. The analysis begins in Section 2 which describes the TOAR dataset applied here and briefly summarizes preliminary trend analyses: the approach of fitting a separate regression model for each station time series. We discuss the narrowness of this approach as it ignores the spatial dependence between sites. Section 3 explains how the GAMM can be used to represent ozone's systematic regional variations (i.e. dependence of the mean ozone level on space and time) in terms of the random adjustments of station-specific effects over time. In Section 4, we demonstrate our approach to determine the monthly systematic regional variations of summertime (April–September) ozone across eastern North America, including the regional trends in rural and urban sites, and investigate the change of spatial patterns by year. We then expand our analysis to Europe and East Asia. In Section 5 we conduct a trend analysis using summertime means rather than monthly means in order to efficiently investigate additional surface ozone metrics. Finally, to demonstrate the useful information on regional ozone trends afforded by sites with relatively weak trends, we illustrate the ability of GAMM to quantify regional ozone trends even when sites with the most robust trends are removed from the analysis. In Section 6, we provide a summary of our trend analysis.

2. TOAR dataset and preliminary analysis

We use two surface ozone metrics in our monthly trend analysis: (1) Monthly mean of the daytime average: defined as an average of hourly values for the 12-h period from 08:00 h to 19:59 h solar time; (2) Monthly mean of the daily maximum 8-hour average (DMA8): according to the new US EPA (Environmental Protection Agency) definition. 8-hour averages are calculated from 7 h local time to 23 h local time (U.S. Environmental Protection Agency, 2013). Note that if less than 75% of data are present (i.e. less than 9 hours for daytime average or 6 hours for DMA8), the value is considered missing. These metrics are volumetric mixing ratios in units of ppb (i.e. parts per billion by volume) and retrieved from April to September over 2000–2014 from the TOAR database. For the summertime mean trend analysis in Section 5, these metrics are averaged across the months of April–September and are also retrieved from the TOAR database (Schultz et al., 2017). Note that when extracting DMA8 for the 6-month summertime period the value returned is the 4th highest daily 8-hour maximum of the April–September aggregation period, which aligns with the US EPA National Ambient Air Quality Standard for ozone (https://www.epa.gov/criteria-air-pollutants/naaqs-table); but when extracting DMA8 for individual months the value is simply the monthly mean.

In this study we consider four explanatory variables at the location of each ozone monitoring site: (1) Station elevation: the value of the station elevation in meters above sea level, as obtained from the google maps API (Application Programming Interface); (2) Population density: the population density in a 5 km radius around the station location, the unit is people per km²; (3) NO_x emissions: annual anthropogenic surface NO_x emissions for the year 2010 from the HTAP_v2.2 (Hemispheric Transport of Air Pollution) global emissions inventory (Janssens-Maenhout et al., 2015) (gridded data in

0.1 degree resolution and in units of grams of NO_2 m^{-2} yr^{-1}); values range from 0 to 1000; (4) OMI (Ozone Monitoring Instrument) tropospheric column NO_2: 5-year average (2011–2015) high-resolution NO_2 column value from the OMI satellite instrument in units of 10^{15} molecules cm^{-2}. Values are in the range of 0 to 20.80. High values are indicative of regions with fresh emissions of nitrogen oxides, an important ozone precursor primarily emitted by fossil fuel combustion. All of these variables were made available through the TOAR database and further details on their sources are provided by Schultz et al. (2017). The TOAR dataset also identifies stations that are "rural, low elevation", "rural, high elevation or mountain", and "urban" sites using an objective methodology based on satellite-detected nighttime lights, OMI tropospheric column NO_2 and population density. Roughly one half of all stations in the database are characterized by one of these labels. For the other half, the categorization is not robust, therefore these stations are labeled as "unclassified" (Schultz et al., 2017).

The trend analysis provided by the TOAR database records the results of regression analyses of ozone time series for each selected station during 2000–2014. The analysis fits the time series itself without including any explanatory variables. The assessment method is the Sen-Theil estimator and p-values were derived from Mann-Kendall tests (Theil, 1950; Sen, 1968; Kendall, 1975). We provide an illustration of summaries from many individual trends across eastern North America in the supplemental material for the interested reader (see Figures S1–S2 and Table S1. DOI: https://doi.org/10.1525/elementa.243.s1).

Separate modeling of single surface ozone time series is the simplest approach for a trend analysis. However, for assessing regional trends this approach has been criticized because it does not account for the representativeness of a site in the ozone monitoring network, and ignores spatial dependency (e.g. ozone can change at neighboring stations in a similar manner due to changes in meteorology) and may thus cause a statistically less powerful and possibly misleading analysis for the assessment of regional trends (Thompson et al., 2001). This approach can also be severely biased by failing to account for the spatial dependency or irregularity (i.e. sub-regions of the network are more heavily weighted) of the station network. Therefore an advanced technique is required to quantify the systematic regional variations.

3. Methods: Generalized additive mixed model

The generalized linear model (GLM) is a mathematical extension of the classical linear regression model, which assumes a specific relationship (presumed linear dependence, but a polynomial relationship is allowed) between response and covariates via a link function (e.g. identical, log or logit). The GLM framework is widely used in the study of environmental time series (Chandler and Wheater, 2002; Yan et al., 2002; Chandler, 2005; Yang et al., 2005). In order to alleviate the linear constraints in the GLM, the generalized additive model (GAM) allows that one or more covariates depend on nonparametric smooth functions. Each unknown smooth function is represented by a linear combination of spline basis functions, i.e. linear covariates in a GLM are partly replaced by nonparametric spline functions in a GAM (Hastie and Tibshirani, 1990).

We consider an ozone time series where observations were averaged to monthly or seasonal means over a number of years. Nonlinear seasonal and interannual variations can thus be assessed by using the spline basis representation within the GAM framework. The GAM is also widely applied to the spatial analysis of regular or irregular data in order to account for geographical variability (Wood and Augustin, 2002; Wood, 2004). The generalized additive mixed model (GAMM) is an extension of GAM and allows for the incorporation of complex autocorrelation, and is therefore flexible for multi-site modeling (Carslaw et al., 2007; Ambrosino and Chandler, 2013; Park et al., 2013).

To describe the general framework in multi-site modeling, let $y(i, t)$ be the ozone value at station i and time t, then the GAMM can be expressed as:

$$y(i,t) = x\beta + f_1(\text{seasonal}) + f_2(\text{interannual}) + f_3(\text{spatial}) + b(i,t) + \epsilon(i,t).$$

It decomposes the observations into the following additive components:

1) Linear terms $x\beta$: x denotes the vector of explanatory variables listed in the previous section (including an intercept μ, representing the overall mean), β is the coefficient vector. Note that the demographic and physiographic information (e.g. station elevation) for each station remain unchanged over time in the TOAR dataset.

2) Smooth terms $f(\cdot)$: the covariates that are considered with a functional nature and thus modeled as nonlinear functions. The systematic regional variation can be regarded as a function of space and time. An explicit parameterization for this space-time variation is generally impractical, but in many cases it would be feasible by using spline smoothing controlled by the dimension of spline basis functions (Wood et al., 2016). In this study we consider three smooth components: seasonal (within-year), interannual (between-year) and purely spatial effects (i.e. not varying with time); each term is represented by a linear combination of spline basis functions. We refer to the interannual effect as the (deseasonalized) regional trend. We explain at the end of this section how to choose the type of spline basis function for representing underlying structure of these smooth terms.

3) Station-specific effects $b(i, t)$: statistical models often assume independent observations. The model does not recognize that a series of observations is produced from the same station with a particular instrument. The random effects are introduced to avoid violating this assumption and therefore permit the clustering of observations by stations.

The residual noise series, $\epsilon(i, t)$, is modeled as an autoregressive process of the order 1. In the regional trend analysis we do not intend to estimate the spatial variations for individual months (i.e. the temporally varying spatial patterns). Instead, we add the station-specific random effects, $b(i, t)$ (including random adjustments to the ozone baseline as well as random adjustments to the slope of the trend in each station and each year), to account for unobserved heterogeneity or correlation. These random effects enable the self-adjustment of the difference of an individual trend against the regional trend. Indeed, since we assume there is an "overall and averaged" regional trend, the measurements from each station should reveal at least some deviations from the average. As a result, the estimates for the explanatory variables may become less confident, but the autocorrelation of the residuals will be reduced. Moreover, since f_2(interannual) is representing the overall and averaged regional trend in the study region, the individual trend for a station i can thus be represented by f_2(interannual) + $b(i, t)$.

For the model implementations, we choose spline basis functions for each smooth term and station-specific effects. Spline functions are known to provide an efficient approach for numerical computation. Each spline function is evaluated at knots and so we need to choose the number and locations of these knots in order to create a flexible and appropriate smooth system. The degree of smoothness can be controlled by the maximum number of knots K: the number is required to be a good compromise between computational feasibility and fidelity to the data (Wood, 2006). The seasonal cycle and interannual trend can be represented via basis expansions:

$$f_1(\text{seasonal}) = \sum_{k=1}^{K_1} w_{1k}\phi_{1k}(\text{Month}) \quad \text{and}$$

$$f_2(\text{interannual}) = \sum_{k=1}^{K_2} w_{2k}\phi_{2k}(\text{Year}),$$

where $\{w_1\}$ and $\{w_2\}$ are associated coefficients to be estimated, $\{\phi_1\}$ is the penalized regression cyclic cubic splines (assumed with periodic nature) and places the knots at each month ($K_1 = 6$); and $\{\phi_2\}$ is the penalized regression cubic splines (provided a convenient basis for computational efficiency) and places the knots at each year ($K_2 = 15$). The same basis functions are also employed for the station-specific effects.

A large number of knots is required for the spatial variations to minimize complications from the irregular distribution of the network stations. For the spatial variations, the smooth function can be expressed as:

$$f_3(\text{spatial}) = \sum_{k=1}^{K_3} w_{3k}\phi_{3k}(L, l),$$

where (L, l) is the collection of latitude and longitude from each site, $\{\phi_{3k}(L, l)\}$ is the collection of spatial spline basis functions evaluated at each location (L, l), and $\{w_{3k}\}$

is the collection of associated coefficients. This procedure is a method of spatial interpolation/kriging for which the interpolated values are modeled by the Gaussian process penalized regression splines (Kammann and Wand, 2003), based on the Matérn covariance. A greater number of basis functions allows the fitted surface to be more complex and have a higher spatial resolution. If the value of K_3 is too small, the basis representation will not have enough degrees of freedom to represent local variability. Note that the irregularity also exists in the temporal domain, the so called time sampling issue (Tiao et al., 1990), but it is difficult to address explicitly if the observations are aggregated into a (regular) monthly average. The irregularity in the spatial domain, however, cannot be ignored in our analysis, therefore a higher dimension of basis functions is required for the capture of local variations (whenever appropriate). The choice of the number of knots is currently post hoc. The number of knots for each smooth term is chosen so that a further increase of knots would have negligible impact on the result (here $K_3 = 80$). More details about spline functions in the trend analysis can be found in Park et al. (2013) and Wood et al. (2016). The estimation is implemented in R package *mgcv*.

4. Summertime ozone trend analysis

In this section, we show how the systematic regional variations can be determined by employing the GAMM technique. We focus on the summertime period, which TOAR defines as the 6-month warm season (April–September

in the Northern Hemisphere and October–March in the Southern Hemisphere). There are two reasons to take this approach: (1) many sites in the USA only measure ozone in the warm season, therefore our analysis for the USA will have less data interruptions; (2) Emissions have different effects on ozone in the warm and cold season. In the warm season emissions tend to produce ozone, while in the cold season fresh emissions tend to destroy ozone in urban areas. By focusing on the warm season it will be easier to interpret the results. We first provide a detailed demonstration of our approach as applied to a temporal trend analysis for rural and urban ozone measurements in eastern North America, and extend to a spatial and temporal trend analysis for accommodating all categories of monitoring sites. After demonstrating the methodology, we then expand our analysis to Europe and East Asia, two other regions of the world with enough stations to perform a reliable trend analysis.

4.1 Eastern North America

4.1.1 Surface ozone in rural and urban sites

A total of 64,567 observations from 756 stations is used to construct the summertime ozone regional trend over eastern North America, including urban (140 sites), rural (273 sites) and unclassified (343 sites). We do not account for the spatial variations in this section as rural and urban sites are non-separable in space.

Figure 2 shows the estimated summertime cycles and long-term changes for daytime average (blue) and DMA8 (red) ozone, separated by rural and urban sites. The

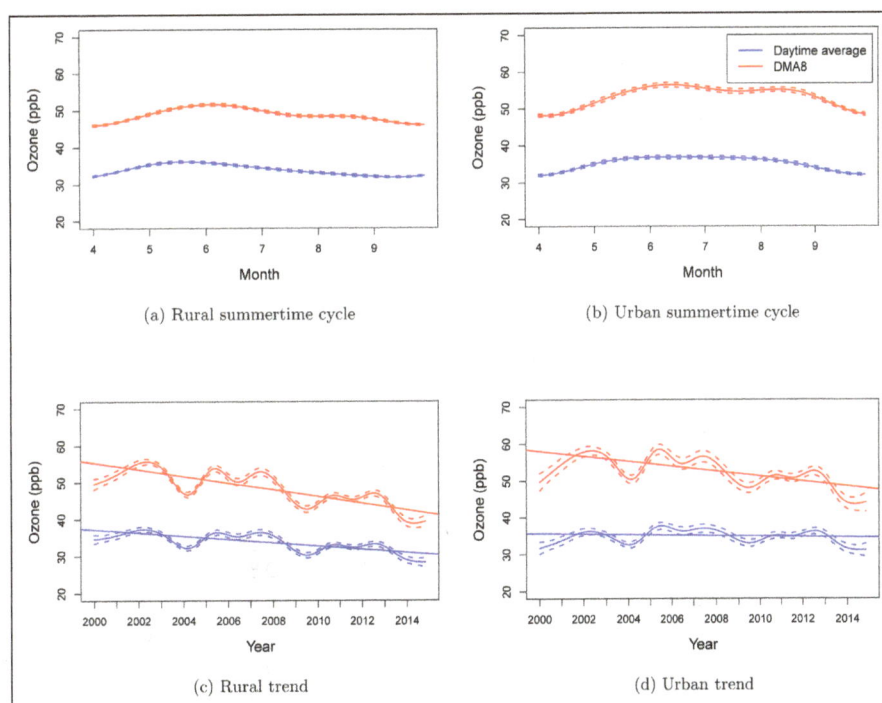

(a) Rural summertime cycle   (b) Urban summertime cycle

(c) Rural trend   (d) Urban trend

Figure 2: Seasonal cycles and interannual trends of summertime ozone at rural and urban sites. Trend analysis result of monthly mean of daytime average (blue) and DMA8 (red) from rural and urban sites in eastern North America. The dashed lines represent ±1 standard error of the mean of seasonal cycle or trend. The linear lines are linear regression fits. The same presentations are applied to all the trend and seasonal cycle plots. DOI: https://doi.org/10.1525/elementa.243.f2

Sen-Theil estimator is also fit to illustrate the tendency. The curvature in the estimated trend is generally slight, therefore the linearity would not be inappropriate in this case. We emphasize that the regional trend is referenced to the curve directly derived by the GAMM, the regression line merely enables us to summarize the overall tendency. It should be noted that autocorrelation is not taken into account for the Theil-Sen estimator and Mann-Kendall test in the TOAR dataset (though could be by using bootstrap simulations (Kunsch, 1989) or by incorporating an autoregressive process (Hamed and Rao, 1998)). The positive autocorrelation of the residuals can result in the underestimation of the uncertainty for the slope. In our approach the GAMM accounts for the autocorrelation by incorporating an AR(1) model and the autocorrelation is generally negligible for the resulting regional trend, it is therefore reasonable to use the Sen-Theil method.

Note that all the nonlinear smooth terms can be regarded as "anomalies" (i.e. departures from an overall mean level, adjusted by explanatory variables in some cases). The estimation of means from which to calculate such anomalies introduces uncertainty, which is displayed within the band between dashed lines. The model intercept, μ, in each scenario was added back to these anomalies in order to compare the results over the original scale (we are doing so throughout the paper).

The Sen-Theil estimators (and 95% confidence interval for the slope estimated by a bootstrap method) for the regional trends from daytime average and DMA8, with p-values from the Mann-Kendall test to detect tendency, are shown in part of **Table 1**. The intercept and slope values in the Table are referenced to the year 2000. The results show that rural ozone decreased relatively faster than urban ozone in both metrics; daytime average ozone in urban sites does not reveal substantial changes over 15 years. The DMA8 reveals a larger decline than the daytime average in both rural and urban sites.

The advantage of using the GAMM over the simple method described in the previous section is that it enables us to learn about associations in the environmental system by the visualization of uncertainty as well as the explanatory variable analysis. However, a practical issue is that when the number of observations is very large, standard errors (SEs) of estimates in the regression model become very small. As a consequence, most p-values for explanatory variables turned out to be highly statistically significant (i.e. very small p-value). In the large dataset statistical significance tends to diverge from practical significance. Hence we are conservative with the results even when the coefficients reach statistical significance. The results should not be over-interpreted.

The detailed summary statistics (mean, SE and p-value) of the fixed effects, i.e. the β covariate coefficients,

Table 1: The Sen-Theil estimators (with 95% confidence interval for the slope) for the regional trends from monthly mean of daytime average and DMA8, p-values are derived from Mann-Kendall tests. The overall statistics include the TOAR unclassified category (monthly mean in different regions). DOI: https://doi.org/10.1525/elementa.243.t1

Monthly mean of daytime average					
Region		**Intercept (ppb)**	**Slope (ppb yr^{-1})**	**Lower CI (ppb yr^{-1})**	**Higher CI (ppb yr^{-1})**	**p-value**
Eastern N America	Overall	43.72	−0.28	−0.30	−0.26	<0.01
	Rural	37.27	−0.46	−0.48	−0.43	<0.01
	Urban	35.67	−0.09	−0.10	−0.06	0.16
Europe	Overall	37.31	−0.06	−0.07	−0.05	0.01
	Rural	39.39	−0.17	−0.17	−0.15	<0.01
	Urban	38.79	0.05	0.02	0.06	0.19
E Asia	Overall	40.91	0.45	0.43	0.46	<0.01
	Rural	44.00	0.21	0.19	0.22	<0.01
	Urban	36.42	0.45	0.42	0.48	<0.01
SE Asia	Overall	26.67	0.20	0.20	0.20	<0.01
Monthly mean of DMA8						
Region		**Intercept (ppb)**	**Slope (ppb yr^{-1})**	**Lower CI (ppb yr^{-1})**	**Higher CI (ppb yr^{-1})**	**p-value**
Eastern N America	Overall	65.31	−0.81	−0.85	−0.79	<0.01
	Rural	55.74	−0.96	−1.01	−0.91	<0.01
	Urban	58.03	−0.69	−0.73	−0.65	<0.01
Europe	Overall	55.40	−0.30	−0.32	−0.28	<0.01
	Rural	53.47	−0.39	−0.41	−0.36	<0.01
	Urban	54.81	−0.13	−0.15	−0.12	0.01
E Asia	Overall	69.13	0.35	0.32	0.37	<0.01
	Rural	60.11	0.14	0.12	0.16	<0.01
	Urban	62.05	0.30	0.24	0.37	0.02
SE Asia	Overall	41.19	0.45	0.45	0.45	<0.01

separated by rural and urban sites with different metrics, are provided in Table S2. We present the main finding as follows: due to the current version of the TOAR database has only 1 year average of OMI NO_2 and NO_x emission data, the correlation should be interpreted in a geographical sense. For example, similar to a higher ozone level typically observed at site located in a higher elevation (Vingarzan, 2004), NO_2 column reveals a strong positive correlation with rural ozone (i.e. higher correlations between ozone and NO_2 are observed in rural sites than urban sites, which is consistent to the result of Safieddine et al. (2013)). NO_x emissions reveal a significant factor for both rural and urban sites with an opposite correlation. The emissions tend to be negatively correlated with rural ozone and positively correlated with urban ozone. Population density is less crucial for rural ozone.

4.1.2 Regional trend analysis

Figure 3(a) and **(b)** display the estimated summertime cycles and long-term changes in eastern North America. The model here is fit to data from all 756 available sites. **Figure 3(c)** and **(d)** show the estimated spatial distribution of the ozone level averaged over 2000–2014. After partitioning out the seasonal cycle and spatial variation, the regional trend from DMA8 ozone shows a faster decline than for the daytime average, as in the previous analysis (see **Figure 2(c)** and **(d)**).

The map presented here is the spatial prediction from statistical interpolation across the gap between each site,

based on the GAMM fitting result. Any area more than 5% of the regional width from the nearest ozone monitoring site is left blank on the map (i.e. we only interpolate any gap less than 5% regional width, as extrapolation or interpolation across too large a distance tends to cause greater uncertainty). The interpolation method is assumed to be realizations of a Gaussian random spatial process from the GAMM estimation. It spatially interpolates values as linear combinations of the original observations (a weighted average of the observations in the neighborhood of the location), and this constitutes the spatial inference of quantities in unobserved locations. The GAMM itself accounts for the spatial weight estimated from each site that best describes the set of observed data. Therefore the spatial interpolation is independent from the estimation of interannual trends and seasonal cycles. The spatial distribution shows a lower mean level in the north and south of the region and a higher mean level in the middle area for both metrics. Since the regional trend is the primary focus of this work, the estimation of station-specific effects is relegated to the supplemental material (see Figure S3).

Summarizing the explanatory variables analysis indicates that a similar pattern can be found in station elevation and population density in both metrics (please refer to Table S2 for detailed numbers); elevation has a significant positive correlation with mean ozone level. A significant and negative relationship is observed between population density and ozone level. NO_x emissions reveal an insignificant

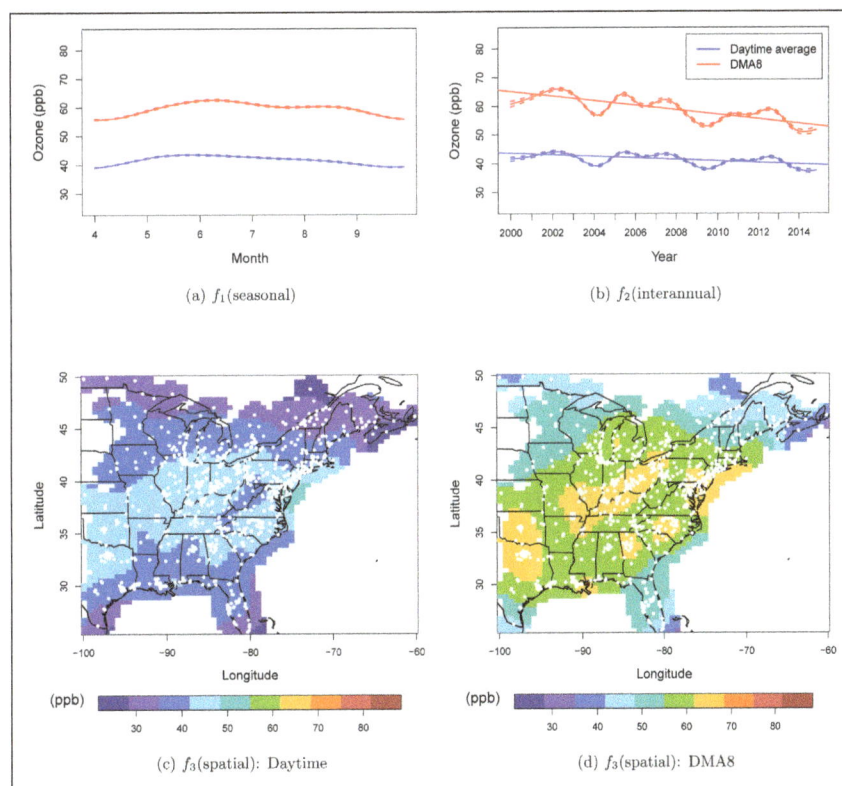

(a) f_1(seasonal)   (b) f_2(interannual)   (c) f_3(spatial): Daytime   (d) f_3(spatial): DMA8

Figure 3: Seasonal cycles, interannual trends and spatial distributions of ozone in eastern North America. Each curve or map represents a smooth component, i.e. f_1: seasonal cycle, f_2: interannual trend and f_3: spatial distribution, in the GAMM. The results are obtained from monthly mean of daytime average (blue) and DMA8 (red). The white points indicate the locations of stations. DOI: https://doi.org/10.1525/elementa.243.f3

contribution due to an opposite effect in urban and rural sites (as in the previous section), and the significance could be neutralized in the whole regional analysis. The results from linear regression in **Table 1** suggest a 4.2 and 12.2 ppb reduction of daytime and DMA8 ozone, respectively, in summertime over 2000–2014.

4.1.3 Investigation of spatial patterns

In order to investigate the changes of the spatial structure of the summertime mean of daytime average ozone and the 4th highest DMA8 (one summertime value per year per site, as opposed to using monthly means) over the same period, we interpolate the summertime distributions over the study region each year with a statistical method and then carry out the regression analysis to the regional summertime means from the interpolation. The analysis step is firstly fitting a surrogate statistical model based on the irregularly spaced observations (L_i, l_i), $i = 1$, ..., 756 sites, then smoothing out the irregularity by predicting the values over a regular network (L', l') (here we use $0.5° \times 0.5°$ regular grid), and finally averaging the predicted values over all grid points in the region (i.e. spatial aggregation). This model projection is particularly useful if we aim to compare the spatial distribution from observations to satellite data or global atmospheric chemistry model output, because we can project the

interpolation from the surrogate statistical model onto the exact same grid point as the satellite data or atmospheric chemistry model output (Chang et al., 2015).

We separate the spatial interpolations from the trend analysis (i.e. only spatial variations are evaluated), thus we can investigate the spatial patterns in different years. The station network in eastern North America is dense enough to allow us to do so. The spatial aggregation approach aims to estimate the ozone distribution in each year; the long-term regional mean changes are based on the results of 15 summertime averages of ozone distribution over a designed regular grid within the monitoring network. This approach implicitly assumes that the regional change can be represented by a series of estimated summertime means and the rate of change is the same for each site. This technique would be intuitive and straightforward under this assumption. However, the spatial aggregation approach may not have enough degrees of freedom to capture all of the temporal variability in the observations. In addition, the spatial coverage of the station network might change due to time series interruptions (e.g. the records at the Canadian sites are only available through 2013). Hence we only use this approach to investigate the spatial patterns.

Figure 4 displays the approximated regional daytime average surface ozone distributions in 2000, 2004, 2009

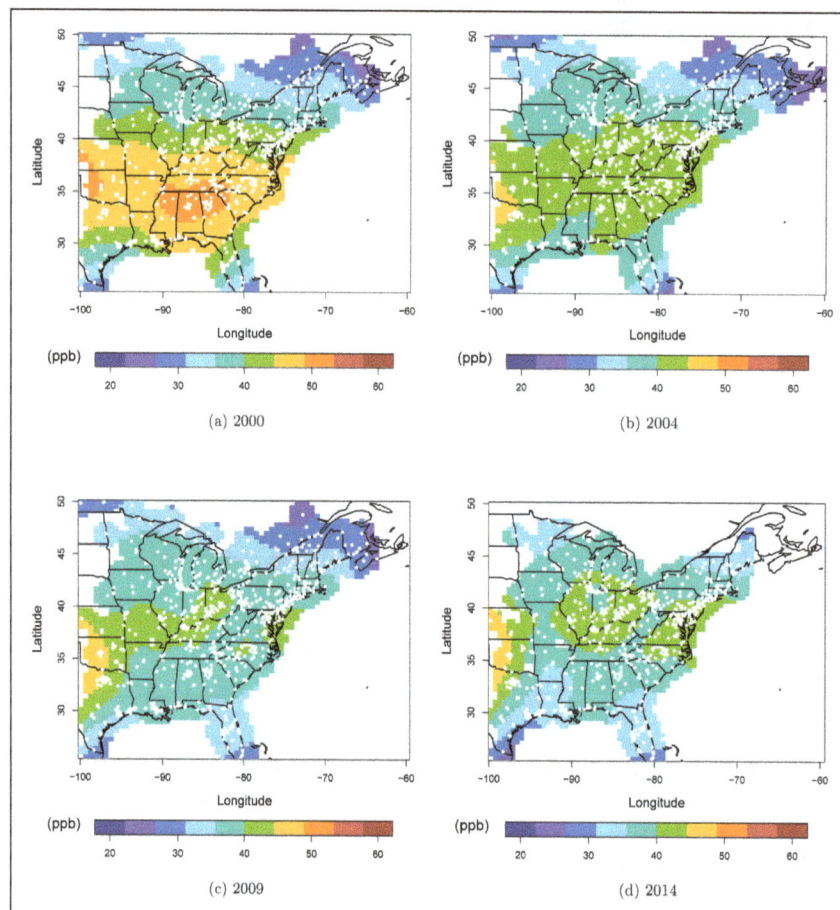

Figure 4: Spatial distributions for summertime mean of daytime average over eastern North America. Statistical estimations of summertime spatial mean distributions of daytime average in four different years. The white points indicate the locations of stations. DOI: https://doi.org/10.1525/elementa.243.f4

and 2014. An ozone reduction can be found over the central area of the region. The ozone distributions of the 4th highest DMA8 in the selected years are shown in **Figure 5**. The 4th highest DMA8 shows a relatively clear decline in the whole region. There are two reasons for the low ozone in 2004. One is that summer 2004 was unusually cool, associated with meteorological conditions that were not conducive for stagnant air pollution events. This was was also the year that power plants across the eastern USA began using scrubbers to reduce NO_x emissions, which is one of the main reasons why ozone has decreased across the eastern US over the past decade (Frost et al., 2006; Kim et al., 2006).

The regression intercepts (and slopes) for the regional mean changes of daytime average ozone and the 4th highest DMA8 are 41.20(−0.22) and 77.11(−1.01), respectively. All three approaches discussed in this paper provide similar results for the daytime average ozone trend. However, the results cannot be directly compared for the DMA8. For the simple method described in Section 2 and this section, we used the summertime 4th highest DMA8. For the GAMM approach in the previous section (**Figure 3**), we used the monthly mean of DMA8. Therefore the decrease here of 15.2 ppb cannot be directly compared to the decrease of 12.2 ppb in **Figure 3**. All the results still indicate that surface ozone decreased over 2000–2014 in both metrics.

4.2 Europe

We select the sites located in Europe up to 66°N (North of 66°N is the Arctic circle according to the European region defined by the Task Force on HTAP). As a result, a total 76,520 observations from 1,007 stations are used in Europe, including urban (260 sites), rural (290 sites) and unclassified (457 sites).

Figure 6(a) and **(b)** display the estimated summertime cycles and long-term changes. The spatial variations reveal similar patterns in **Figure 3(c)** and **(d)**: lower values in western and northern Europe and higher values in southeast Europe. Both metrics indicate that a large spike occurred in 2003 (a well-known event associated with an extreme heatwave), followed by a small spike in 2006. The overall tendency of the regional trend for daytime average ozone is slightly decreasing over 2000–2014. A small amount of reduction can be observed in the DMA8.

The Sen-Theil estimators for summertime regional trends from daytime average and DMA8, with p-values from the Mann-Kendall test, are shown in part of **Table 1**. One important finding is that decreasing rural ozone can be observed in both metrics. Another finding suggests that decreasing DMA8 is detected in urban sites.

The detailed summary statistics of the fixed effects can be found in Table S3. Our main result is provided as follows: elevation has a significant positive correlation

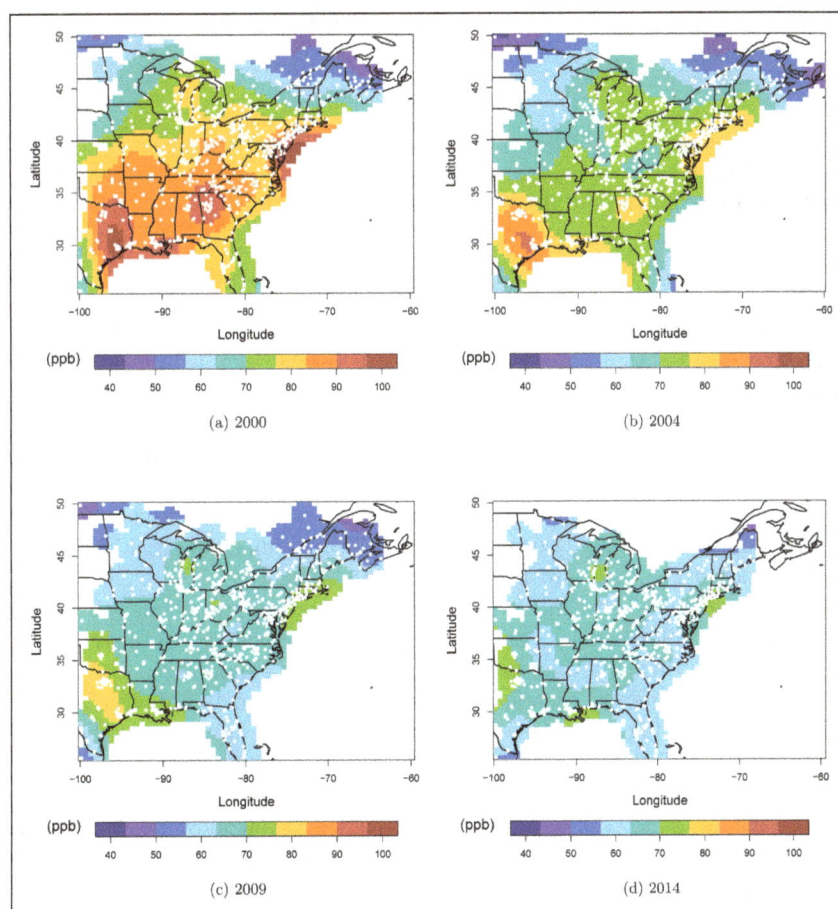

Figure 5: Spatial distributions for summertime mean of the 4th highest DMA8 over eastern North America. Statistical estimations of summertime spatial mean distributions from the 4th highest DMA8 in four different years. The white points indicate the locations of stations. DOI: https://doi.org/10.1525/elementa.243.f5

186

Human Impacts on the Environment: Past, Present and Future

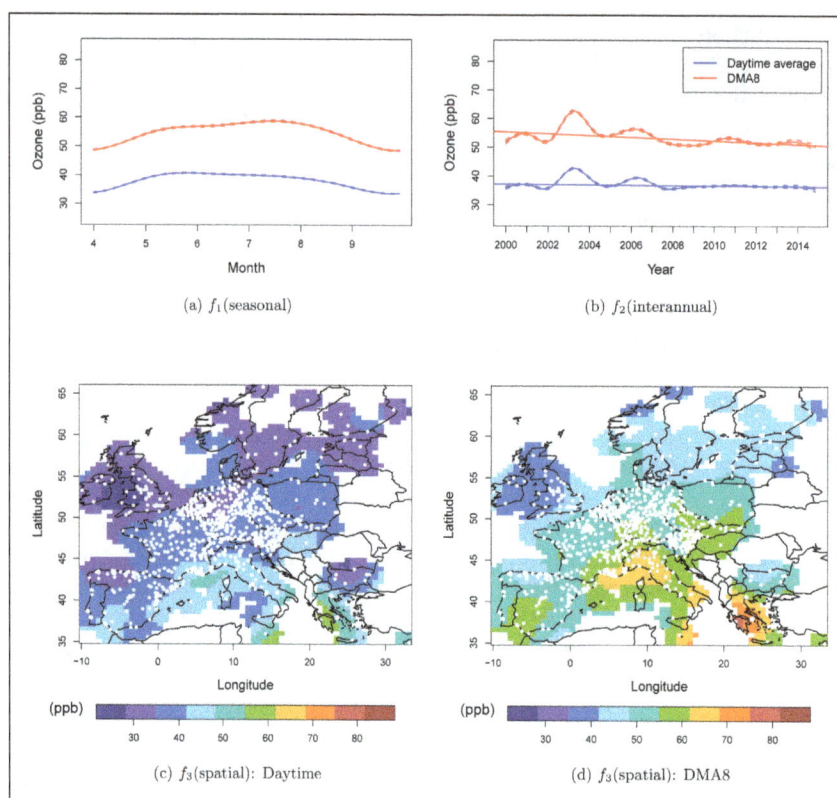

Figure 6: Seasonal cycles, interannual trends and spatial distributions of ozone in Europe. Each curve or map represents a smooth component, i.e. f_1: seasonal cycle, f_2: interannual trend and f_3: spatial distribution, in the GAMM. The results are obtained from monthly mean of daytime average (blue) and DMA8 (red). The white points indicate the locations of stations. DOI: https://doi.org/10.1525/elementa.243.f6

with mean ozone level. Population density shows that the correlation with ozone level is negative in urban sites and positive for rural sites (only for DMA8); the overall statistics show a negative impact (this finding is the same as the results in eastern North America). NO_x emissions reveal an insignificant contribution to DMA8 in urban sites. NO_2 column tends to negatively correlate with ozone in urban sites and daytime ozone in rural sites, whereas the DMA8 has a positive correlation in rural sites.

4.3 East Asia

A unique overall trend is difficult to determine in East and Southeast Asia due to a large spatial gap between Japan and Taiwan. The results will be highly uncertain if we interpolate across such a gap, and also because there is a strong network asymmetry between East Asia (557 stations) and Southeast Asia (19 stations). In addition, an analysis indicates that the systematic variations in East and Southeast Asia behave differently. Therefore we separate the analysis over East Asia (including Japan and South Korea) and Southeast Asia (including Taiwan and Hong Kong).

A total of 42,792 observations from 557 stations are used in East Asia, including urban (217 sites), rural (39 sites) and unclassified (217 sites). **Figure 7** displays the estimated smooth terms in the model. The estimated regional trends are increasing for both metrics. A high mean level of ozone can be observed in south Japan. The Sen-Theil estimators

for summertime regional trends and the Mann-Kendall test statistics are shown in **Table 1**. Urban ozone in East Asia shows a higher increasing rate than rural ozone. These results suggest a 6.8 and 4.5 ppb growth of the daytime average and DMA8 in urban ozone, respectively. The explanatory variable analysis shows that elevation and population density have a significant positive and negative correlation with mean ozone level, respectively. NO_x emissions reveal an insignificant contribution to the DMA8, and have a significant contribution to daytime average with opposite correlation in rural and urban sites. NO_2 column tends to negatively correlate with overall statistics, whereas it shows a positive correlation with DMA8 when rural and urban sites are separated (see also Table S4).

There are too few stations to estimate accurately the spatial variations in Southeast Asia, thus the spatial variations will not be evaluated. A total of 1,632 observations from 19 stations are used in Southeast Asia, including urban (11 sites), rural (3 sites) and unclassified (5 sites). We do not separate the urban and rural sites here as the results would not be robust for only 3 rural sites available. We only show the results on the seasonal cycles, long-term changes and the explanatory variables analysis.

Figure 8 displays the estimated summertime cycles and long-term changes. A different seasonal effect is observed in contrast to the other regions: a marked drop from May–July can be found in both metrics (about 20 ppb

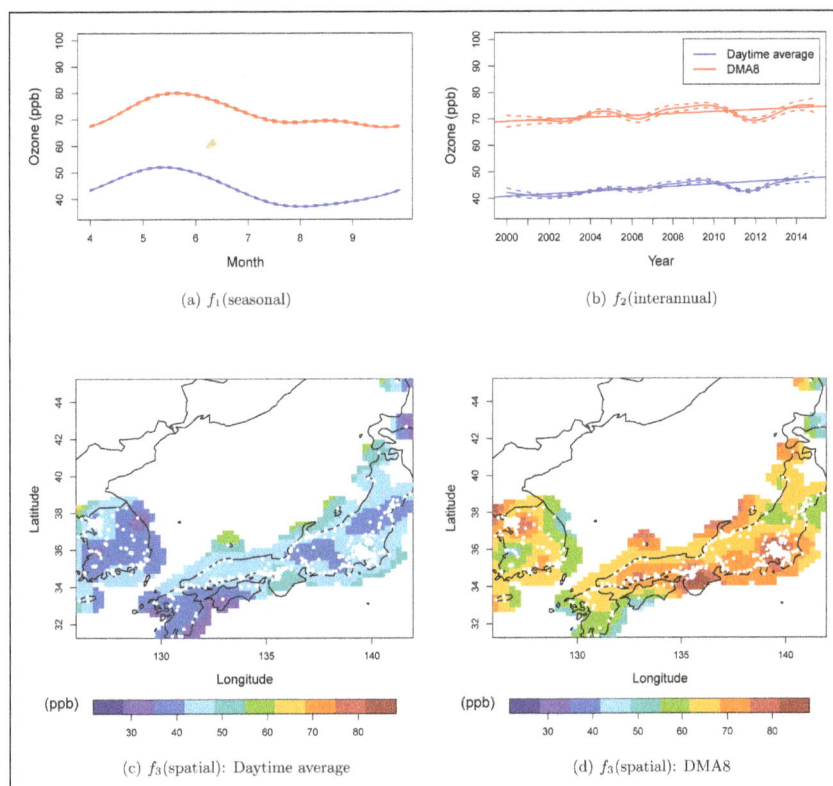

Figure 7: Seasonal cycles, interannual trends and spatial distributions of ozone in East Asia. Each curve or map represents a smooth component, i.e. f_1: seasonal cycle, f_2: interannual trend and f_3: spatial distribution, in the GAMM. The results are obtained from monthly mean of daytime average (blue) and DMA8 (red). The white points indicate the locations of stations. DOI: https://doi.org/10.1525/elementa.243.f7

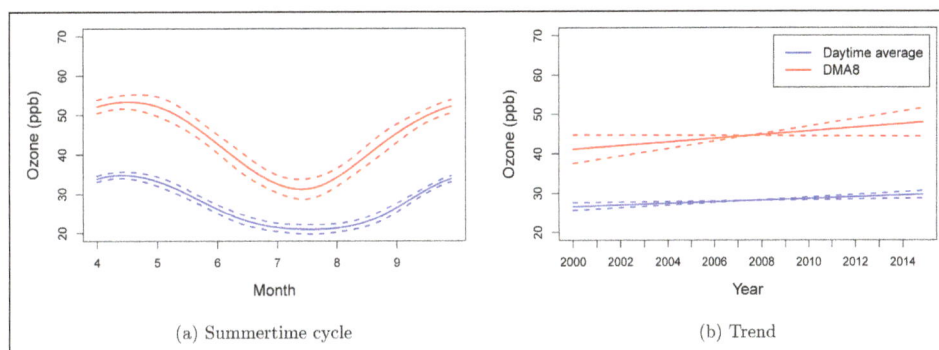

Figure 8: Seasonal cycles and interannual trends of ozone in Southeast Asia. Trend analysis result of monthly mean of daytime average (blue) and DMA8 (red). DOI: https://doi.org/10.1525/elementa.243.f8

decline in DMA8). The regional trend turns out to be linear increasing for both metrics. The increasing rate of the DMA8 is more than twice as great as the daytime average (see also **Table 1**). The result also shows that mean ozone level reveals a significant positive correlation with elevation and NO_x emissions, and a negative correlation with population density. Tropospheric column NO_2 reveals a negative contribution to daytime average and a positive contribution to DMA8 in southeast Asia (See Table S5).

5. Summertime means trend analysis

In order to explore the regional trends of other ozone metrics, we carry out the trend analysis using annual summertime values (i.e. one value per site per year, calculated for the period from April to September) of the following four metrics: (1) daytime average: this metric has already been discussed and it will be relevant to the climate community, especially at rural sites because it gives a broad overview as to how the mid-range ozone values are changing, which can be compared to global models with relatively coarse horizontal resolution; (2) summertime mean of all daily 8-hour maximum values (avgdma8epax). This metric is used to determine the mortality due to long-term ozone exposure and is of great interest to researchers who study the impact of ozone on human health; (3) AOT40 is defined as cumulative ozone exposure over a threshold of 40 ppb. This is a metric designed to study the impacts of ozone exposure on vegetation; (4) A useful metric for the human health community is the number of days per summertime period in which the maximum daily 8-hour average exceeds 70 ppb (NVGT070). A potential complication for this metric is that some sites never exceed 70 ppb so their value is always zero.

Due to different data characteristics, additional treatments are required for AOT40 and NVGT070. Therefore in this section the analysis is laid out by metric, rather than by region. This arrangement also enables us to directly compare the ozone pollution in the three regions with the most extensive ozone monitoring networks: eastern North America, Europe and East Asia. In order to assess the quality of spatial coverage of monitoring sites in different regions, we compare our approach to the results from the simple method of averaging all individual trends (as described in Section 2). Similar results are expected if the monitoring network is well developed with good spatial coverage. Since we use the summertime data in this section, no seasonal cycle will be evaluated, and we only focus on the estimation of the regional trends and spatial variations (without going through the details of the station-specific effects).

In this analysis we quantified regional ozone trends using all available stations regardless of the strength and statistical significance of the trend at each site because even sites with weak trends provide useful information that can be considered for the calculation of the regional trend (Chandler and Scott, 2011). To explore the contribution to the regional trend by sites that have statistically significant trends ($p < 0.05$) and sites that have insignificant trends ($p > 0.05$), we remove sites from the analysis sequentially according to p-value, beginning with the lowest p-values. This analysis is applied to eastern North America using the summertime mean of daytime average and DMA8 in the final step.

5.1 Daytime average
We first present the summertime daytime average trend analysis across eastern North America, Europe and East Asia. **Figure 9** shows the estimated interannual trends and spatial variations based on summertime means over 2000–2014. The Sen-Theil estimator is less sensitive to the extreme event in 2003 and the ozone mean level across Europe remains steady over the study period. The mean daytime ozone reveals that there has been a gradual decline in eastern North America, while ozone shows a

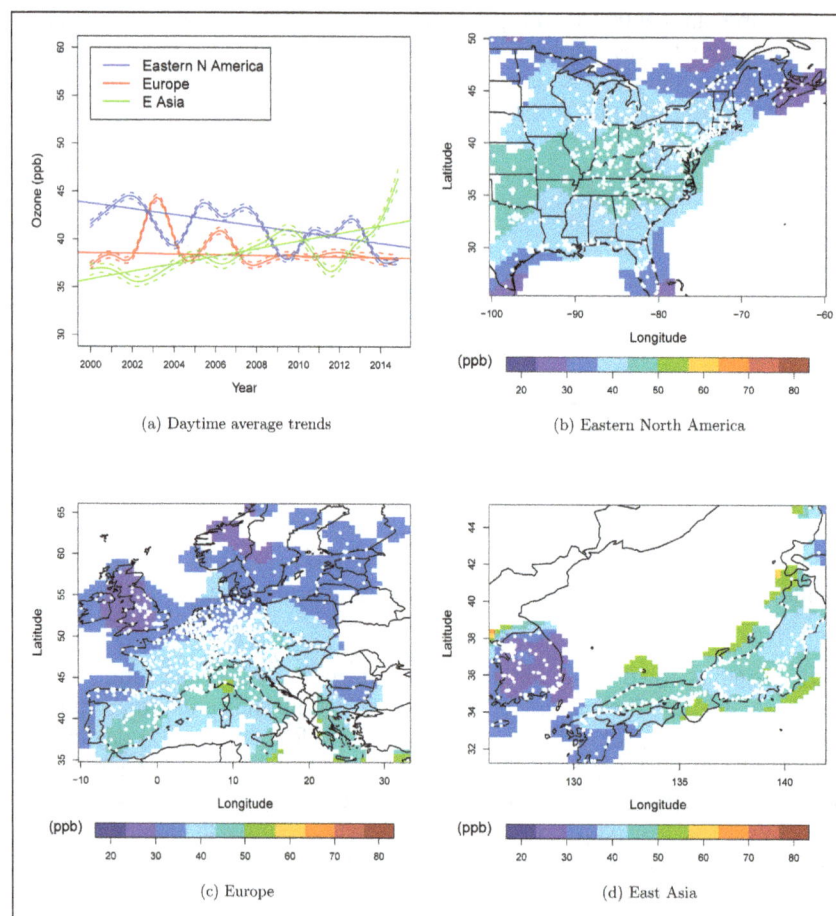

Figure 9: Regional trends and spatial distributions for summertime mean of daytime average in different regions. Estimated daytime average (ppb) long-term changes in eastern North America (blue), Europe (red) and East Asia (green), along with the spatial mean distributions in each region. The white points indicate the locations of stations. DOI: https://doi.org/10.1525/elementa.243.f9

sharp rise in East Asia after 2011 (>10 ppb). No extreme ozone levels are found over eastern North America on the regional scale. A lower mean concentration can be found in northern and western Europe. In East Asia, the mean ozone level in the Republic of Korea is lower than Japan. The area with the highest mean level in Japan is Wakayama.

The first half of **Table 2** shows the Sen-Theil estimators for the detrended line by different regions. Rural ozone reveals a relatively higher baseline than urban ozone in all regions. Rural ozone in eastern North America reveals a steeper decline than in Europe. There is no significant trend for urban ozone in Europe. A slight decline of urban ozone can be detected in eastern North America. Both rural and urban ozone are increasing in East Asia, with urban ozone increasing faster than rural ozone.

The last two columns in **Table 2** report the mean and standard deviation (SD) of all available individual trend estimates (i.e. "regional mean approach"). In cases when the slopes from the GAMM and regional mean approaches are similar, we conclude that the station network is well covered in this region, and a sophisticated statistical approach might not be required to assess the regional trend. This is the case for eastern North America and East Asia. A discrepancy is expected for the results in Europe due to the network being more scattered across northern and eastern Europe (the same scenario can be observed in

the analysis of the summertime mean of DMA8, described below).

5.2 Summertime mean of DMA8

Figure 10 shows the estimated interannual trends and spatial variations based on the summertime mean of DMA8 over 2000–2014. The trend and spatial features are similar to the summertime daytime average in the previous section. The second half of Table 10 displays the Sen-Theil estimators of the detrended line. The difference from the previous analysis is that the DMA8 reveals a larger decline than daytime average for both rural and urban ozone in eastern North America and Europe, while urban ozone shows insignificant changes in Europe. DMA8 at rural and urban sites in east Asia shows similar levels of increment as daytime average.

5.3 AOT40

The AOT40 values are summertime cumulative values and the range of values is relatively more wide spread than daytime average and DMA8. For instance, the range of AOT40 values in eastern North America over 2000–2014 is 8 to 46,234 ppb hr. In order to improve the linearity and the fit, we transform the AOT40 values by using the natural logarithm. Therefore the results should be interpreted by their exponential values. **Figure 11**

Table 2: The Sen-Theil estimators for summertime mean of daytime average and DMA8 regional trends, p-values are derived from Mann-Kendall test (GAMM approach), along with the means (SDs) of all individual intercepts and slopes (Regional mean approach). The overall statistics include the TOAR unclassified category. DOI: https://doi.org/10.1525/elementa.243.t2

Region		Summertime mean of daytime average				
		GAMM approach			Regional mean approach	
		Intercept (ppb)	Slope (ppb yr⁻¹)	p-value	Intercept (ppb)	Slope (ppb yr⁻¹)
Eastern N America	Overall	43.73	−0.30	<0.01	43.13(6.04)	−0.30(0.29)
	Rural	45.06	−0.42	<0.01	44.40(6.05)	−0.42(0.24)
	Urban	39.79	−0.10	0.01	39.15(6.19)	−0.07(0.30)
Europe	Overall	38.56	−0.04	0.09	39.17(7.18)	−0.08(0.33)
	Rural	42.35	−0.17	<0.01	42.72(6.21)	−0.21(0.30)
	Urban	35.66	0.01	0.78	35.57(7.40)	0.05(0.31)
E Asia	Overall	35.78	0.40	<0.01	35.40(8.42)	0.41(0.56)
	Rural	40.42	0.23	<0.01	39.95(5.67)	0.22(0.62)
	Urban	34.05	0.51	<0.01	33.74(9.35)	0.52(0.51)

Region		Summertime mean of DMA8				
		GAMM approach			Regional mean approach	
		Intercept (ppb)	Slope (ppb yr⁻¹)	p-value	Intercept (ppb)	Slope (ppb yr⁻¹)
Eastern N America	Overall	50.15	−0.43	<0.01	49.37(6.63)	−0.43(0.30)
	Rural	50.85	−0.52	<0.01	50.10(6.66)	−0.52(0.26)
	Urban	46.57	−0.25	<0.01	45.79(6.99)	−0.21(0.33)
Europe	Overall	43.87	−0.08	<0.01	44.63(7.34)	−0.14(0.34)
	Rural	47.14	−0.21	<0.01	47.51(6.29)	−0.25(0.31)
	Urban	41.17	−0.05	0.13	41.25(7.92)	−0.03(0.32)
E Asia	Overall	43.72	0.37	<0.01	43.38(9.50)	0.37(0.66)
	Rural	46.83	0.23	<0.01	46.25(6.01)	0.20(0.66)
	Urban	42.37	0.48	<0.01	42.11(10.81)	0.49(0.63)

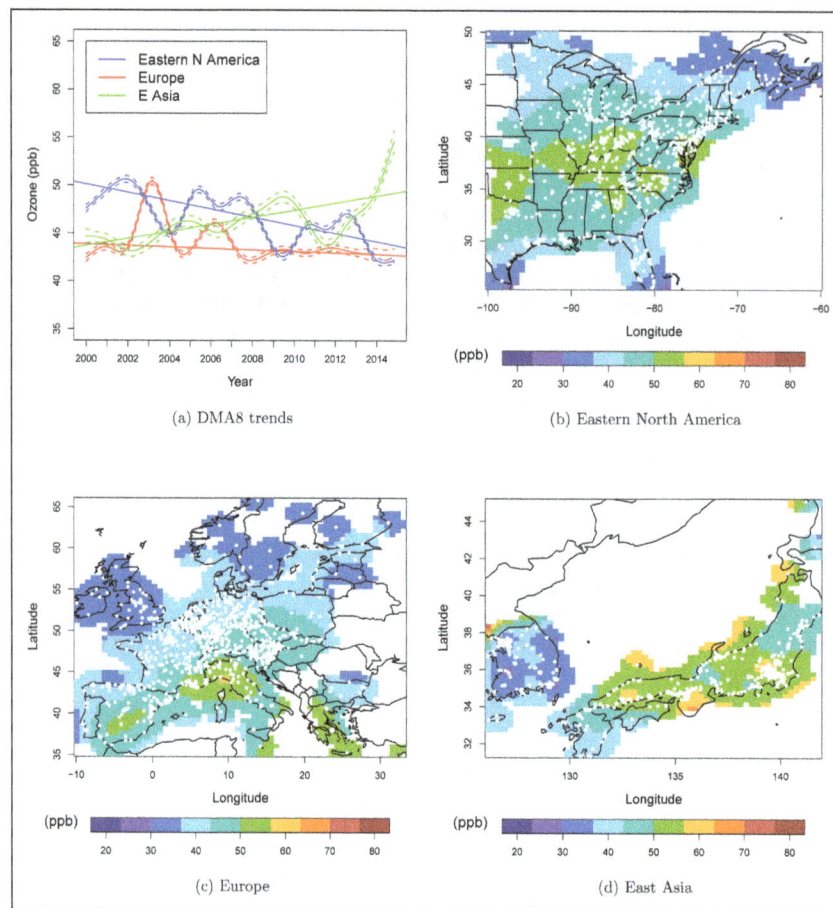

Figure 10: Regional trends and spatial distributions for summertime mean of DMA8 in different regions. Estimated DMA8 (ppb) long-term changes in eastern North America (blue), Europe (red) and East Asia (green), along with the spatial mean distributions in each region. The white points indicate the locations of stations. DOI: https://doi.org/10.1525/elementa.243.f10

displays the estimated trends and spatial distributions for different regions. The highest AOT40 mean concentration was found in Wakayama, on the southern side of Japan, corresponding to roughly 34,000 ppb hr. There is a cluster of high AOT40 values in southern Europe, corresponding to roughly maximal 32,000 ppb hr over land. The highest AOT40 value in eastern North America is about 22,000 ppb hr.

The first half of **Table 3** reports the Sen-Theil intercepts and slopes of the estimated regional trends in these three areas. The linear regression results suggest AOT40 decreased by half (from ~20,800 to ~11,200 ppb hr) in eastern North America over 2000–2014. In the same period AOT40 decreased from ~12,900 to ~10,200 ppb hr in Europe and increased from ~15,700 to ~19,600 ppb hr in East Asia. Most of the increase in East Asia is driven by the years after 2011, see **Figure 11(a)**.

5.4 NVGT070

NVGT070 is the cumulative number of days per summertime period in which the maximum daily 8-hour average exceeds 70 ppb, and the values are treated as non-negative integer values, in contrast to daytime average and DMA8 which are treated as continuous values. Therefore the statistical model assumption needs to be adjusted. Poisson regression is a

generalized form of regression analysis used to model count data. It assumes the response has a Poisson distribution, and assumes the logarithm of its expected value can be modeled by a linear combination of unknown covariates. There are two major issues in employing the Poisson regression: (1) A common problem with Poisson regression is excess zeros, for instance, 1,677 zeros out of 10,949 (15.3%) observations in eastern North America. The high proportion of zeros is often used to justify the use of zero-inflated models, although this sort of model is only appropriate when none of the covariates help to explain the zeros in the data (Wood et al., 2016). In our investigation the zero NVGT070 values are highly clustered in space, suggesting the need for a process with a spatially varying structure, rather than zero inflation; (2) A characteristic of the Poisson distribution is that its mean is equal to its variance. In certain circumstances, it will be found that the observed variance is greater than the mean; this is known as overdispersion and indicates that the model is not appropriate. There are several approaches to tackle this issue, here we adopt a negative binomial regression instead.

Negative binomial regression can be considered as a generalization of Poisson regression since it has the same mean structure as Poisson regression and it has an extra parameter to model the overdispersion. We briefly

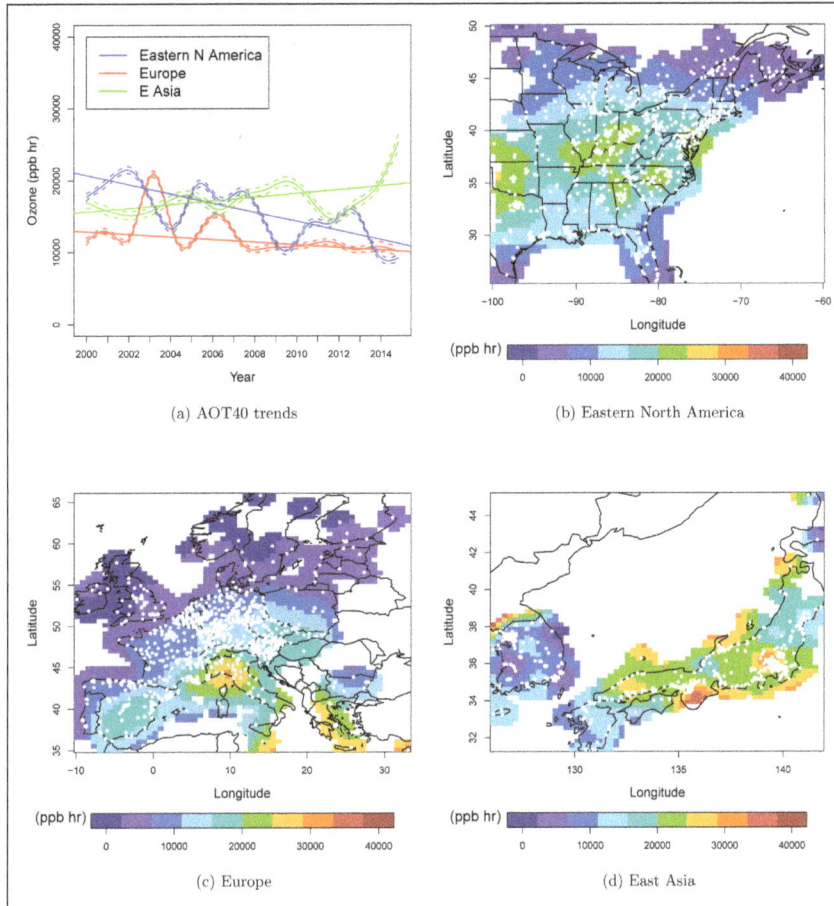

Figure 11: Regional trends and spatial distributions for summertime mean of AOT40 in different regions. Estimated AOT40 (ppb hr) long-term changes in eastern North America (blue), Europe (red) and East Asia (green), along with the spatial mean distributions in each region. The white points indicate the locations of stations. DOI: https://doi.org/10.1525/elementa.243.f11

introduce the structure of negative binomial regression as follows. Let $\tilde{y}(i, t)$ be the NVGT070 value at station i and year t, then the probability mass function is modeled as:

$$p\left(\tilde{y}(i,t)\right) = \frac{\Gamma\left(\tilde{y}(i,t)+1/\theta\right)}{\Gamma\left(\tilde{y}(i,t)+1\right)\Gamma(1/\theta)}\left(\frac{1}{1+\theta\lambda(i,t)}\right)^{1/\theta}$$

$$\left(\frac{\theta\lambda(i,t)}{1+\theta\lambda(i,t)}\right)^{\tilde{y}(i,t)},$$

where

$$\log\lambda(i,t) = f_2(\text{interannual}) + f_3(\text{spatial}) + b(i,t),$$

where $\Gamma(\cdot)$ is the gamma function and $\theta > 0$ is the heterogeneity parameter. This structure has a property $\text{Var}\left(\tilde{y}(i,t)\right) = \lambda(i,t) + \theta\lambda^2(i,t) > \text{E}\left(\tilde{y}(i,t)\right) = \lambda(i,t)$ to accommodate the overdispersion.

Figure 12 shows the NVGT070 summertime mean trends and spatial distributions over different regions. A marked difference from previous analyses on different metrics is that a larger coverage of high values in mean NVGT070 can be found over southern Japan,

corresponding to at least 20 days per summertime period when the maximum daily 8-hour average exceeds 70 ppb.

The second half of **Table 3** shows the Sen-Theil estimators of the detrended line. The mean NVGT070 value decreases from 12 days to less than 1 day in eastern North America over 2000–2014 (the overall mean from all sites is 1.47 in eastern US vs. 0.97 by model prediction over eastern North America in 2014, this discrepancy might be due to the lack of Canadian data in the year 2014). In the same period the mean NVGT070 decreases from 6 days to 2 days in Europe (the average of all available sites in 2014 is 3.15 days) and increases from 15 days to over 20 days in East Asia (the average of all available sites in 2014 is 24.33 days).

5.5 Sensitivity analysis of the trend to the representativeness of sites

We use 756 sites located over eastern North America in summertime as an illustration (with the same metrics described in Section 5.1 and 5.2). We conduct a sensitivity analysis for the long-term mean estimations by removing stations sequentially, according to the p-value for the slope of the trend at each site. We only illustrate the

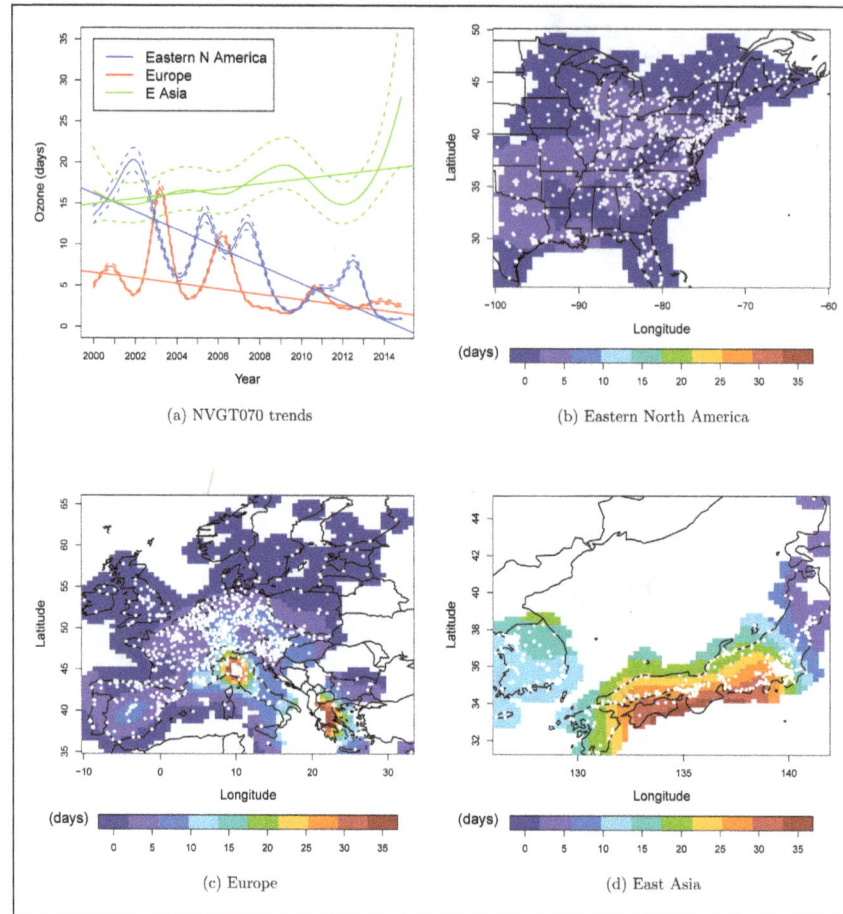

Figure 12: Regional trends and spatial distributions for summertime mean of NVGT070 in different regions. Estimated NVGT070 (days) long-term changes in eastern North America (blue), Europe (red) and East Asia (green), along with the spatial mean distributions in each region. The white points indicate the locations of stations. DOI: https://doi.org/10.1525/elementa.243.f12

impact of removing stations on the trends, the rest of the estimations will not be shown.

Figure 13(a) shows the regional daytime average trend for eastern North America using all 756 available stations (red, labeled as ALL). The slope is statistically significant with a slope of –0.30. We then threw out the 273 stations with p-values less than 0.05 to see the impact on the regional trend. Because these dropped stations have the strongest negative trends, the slope relaxed to –0.20 but it was still statistically significant (orange). We then threw out all stations with p-values less than 0.10 (a total of 312 stations thrown out) for a similar result. The trend relaxed to –0.17 and was still statistically significant (light blue). We then threw out all stations with p-values less than 0.15 (a total of 400 stations thrown out and over half of sites removed at this point) for a similar result. The trend was similar at –0.15 and was statistically significant (dark blue). Using a different metric we found similar results for the DMA8 (see **Figure 13(b)**), however the slope dropped faster due to more stations removed in each iteration.

Table 4 shows the linear regression coefficients for the regional trend for further experiments. The daytime average regional trend remains negative and significant even after removing 551 stations with p-values less than 0.40. Despite the slope dropping faster for DMA8, the

trend remains negative and significant after removing 663 (87.7% of sites) stations with p-value less than 0.50. Beyond this point there are few stations left for the regional analysis and the slope of the regional trend then becomes insignificant. We also provide an example of how the result of spatial kriging can be affected by the similar throwing out procedure in Figures S4 and S5, this example also suggests that useful information can be gleaned from many individual trends with p-values larger than 0.05, 0.10 or even 0.34.

6. Conclusions

This paper provides a trend analysis of summertime surface ozone in eastern North America, Europe and East Asia for several metrics during 2000–2014. Our approach assumes that there is an overall and averaged seasonal cycle and an interannual trend in the study region. The expected achievement in this approach lies in the combination and adjustment of the deviations from each station to the overall regional trend. All of the components in the GAMM are not new techniques, however, this sophisticated incorporation with a focus on overall variations of multiple time series for large and irregular spatial datasets has not been accounted as a whole in previous studies.

Table 3: The Sen-Theil estimators for summertime mean of AOT40 and NVGT070 regional trends, p-values are derived from Mann-Kendall Test (GAMM approach), along with the means (SDs) of all individual intercepts and slopes (Regional mean approach). The overall statistics include the TOAR unclassified category. DOI: https://doi.org/10.1525/elementa.243.t3

		Summertime mean of AOT40				
		GAMM approach			**Regional mean approach**	
Region		**Intercept (ppb hr)**	**Slope (ppb hr yr⁻¹)**	**p-value**	**Intercept (ppb hr)**	**Slope (ppb hr yr⁻¹)**
Eastern N America	Overall	20751	−640	<0.01	19543(7861)	−633(390)
	Rural	20821	−735	<0.01	19801(8490)	−721(407)
	Urban	17282	−439	<0.01	16129(7330)	−412(345)
Europe	Overall	12874	−180	<0.01	13383(8069)	−255(381)
	Rural	15755	−303	<0.01	15882(8602)	−364(406)
	Urban	10245	−130	<0.01	10433(7545)	−138(325)
E Asia	Overall	15663	260	<0.01	15098(9170)	253(678)
	Rural	17033	302	<0.01	15865(6565)	280(763)
	Urban	15152	343	<0.01	14785(10199)	335(627)

		Summertime mean of NVGT070				
		GAMM approach			**Regional mean approach**	
Region		**Intercept (days)**	**Slope (days yr⁻¹)**	**p-value**	**Intercept (days)**	**Slope (days yr⁻¹)**
Eastern N America	Overall	15.37	−0.96	<0.01	17.07(11.41)	−1.16(0.75)
	Rural	15.58	−1.04	<0.01	14.77(10.65)	−1.12(0.76)
	Urban	16.36	−1.03	<0.01	15.19(11.21)	−0.92(0.70)
Europe	Overall	6.52	−0.33	<0.01	7.88(9.60)	−0.42(0.59)
	Rural	9.08	−0.48	<0.01	8.27(10.45)	−0.48(0.70)
	Urban	5.87	−0.26	<0.01	6.15(8.39)	−0.33(0.53)
E Asia	Overall	14.92	0.30	<0.01	16.46(16.38)	0.12(1.24)
	Rural	17.22	0.05	<0.01	13.76(10.30)	0.30(1.30)
	Urban	13.76	0.55	<0.01	17.43(18.42)	0.21(1.21)

Table 4: The Sen-Theil estimators for summertime mean of daytime average and DMA8 regional trends, with p-values derived from Mann-Kendall tests (summertime mean in eastern North America). DOI: https://doi.org/10.1525/elementa.243.t4

	Daytime average			
	Intercept (ppb)	**Slope (ppb yr⁻¹)**	**p-value**	**# Site**
All sites	43.73	−0.30	<0.0001	756
$p = [0.05–1.00]$	42.55	−0.20	<0.0001	483
$p = [0.10–1.00]$	42.39	−0.17	0.0002	414
$p = [0.15–1.00]$	42.07	−0.15	0.0011	356
$p = [0.20–1.00]$	41.87	−0.13	0.0067	304
$p = [0.30–1.00]$	41.69	−0.12	0.0218	251
$p = [0.40–1.00]$	41.32	−0.11	0.0365	205
$p = [0.50–1.00]$	40.86	−0.09	0.0657	158
	DMA8			
	Intercept (ppb)	**Slope (ppb yr⁻¹)**	**p-value**	**# site**
All sites	50.15	−0.43	<0.0001	756
$p = [0.05–1.00]$	48.25	−0.28	<0.0001	382
$p = [0.10–1.00]$	47.58	−0.23	<0.0001	301
$p = [0.15–1.00]$	47.10	−0.21	<0.0001	255
$p = [0.20–1.00]$	46.69	−0.18	0.0007	208
$p = [0.30–1.00]$	46.25	−0.16	0.0038	158
$p = [0.40–1.00]$	45.65	−0.13	0.0185	119
$p = [0.50–1.00]$	45.03	−0.11	0.0432	93

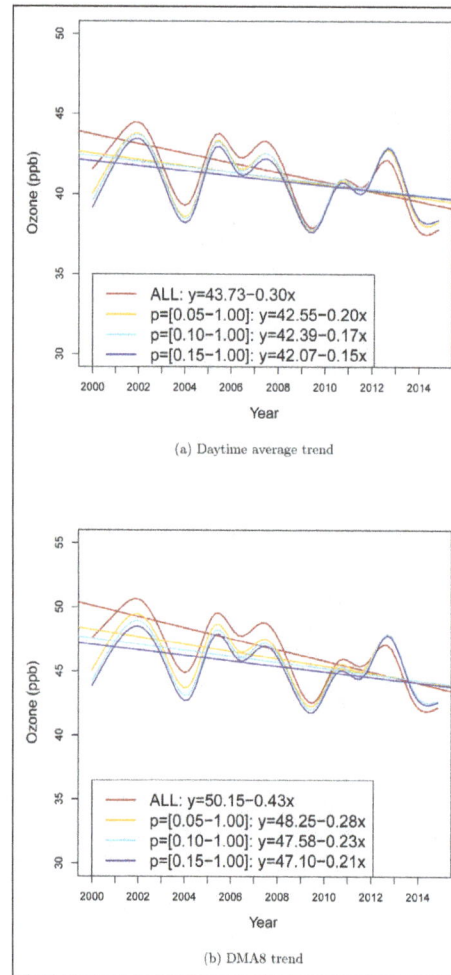

Figure 13: Impact of the representativeness of sites on trends. Estimated long-term changes for summertime mean of **(a)** Daytime average and **(b)** DMA8 using all 756 sites (red), and only the sites with p-value of slope of the trend within the range of [0.05, 1.00] (orange), [0.10, 1.00] (light blue) and [0.15, 1.00] (dark blue). DOI: https://doi.org/10.1525/elementa.243.f13

All of our approaches in this paper are easy to implement under moderate computational costs, and are suitable for application to the TOAR dataset.

The main results are summarized as follows:

1) In eastern North America surface ozone has decreased strongly in summertime (although the daytime average trend at urban sites is less certain). The summertime mean of DMA8 shows a larger decrease than daytime average. AOT40 is reduced by roughly half (from ~20,800 to ~11,200 ppb hr) over the 15-year period. The average modeled value of NVGT070 decreased to less than 1 day in 2014.

2) The regression result of the ozone trends in Europe shows that significant decreases of daytime average and summertime mean of DMA8 are only detecTable in rural sites. AOT40 and NVGT070 decreased significantly in both rural and urban sites. The spatial distributions estimated from different metrics display a similar result: lower values in western and northern Europe and higher values in southern Europe.

3) All the metrics indicate that surface ozone increased over East Asia, with statistically significant trends of 0.40 and 0.37 ppb yr^{-1} estimated for summertime mean of daytime average and DMA8, respectively. AOT40 also reveals a significant increase of 260 ppb hr yr^{-1}. The linear regression predicts the NVGT070 value reached 20 days in summertime 2014. All the trends show a steep increase from 2011–2014.

4) The monitoring network is well covered and developed in eastern North American and East Asia, assessed by several metrics. A consistent result in Europe is difficult to achieve due to relatively limited monitoring sites over northern and eastern Europe.

5) The results from the sensitivity analysis clearly demonstrate that regional trends calculated from just the sites with relatively weak trends are spatially consistent with the regional trends calculated from all sites.

The GAMM has been shown to be applicable to an analysis of the TOAR dataset. It properly accounts for relevant covariate information before producing spatial distributions and regional trends. The GAMM can also take into account other factors known to affect surface ozone. For example the present study did not consider the well-known association between weather and ozone due to lack of meteorological data in the TOAR database. In some regions, ozone is highly correlated with temperature as warm temperatures not only affect reaction rates but they are also associated with stagnant conditions conducive to boundary layer accumulation of ozone precursors (Porter et al., 2015; Pusede et al., 2015; Shen et al., 2016). Continuing development of the TOAR database will permit the inclusion of meteorological variables at all stations (from observations or reanalysis). These data will allow future studies to account for meteorological adjustment of ozone concentrations to provide additional insight, and a more elaborated interpretation, into regional or global scale surface ozone variability (Camalier et al., 2007).

Supplemental Files

The supplemental files for this article can be found as follows:

- **Figure S1.** Locations of stations in eastern North America.

- **Figure S2.** Boxplots of intercepts (ppb) and slopes (ppb yr^{-1}) for trends in eastern North America.

- **Figure S3.** Station-specific effects in different regions.

- **Figure S4.** Impact of the representativeness of sites on spatial interpolation of ozone distribution in summertime 2000.

- **Figure S5.** Impact of the representativeness of sites on spatial interpolation of ozone distribution in summertime 2013.

- **Table S1.** Number of stations with significantly decreasing (D) or increasing (I) summertime mean of daytime average and summertime 4th highest DMA8 based on p-value in eastern North America over 2000–2014, along with the means (standard deviations) of the Sen-Theil intercepts and slopes, separated by site category[a].

- **Table S2.** Numerical output for the explanatory variables from the GAMM divided by rural and urban sites (monthly mean in eastern North America).

- **Table S3.** Numerical output for the explanatory variables from the GAMM divided by rural and urban sites (monthly mean in Europe).

- **Table S4.** Numerical output for the explanatory variables from the GAMM divided by rural and urban sites (monthly mean in East Asia).

- **Table S5.** Numerical output for the explanatory variables from the GAMM (monthly mean in Southeast Asia).

Funding information

TW acknowledges support from the Hong Kong Polytechnic University (G-S023) and The Hong Kong Research Grants Council (PolyU 153042/15E).

Competing interests

The authors have no competing interests to declare.

Author contributions

- Contributed to conception and design: all authors.
- Contributed to acquisition of data: ORC, MGS and TW.
- Contributed to analysis and interpretation of data: all authors.
- Drafted and/or revised the article: KLC and ORC drafted the article while IP, MGS and TW helped with the revision.
- Approved the submitted and revised versions for publication: all authors.

References

Ambrosino, C and **Chandler, RE** 2013 A nonparametric approach to the removal of documented inhomogeneities in climate time series. *Journal of Applied Meteorology and Climatology* **52**(5): 1139–1146. DOI: https://doi.org/10.1175/JAMC-D-12-0166.1

Camalier, L, Cox, W and **Dolwick, P** 2007 The effects of meteorology on ozone in urban areas and their use in assessing ozone trends. *Atmospheric Environment* **41**(33): 7127–7137. DOI: https://doi.org/10.1016/j.atmosenv.2007.04.061

Carslaw, DC, Beevers, SD and **Tate, JE** 2007 Modelling and assessing trends in traffic-related emissions using a generalised additive modelling approach. *Atmospheric Environment* **41**(26): 5289–5299. DOI: https://doi.org/10.1016/j.atmosenv.2007.02.032

Chandler, RE 2005 On the use of generalized linear models for interpreting climate variability. *Environmetrics* **16**(7): 699–715. DOI: https://doi.org/10.1002/env.731

Chandler, RE and **Scott, M** 2011 *Statistical methods for trend detection and analysis in the environmental sciences.* John Wiley & Sons. DOI: https://doi.org/10.1002/9781119991571

Chandler, RE and **Wheater, HS** 2002 Analysis of rainfall variability using generalized linear models: a case study from the west of Ireland. *Water Resources Research* **38**(10). DOI: https://doi.org/10.1029/2001WR000906

Chang, KL, Guillas, S and **Fioletov, VE** 2015 Spatial mapping of ground-based observations of total ozone. *Atmospheric Measurement Techniques*

8(10): 4487–4505. DOI: https://doi.org/10.5194/amt-8-4487-2015

Cooper, OR, Gao, RS, Tarasick, D, Leblanc, T and **Sweeney, C** 2012 Long-term ozone trends at rural ozone monitoring sites across the United States, 1990–2010. *Journal of Geophysical Research: Atmospheres* **117**(D22). DOI: https://doi.org/10.1029/2012JD018261

Cooper, OR, Parrish, DD, Ziemke, J, Balashov, NV, Cupeiro, M, et al. 2014 Global distribution and trends of tropospheric ozone: An observation-based review. *Elem Sci Anth.* **2**: 29. DOI: https://doi.org/10.12952/journal.elementa.000029

Derwent, RG, Witham, CS, Utembe, SR, Jenkin, ME and **Passant, NR** 2010 Ozone in Central England: the impact of 20 years of precursor emission controls in Europe. *environmental science & policy* **13**(3): 195–204.

Duncan, BN, Lamsal, LN, Thompson, AM, Yoshida, Y, Lu, Z, et al. 2016 A space-based, high-resolution view of notable changes in urban NO_x pollution around the world (2005–2014). *Journal of Geophysical Research: Atmospheres.*

European Environment Agency 2016 Air quality in Europe 2016 report, EEA Report No 28/2016. http://www.eea.europa.eu/publications/air-quality-in-europe-2016.

Frost, G, McKeen, S, Trainer, M, Ryerson, T, Neuman, J, et al. 2006 Effects of changing power plant NO_x emissions on ozone in the eastern United States: Proof of concept. *Journal of Geophysical Research: Atmospheres* **111**(D12). DOI: https://doi.org/10.1029/2005JD006354

Granier, C, Bessagnet, B, Bond, T, D'Angiola, A, van Der Gon, HD, et al. 2011 Evolution of anthropogenic and biomass burning emissions of air pollutants at global and regional scales during the 1980–2010 period. *Climatic Change* **109**(1–2): 163. DOI: https://doi.org/10.1007/s10584-011-0154-1

Hamed, KH and **Rao, AR** 1998 A modified Mann-Kendall trend test for auto-correlated data. *Journal of Hydrology* **204**(1–4): 182–196. DOI: https://doi.org/10.1016/S0022-1694(97)00125-X

Hastie, TJ and **Tibshirani, RJ** 1990 *Generalized additive models* **43**. CRC press.

Janssens-Maenhout, G, Crippa, M, Guizzardi, D, Dentener, F, Muntean, M, et al. 2015 HTAP_v2.2: a mosaic of regional and global emission grid maps for 2008 and 2010 to study hemispheric transport of air pollution. *Atmospheric Chemistry and Physics* **15**(19): 11411–11432. DOI: https://doi.org/10.5194/acp-15-11411-2015

Kammann, E and **Wand, MP** 2003 Geoadditive models. *Journal of the Royal Statistical Society: Series C (Applied Statistics)* **52**(1): 1–18. DOI: https://doi.org/10.1111/1467-9876.00385

Kendall, M 1975 *Rank correlation methods.* Charles Griffin: London.

Kim, SW, Heckel, A, McKeen, S, Frost, G, Hsie, EY, et al. 2006 Satellite-observed US power plant NO_x emission reductions and their impact on air quality. *Geophysical Research Letters* **33**(22). DOI: https://doi.org/10.1029/2006GL027749

Krotkov, NA, McLinden, CA, Li, C, Lamsal, LN, Celarier, EA, et al. 2016 Aura OMI observations of regional SO_2 and NO_2 pollution changes from 2005 to 2015. *Atmospheric Chemistry and Physics* **16**(7): 4605–4629. DOI: https://doi.org/10.5194/acp-16-4605-2016

Kunsch, HR 1989 The jackknife and the bootstrap for general stationary observations. *The annals of Statistics*, 1217–1241. DOI: https://doi.org/10.1214/aos/1176347265

Lefohn, AS, Shadwick, D and **Oltmans, SJ** 2010 Characterizing changes in surface ozone levels in metropolitan and rural areas in the United States for 1980–2008 and 1994–2008. *Atmospheric Environment* **44**(39): 5199–5210. DOI: https://doi.org/10.1016/j.atmosenv.2010.08.049

Li, G, Bei, N, Cao, J, Wu, J, Long, X, et al. 2017 Widespread and persistent ozone pollution in eastern China during the non-winter season of 2015: observations and source attributions. *Atmospheric Chemistry and Physics* **17**(4): 2759–2774. DOI: https://doi.org/10.5194/acp-17-2759-2017

Lin, M, Horowitz, LW, Payton, R, Fiore, AM and **Tonnesen, G** 2017 US surface ozone trends and extremes from 1980 to 2014: quantifying the roles of rising Asian emissions, domestic controls, wildfires, and climate. *Atmospheric Chemistry and Physics* **17**(4): 2943–2970. DOI: https://doi.org/10.5194/acp-17-2943-2017

Liu, F, Zhang, Q, Zheng, B, Tong, D, Yan, L, et al. 2016 Recent reduction in NO_x emissions over China: synthesis of satellite observations and emission inventories. *Environmental Research Letters* **11**(11): 114002. DOI: https://doi.org/10.1088/1748-9326/11/11/114002

LRTAP Convention 2015 Draft Chapter III: Mapping critical levels for vegetation, of the manual on methodologies and criteria for modelling and mapping critical loads and levels and air pollution effects, risks and trends. http://icpmapping.org/Mapping_Manual.

Ma, Z, Xu, J, Quan, W, Zhang, Z, Lin, W, et al. 2016 Significant increase of surface ozone at a rural site, north of eastern China. *Atmospheric Chemistry and Physics* **16**(6): 3969–3977. DOI: https://doi.org/10.5194/acp-16-3969-2016

Miyazaki, K, Eskes, H, Sudo, K, Boersma, KF, Bowman, K, et al. 2017 Decadal changes in global surface NO_x emissions from multi-constituent satellite data assimilation. *Atmospheric Chemistry and Physics* **17**(4): 807–837. DOI: https://doi.org/10.5194/acp-17-807-2017

Monks, PS, Archibald, A, Colette, A, Cooper, OR, Coyle, M, et al. 2015 Tropospheric ozone and its precursors from the urban to the global scale from air quality to short-lived climate forcer. *Atmospheric Chemistry and Physics* **15**(15): 8889–8973. DOI: https://doi.org/10.5194/acp-15-8889-2015

Paciorek, CJ, Yanosky, JD, Puett, RC, Laden, F and Suh, HH 2009 Practical large-scale spatio-temporal modeling of particulate matter concentrations. *The Annals of Applied Statistics* **3**(1): 370–397. DOI: https://doi.org/10.1214/08-AOAS204

Park, A, Guillas, S and Petropavlovskikh, I 2013 Trends in stratospheric ozone profiles using functional mixed models. *Atmospheric Chemistry and Physics* **13**(22): 11473–11501. DOI: https://doi.org/10.5194/acp-13-11473-2013

Porter, WC, Heald, CL, Cooley, D and Russell, B 2015 Investigating the observed sensitivities of airquality extremes to meteorological drivers via quantile regression. *Atmospheric Chemistry and Physics* **15**(18): 10349–10366. DOI: https://doi.org/10.5194/acp-15-10349-2015

Pusede, SE, Steiner, AL and Cohen, RC 2015 Temperature and recent trends in the chemistry of continental surface ozone. *Chemical reviews* **115**(10): 3898–3918. DOI: https://doi.org/10.1021/cr5006815

REVIHAAP 2013 Review of evidence on health aspects of air pollution–REVIHAAP Project. World Health Organization (WHO) Regional Office for Europe Bonn, http://www.euro.who.int/data/assets/pdf_file/0004/193108/REVIHAAP-Final-technical-report-final-version.pdf.

Safieddine, S, Clerbaux, C, George, M, Hadji-Lazaro, J, Hurtmans, D, et al. 2013 Tropospheric ozone and nitrogen dioxide measurements in urban and rural regions as seen by IASI and GOME-2. *Journal of Geophysical Research: Atmospheres* **118**(18). DOI: https://doi.org/10.1002/jgrd.50669

Schultz, MG, Schroeder, S, Lyapina, O, Cooper, OR, Galbal, I, et al. 2017 Tropospheric Ozone Assessment Report: Database and metrics data of global surface ozone observations. *Elem Sci Anth*. In press for TOAR Special Feature.

Sen, PK 1968 Estimates of the regression coefficient based on Kendall's tau. *Journal of the American Statistical Association* **63**(324): 1379–1389. DOI: https://doi.org/10.1080/01621459.1968.10480934

Shen, L, Mickley, LJ and Gilleland, E 2016 Impact of increasing heat waves on US ozone episodes in the 2050s: Results from a multimodel analysis using extreme value theory. *Geophysical Research Letters* **43**(8): 4017–4025. DOI: https://doi.org/10.1002/2016GL068432

Simon, H, Reff, A, Wells, B, Xing, J and Frank, N 2015 Ozone trends across the United States over a period of decreasing NO_x and VOC emissions. *Environmental science & technology* **49**(1): 186–195. DOI: https://doi.org/10.1021/es504514z

Simpson, D, Arneth, A, Mills, G, Solberg, S and Uddling, J 2014 Ozone- the persistent menace: interactions with the N cycle and climate change. *Current Opinion in Environmental Sustainability* **9**: 9–19. DOI: https://doi.org/10.1016/j.cosust.2014.07.008

Sun, L, Xue, L, Wang, T, Gao, J, Ding, A, et al. 2016 Significant increase of summertime ozone at Mount Tai in Central Eastern China. *Atmospheric Chemistry and Physics* **16**(16): 10637–10650. DOI: https://doi.org/10.5194/acp-16-10637-2016

Theil, H 1950 A rank-invariant method of linear and polynomial regression analysis, 3; confidence regions for the parameters of polynomial regression equations. *Stichting Mathematisch Centrum Statistische Afdeling* (SP 5a/50/R), 1–16.

Thompson, ML, Reynolds, J, Cox, LH, Guttorp, P and Sampson, PD 2001 A review of statistical methods for the meteorological adjustment of tropospheric ozone. *Atmospheric environment* **35**(3): 617–630. DOI: https://doi.org/10.1016/S1352-2310(00)00261-2

Tiao, G, Reinsel, G, Xu, D, Pedrick, J, Zhu, X, et al. 1990 Effects of autocorrelation and temporal sampling schemes on estimates of trend and spatial correlation. *Journal of Geophysical Research: Atmospheres* **95**(D12): 20507–20517. DOI: https://doi.org/10.1029/JD095iD12p20507

US Environmental Protection Agency 2013 Integrated Science Assessment for Ozone and Related Photochemical Oxidants. EPA 600/R-10/076F.

Van der A, RJ, Mijling, B, Ding, J, Elissavet Koukouli, M, Liu, F, et al. 2017 Cleaning up the air: effectiveness of air quality policy for SO_2 and NO_x emissions in China. *Atmospheric Chemistry & Physics* **17**: 1775–1789. DOI: https://doi.org/10.5194/acp-17-1775-2017

Vingarzan, R 2004 A review of surface ozone background levels and trends. *Atmospheric Environment* **38**(21): 3431–3442. DOI: https://doi.org/10.1016/j.atmosenv.2004.03.030

Wagner, HH and Fortin, MJ 2005 Spatial analysis of landscapes: concepts and statistics. *Ecology* **86**(8): 1975–1987. DOI: https://doi.org/10.1890/04-0914

Wang, T, Xue, L, Brimblecombe, P, Lam, YF, Li, L, et al. 2017 Ozone pollution in China: A review of concentrations, meteorological influences, chemical precursors, and effects. *Science of The Total Environment* **575**: 1582–1596. DOI: https://doi.org/10.1016/j.scitotenv.2016.10.081

Wood, SN 2004 Stable and efficient multiple smoothing parameter estimation for generalized additive models. *Journal of the American Statistical Association* **99**(467): 673–686. DOI: https://doi.org/10.1198/016214504000000980

Wood, SN 2006 *Generalized additive models: an introduction with R*. CRC press.

Wood, SN and Augustin, NH 2002 GAMs with integrated model selection using penalized regression splines and applications to environmental modelling. *Ecological modelling* **157**(2): 157–177. DOI: https://doi.org/10.1016/S0304-3800(02)00193-X

Wood, SN, Pya, N and Säfken, B 2016 Smoothing parameter and model selection for general smooth models. *Journal of the American Statistical Association* **111**(516): 1548–1563. DOI: https://doi.org/10.1080/01621459.2016.1180986

Xu, W, Lin, W, Xu, X, Tang, J, Huang, J, et al. 2016 Longterm trends of surface ozone and its influencing

factors at the Mt Waliguan GAW station, China–Part
1: Overall trends and characteristics. *Atmospheric
Chemistry and Physics* **16**(10): 6191–6205. DOI:
https://doi.org/10.5194/acp-16-6191-2016

Yan, Z, Bate, S, Chandler, RE, Isham, V and **Wheater, H**
2002 An analysis of daily maximum wind speed in
northwestern Europe using generalized linear models.
Journal of Climate **15**(15): 2073–2088. DOI: https://
doi.org/10.1175/1520-0442(2002)015<2073:AAO
DMW>2.0.CO;2

Yang, C, Chandler, RE, Isham, V and **Wheater, H** 2005
Spatial-temporal rainfall simulation using generalized

linear models. *Water Resources Research* **41**(11). DOI:
https://doi.org/10.1029/2004WR003739

**Zhang, Y, Cooper, OR, Gaudel, A, Thompson, AM,
Nédélec, P,** et al. 2016 Tropospheric ozone change
from 1980 to 2010 dominated by equatorward
redistribution of emissions. *Nature Geoscience.* DOI:
https://doi.org/10.1038/ngeo2827

Zhao, B, Wang, S, Liu, H, Xu, J, Fu, K, et al. 2013 NO_x
emissions in China: historical trends and future
perspectives. *Atmospheric Chemistry and Physics*
13(19): 9869–9897. DOI: https://doi.org/10.5194/
acp-13-9869-2013

PERMISSIONS

All chapters in this book were first published in ESTA, by University of California Press; hereby published with permission under the Creative Commons Attribution License or equivalent. Every chapter published in this book has been scrutinized by our experts. Their significance has been extensively debated. The topics covered herein carry significant findings which will fuel the growth of the discipline. They may even be implemented as practical applications or may be referred to as a beginning point for another development.

The contributors of this book come from diverse backgrounds, making this book a truly international effort. This book will bring forth new frontiers with its revolutionizing research information and detailed analysis of the nascent developments around the world.

We would like to thank all the contributing authors for lending their expertise to make the book truly unique. They have played a crucial role in the development of this book. Without their invaluable contributions this book wouldn't have been possible. They have made vital efforts to compile up to date information on the varied aspects of this subject to make this book a valuable addition to the collection of many professionals and students.

This book was conceptualized with the vision of imparting up-to-date information and advanced data in this field. To ensure the same, a matchless editorial board was set up. Every individual on the board went through rigorous rounds of assessment to prove their worth. After which they invested a large part of their time researching and compiling the most relevant data for our readers.

The editorial board has been involved in producing this book since its inception. They have spent rigorous hours researching and exploring the diverse topics which have resulted in the successful publishing of this book. They have passed on their knowledge of decades through this book. To expedite this challenging task, the publisher supported the team at every step. A small team of assistant editors was also appointed to further simplify the editing procedure and attain best results for the readers.

Apart from the editorial board, the designing team has also invested a significant amount of their time in understanding the subject and creating the most relevant covers. They scrutinized every image to scout for the most suitable representation of the subject and create an appropriate cover for the book.

The publishing team has been an ardent support to the editorial, designing and production team. Their endless efforts to recruit the best for this project, has resulted in the accomplishment of this book. They are a veteran in the field of academics and their pool of knowledge is as vast as their experience in printing. Their expertise and guidance has proved useful at every step. Their uncompromising quality standards have made this book an exceptional effort. Their encouragement from time to time has been an inspiration for everyone.

The publisher and the editorial board hope that this book will prove to be a valuable piece of knowledge for researchers, students, practitioners and scholars across the globe.

LIST OF CONTRIBUTORS

P. Tuccella
LATMOS/IPSL, UPMC Univ. Paris 06 Sorbonne Université, UVSQ, CNRS, Paris, FR
NUMTECH, 6 allée Alan Turing, CS 60242, 63178 Aubiere, FR
Laboratoire de Météorologie Dynamique, Ecole Polytechnique, 91128 Palaiseau, FR

J. L. Thomas, K. S. Law, J.-C. Raut and T. Onishi
LATMOS/IPSL, UPMC Univ. Paris 06 Sorbonne Université, UVSQ, CNRS, Paris, FR

L. Marelle
LATMOS/IPSL, UPMC Univ. Paris 06 Sorbonne Université, UVSQ, CNRS, Paris, FR
Center for International Climate and Environmental Research, Oslo, NO

A. Roiger and H. Schlager
Deutsches Zentrum für Luft- und Raumfahrt (DLR), Institut für Physik der Atmosphäre, 10 Oberpfaffenhofen, DE

B. Weinzierl
University of Vienna, Faculty of Physics, Aerosol Physics and Environmental Physics, Boltzmanngasse 5, A-1090 Vienna, AT

H. A. C. Denier van der Gon
TNO Climate, Air and Sustainability, Princetonlaan 6, 3584 CB Utrecht, NL

Natasha L. Miles, Scott J. Richardson, Thomas Lauvaux, Kenneth J. Davis, Nikolay V. Balashov and Aijun Deng
Department of Meteorology, The Pennsylvania State University, University Park, Pennsylvania, US

Jocelyn C. Turnbull
National Isotope Centre, GNS Science, Lower Hutt, NZ
National Oceanic and Atmospheric Administration/University of Colorado, Boulder, Colorado, US

Colm Sweeney
National Oceanic and Atmospheric Administration/University of Colorado, Boulder, Colorado, US

Kevin R. Gurney, Risa Patarasuk and Igor Razlivanov
Arizona State University, Tempe, Arizona, US

Maria Obiminda L. Cambaliza
Purdue University, West Lafayette, Indiana, US
Ateneo de Manila University, Katipunan Ave, Quezon City, Metro Manila, Philippines 1108, PH

Paul B. Shepson
Purdue University, West Lafayette, Indiana, US

Aijun Deng
Utopus Insights, Inc., New York, US

Thomas Lauvaux, Kenneth J. Davis, Brian J. Gaudet, Natasha Miles, Scott J. Richardson, Kai Wu and Daniel P. Sarmiento
The Pennsylvania State University, Pennsylvania, US

R. Michael Hardesty, Timothy A. Bonin and W. Alan Brewer
Cooperative Institute for Research in Environmental Sciences, University of Colorado/NOAA, Chemical Sciences Division, Colorado, US

Kevin R. Gurney
Arizona State University, Arizona, US

Alexie M. F. Heimburger, Rebecca M. Harvey, Chloe Gore, Olivia E. Salmon, Anna-Elodie M. Kerlo and Tegan N. Lavoie
Department of Chemistry, Purdue University, West Lafayette, Indiana, US

Paul B. Shepson
Department of Chemistry, Purdue University, West Lafayette, Indiana, US
Department of Earth, Atmospheric and Planetary Science and Purdue Climate Change Research Center, Purdue University, West Lafayette, Indiana, US

Brian H. Stirm
Department of Aviation and Transportation Technology, Purdue University, West Lafayette, Indiana, US

Jocelyn Turnbull
National Isotope Center, GNS Science, Lower Hutt, NZ

Maria O. L. Cambaliza
Department of Physics, Ateneo de Manila University, Loyola Heights, Quezon City, PH

Kenneth J. Davis and Thomas Lauvaux
The Pennsylvania State University, Department of Meteorology, University Park, PA

Anna Karion
NOAA/ESRL, Colorado, US
CIRES, University of Colorado at Boulder, Boulder, Colorado, US
NIST, Gaithersburg, Maryland, US

Colm Sweeney
NOAA/ESRL, Colorado, US
CIRES, University of Colorado at Boulder, Boulder, Colorado, US

W. Allen Brewer
NOAA/ESRL, Colorado, US

R. Michael Hardesty
CIRES, University of Colorado at Boulder, Boulder, Colorado, US

Kevin R. Gurney
School of Life Sciences, Arizona State University, Tempe, Arizona, US

Kevin R. Gurney, Jianming Liang, Risa Patarasuk, Darragh O'Keeffe, Jianhua Huang and Maya Hutchins
Arizona State University, Tempe, Arizona, US

Thomas Lauvaux
Department of Meteorology, Pennsylvania State University, University Park, Pennsylvania, US

Jocelyn C. Turnbull
GNS Science, Rafter Radiocarbon Laboratory, Lower Hutt, NZ
National Oceanic and Atmospheric Administration/ University of Colorado, Boulder, Colorado, US

Tomohiro Oda
Global Modeling and Assimilation Office, NASA Goddard Space Flight Center, Greenbelt, Maryland, US
Goddard Earth Sciences Technologies and Research, Universities Space Research Association, Columbia, Maryland, US

Thomas Lauvaux, Natasha L. Miles and Scott J. Richardson
Department of Meteorology and Atmospheric Science, The Pennsylvania State University, University Park, Pennsylvania, US

Dengsheng Lu
Michigan State University, East Lansing, Michigan, US

Preeti Rao
NASA Jet Propulsion Laboratory, Pasadena, California, US

N. S. Wagenbrenner
US Forest Service, Rocky Mountain Research Station, Missoula Fire Sciences Laboratory, Missoula, Minnesota, United States

S. H. Chung and B. K. Lamb
Laboratory for Atmospheric Research, Department of Civil and Environmental Engineering, Washington State University, Pullman, Washington, United States

B. Zhang
Atmospheric Sciences, Michigan Technological University, Houghton, US
National Institute of Aerospace, Hampton, US

R. C. Owen
Atmospheric Sciences, Michigan Technological University, Houghton, US
Environmental Protection Agency, Research Triangle Park, US

J. A. Perlinger, L. R. Mazzoleni and C. Mazzoleni
Atmospheric Sciences, Michigan Technological University, Houghton, US

D. Helmig
§ Institute of Arctic and Alpine Research, University of Colorado, Boulder, US

M. Val Martín
|| Chemical and Biological Engineering, University of Sheffield, Sheffield, UK

L. Kramer
Atmospheric Sciences, Michigan Technological University, Houghton, US
University of Birmingham, Birmingham, UK

Le Kuai
Joint Institute for Regional Earth System Science and Engineering, University of California, Los Angeles, California, US

Kevin W. Bowman and Robert L. Herman
Jet Propulsion Laboratory, California Institute of Technology, Pasadena, California, US

Helen M. Worden
National Center for Atmospheric Research, Boulder, Colorado, US

Susan S. Kulawik
Bay Area Environmental Research Institute, Mountain View, California, US
NASA's Ames Research Center, Mountain View, California,US

Kristina A. Dahl
Dahl Scientific, San Francisco, CA, US

Erika Spanger-Siegfried
Union of Concerned Scientists, Cambridge, MA, US

Astrid Caldas and Shana Udvardy
Union of Concerned Scientists, Washington, DC, US

Ingrid Wiedmann and Marit Reigstad
UiT The Arctic University of Norway, Tromsø, NO

Jean-Éric Tremblay
Québec-Océan and Takuvik, Biology Department, Université Laval, Québec City, Québec, CA

Arild Sundfjord
Norwegian Polar Institute, Tromsø, NO

Kai-Lan Chang
NOAA Earth System Research Laboratory, Boulder, US

Irina Petropavlovskikh and Owen R. Cooper
NOAA Earth System Research Laboratory, Boulder, US
Cooperative Institute for Research in Environmental Sciences, University of Colorado, Boulder, US

Martin G. Schultz
Institute for Energy and Climate Research (IEK-8), Forschungszentrum Jülich, Jülich, DE

Tao Wang
Department of Civil and Environmental Engineering, The Hong Kong Polytechnic University, Hong Kong, CN

Index

www.ingramcontent.com/pod-product-compliance
Lightning Source LLC
Chambersburg PA
CBHW080255230326
41458CB00097B/5008

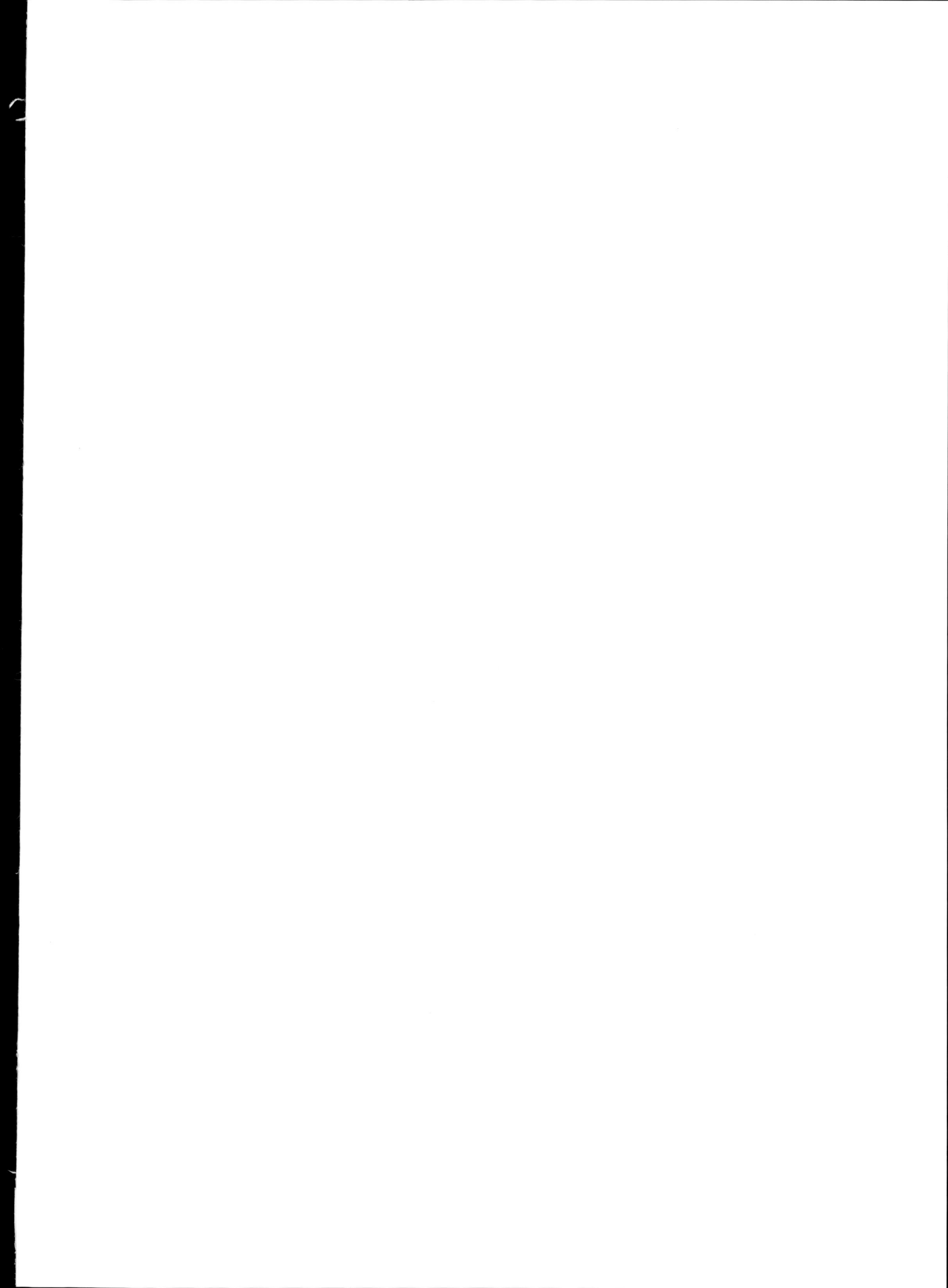